An Int

Mat

Eng

Mecha

An Introduction to
Mathematics for Engineers

Mechanics

Stephen Lee

HODDER
EDUCATION
PART OF HACHETTE LIVRE UK

Orders: please contact Bookpoint Ltd, 130 Milton Park, Abingdon, Oxon OX14 4SB. Telephone: (44) 01235 827720. Fax: (44) 01235 400454. Lines are open from 9.00 - 5.00, Monday to Saturday, with a 24 hour message answering service. You can also order through our website www.hoddereducation.co.uk

> If you have any comments to make about this, or any of our other titles, please send them to educationenquiries@hodder.co.uk

British Library Cataloguing in Publication Data
A catalogue record for this title is available from the British Library

ISBN: 978 0 340 965528

First Edition Published 2008
This Edition Published 2008
Impression number   10 9 8 7 6 5 4 3 2 1
Year   2012 2011 2010 2009 2008

Hachette Livre UK's policy is to use papers that are natural, renewable and recyclable products and made from wood grown in sustainable forests.
The logging and manufacturing processes are expected to conform to the environmental regulations of the country of origin.

Cover photo from Vladimir Godnik/Getty Images and Digital Vision
Typeset by Tech-Set Ltd, Gateshead.
Printed by Martins the Printers, Berwick-upon-Tweed for Hodder Education, a part of Hachette Livre UK, 338 Euston Road, London NW1 3BH

The author and publisher would like to thank the following for permission to reproduce copyright illustrative material:

p.3 Ruth Nossek; p.6 Andrew Ward/Life File Photographic Library Ltd; p.14 David Cumming;Eye Ubiquitous CORBIS; p.37 Topfoto; p.41 DR JEREMY BURGESS / SCIENCE PHOTO LIBRARY; p.42 Tony Henshaw/ActionPlus; p.43 Jeremy Hoare/Life File Photographic Library Ltd; p.48 NASA; p. 53 Jeremy Hoare/Life File Photographic Library Ltd; p.61 Mehau Kulyk/Science Photo Library; p.62 © National Portrait Gallery, London; p.68 John Cox/Life File Photographic Library Ltd; p. 68 (left) Graham Buchan/Life File Photographic Library Ltd; p.68 (right) B&C Alexander/Arctic Photo; p.94 John Greim/Science Photo Library; p. 115 Emma Lee/Life File Photographic Library Ltd; p.123 Jennie Woodcock; Reflections Photo Library/CORBIS; p.155 Northern Counties buses; p.165 Glyn Kirk/Action Plus; p.175 (left) The genius of China, photographer unknown; p.175 (middle) Andrew Ward/Life File Photographic Library Ltd; p.175 (right) Andrew Ward/Life File Photographic Library Ltd; p.181 (left) Emma Lee/Life File Photographic Library Ltd; p.181 (right) Jane Burton/Bruce Coleman collection; p.198 Pat Brydon; p.201 M.C. Escher's "Waterfall" © 2008 The M.C. Escher Company B.V. Holland. All rights reserved. www.mcescher.com; p.214 Neil Tingle/Action Plus; p.221 Colorsport; p.226 Colorsport; p.243 Neil Tingle/Action Plus; p.252 (left) NASA/Roger Ressmeyer/CORBIS; p.252 (middle) Bohemian Nomad Picturemakers / Corbis; p.252 (right) Tim Wright/CORBIS; p.264 © Reuters/CORBIS; p.277 Hodder Picture Library; p.288 Kate Fullerty (left) p.288 Kate Fullerty (right); p.294 Nick Nicholson/Hawkley Studios; p.294 Emma Lee/Life File Photographic Library Ltd; p. 300 Emma Lee/Life File Photographic Library Ltd; p.305 (left) Matthew Mendelsohn/CORBIS; p.305 (right) Gary Braasch/CORBIS; p.305 Emma Lee/Life File Photographic Library Ltd; p.337 Andrew Ward/Life File Photographic Library Ltd; p.342 TopFoto/PA; p.342 ANDREW LAMBERT PHOTOGRAPHY / SCIENCE PHOTO LIBRARY; p.343 AP/AP/PA Photos; p.419 China Photo/Reuters/CORBIS; p.438 © Morey Milbradt/Brand X/Corbis; p.450 Hubert Stadler/CORBIS; p.450 (left) John and Lisa Merrill/CORBIS; P.450 (right) Kim Sayer/CORBIS; p.472 Duomo/CORBIS; p.490 Vince Streano/CORBIS; p.496 (left) Jennie Woodcock, Reflections Photolibrary/CORBIS; p.496 (middle) Richard Cummins/CORBIS; p.496 (right) Bob Battersby/BDI images

Contents

Introduction

This book has been written to meet the needs of students in the early years of engineering and STEM degrees. Recent changes to the structure of A levels mean that an increasing number of those starting such courses in higher education have studied little, if any, mechanics. However, the book will also be valuable to a much wider readership including those on other courses, e.g. engineering diplomas, and those who already have some basic knowledge of mechanics, but wish to explore more advanced topics, such as variable forces, circular motion, simple harmonic motion or stability and small oscillations. This book is based upon the well-respected MEI textbooks.

Throughout the book emphasis is placed on understanding the basic principles of mechanics and the process of modelling the real world, rather than on mere routine calculation. Examples of everyday applications are covered in worked examples and in the various exercises. The book also contains an enclosed CD-ROM with 'Personal Tutor' examples; these are electronic step-by-step solutions to questions at the end of most chapters, with voiceover commentaries. Further exercises, and answers to all exercises, are also included on the CD-ROM

Thanks are due to many people who have given help and advice with the book and also those that have been involved in its preparation, including David Holland, Roger Porkess and all the authors of previous MEI mechanics books.

1 Motion along a straight line

The whole burden of philosophy seems to consist in this – from the phenomena of motions to investigate the forces of nature.

Isaac Newton

1.1 Setting up a mathematical model

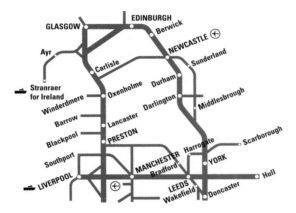

Figure 1.1: Northern Great Britain's Main Railways

QUESTION 1.1 The figure shows part of the map of the main railway lines of northern Great Britain. Which of the following statements can you be sure of just by looking at this map? Which of them might be important to a visitor from abroad?

i) Darlington is on the line from York to Durham.
ii) Carlisle is nearer to Glasgow than it is to Newcastle.
iii) Leeds is due east of Manchester.
iv) The quickest way from Leeds to Glasgow is via York.

This is a *diagrammatic model* of the railway system which gives essential though by no means all the information you need for planning train journeys. You can be sure about the places a line passes through but distances and directions are only approximate and if you compare this map with an ordinary map you will see that statements ii) and iii) are false. You will also need further information from timetables to plan the best way to get from Leeds to Glasgow.

Making simplifying assumptions

When setting up a model, you first need to decide what is essential. For example, what would you take into account and what would you ignore when considering the motion of a car travelling from Bristol to London?

You will need to know the distance and the time taken for parts of the journey, but you might decide to ignore the dimensions of the car and the motion of the wheels. You would then be using the idea of a particle to model the car. *A particle has no dimensions*.

You might also decide to ignore the bends in the road and its width and so treat it as a *straight line with only one dimension*. A length along the line would represent a length along the road in the same way as a piece of thread following a road on a map might be straightened out to measure its length.

You might decide to split the journey up into parts and assume that the speed is constant over these parts.

The process of making decisions like these is called *making simplifying assumptions* and is the first stage of setting up a *mathematical model* of the situation.

Defining the variables and setting up the equations

The next step in setting up a mathematical model is to *define the variables* with suitable units. These will depend on the problem you are trying to solve. Suppose you want to know where you ought to be at certain times in order to maintain a good average speed between Bristol and London. You might define your variables as follows:

● the total time since the car left Bristol is t hours
● the distance from Bristol at time t is x km
● the average speed up to time t is v kmh^{-1}.

Then, at Newbury $t = t_1$ and $x = x_1$; etc.

You can then *set up equations* and go through the mathematics required to solve the problem. Remember to check that your answer is sensible. If it isn't, you might have made a mistake in your mathematics or your simplifying assumptions might need reconsideration.

The theories of mechanics that you will learn about in this book, and indeed any other studies in which mathematics is applied, are based on mathematical models of the real world. When necessary, these models can become more complex as your knowledge increases.

QUESTION 1.2   The simplest form of the Bristol to London model assumes that the speed remains constant over sections of the journey. Is this reasonable?

For a much shorter journey, you might need to take into account changes in the velocity of the car. This chapter develops the mathematics required when an object can be modelled as a *particle moving in a straight line with constant acceleration*. In most real situations this is only the case for part of the motion – you wouldn't expect a car to continue accelerating at the same rate for very long – but it is a very useful model to use as a first approximation over a short time.

1.2 The language of motion

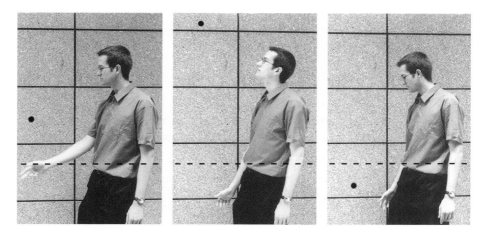

Throw a small object such as a marble straight up in the air and think about the words you could use to describe its motion from the instant just after it leaves your hand to the instant just before it hits the floor. Some of your words might involve the idea of direction. Other words might be to do with the position of the marble, its speed or whether it is slowing down or speeding up. Underlying many of these is time.

Direction

The marble moves as it does because of the gravitational pull of the earth. We understand directional words such as up and down because we experience this pull towards the centre of the earth all the time. The *vertical* direction is along the line towards or away from the centre of the earth.

In mathematics a quantity which has only size, or magnitude, is called a *scalar*. One which has both magnitude and a direction in space is called a *vector*.

Distance, position and displacement

The total *distance* travelled by the marble at any time does not depend on its direction. It is a scalar quantity.

Position and displacement are two vectors related to distance; they have direction as well as magnitude. Here their direction is up or down and you decide which of these is positive. When up is taken to be positive, down is negative.

The *position* of the marble is then its distance above a fixed origin, for example the distance above the place it first left your hand.

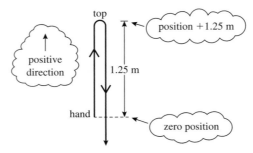

Figure 1.2

When it reaches the top, the marble might have travelled a distance of 1.25 m. Relative to your hand its position is then 1.25 m upwards or +1.25m.

At the instant it returns to the same level as your hand it will have travelled a total distance of 2.5m. Its *position*, however, is zero upwards.

A position is always referred to a fixed origin but a *displacement* can be measured from any position. When the marble returns to the level of your hand, its displacement is zero relative to your hand but −1.25 m relative to the top.

QUESTION 1.3   What are the positions of the particles A, B and C in the diagram below?

Figure 1.3

What is the displacement of B
i) relative to A   ii) relative to C?

Diagrams and graphs

In mathematics, it is important to use words precisely, even though they might be used more loosely in everyday life. In addition, a picture in the form of a diagram or graph can often be used to show the information more clearly.

Figure 1.4 is a *diagram* showing the direction of motion of the marble and relevant distances. The direction of motion is indicated by an arrow. Figure 1.5 is a *graph* showing the position above the level of your hand against the time. Notice that it is *not* the path of the marble.

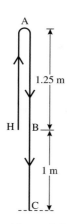

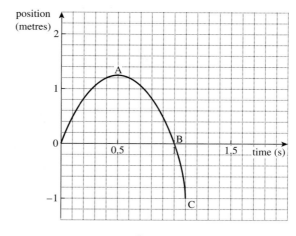

Figure 1.4   Figure 1.5

QUESTION 1.4   The graph in figure 1.5 shows that the position is negative after one second (point B). What does this negative position mean?

Note

When drawing a graph it is very important to specify your axes carefully. Graphs showing motion usually have time along the horizontal axis. Then you have to decide where the origin is and which direction is positive on the vertical axis. In this graph the origin is at hand level and upwards is positive. The time is measured from the instant the marble leaves your hand.

Notation and units

As with most mathematics, you will see in this book that certain letters are commonly used to denote certain quantities. This makes things easier to follow. Here the letters used are:

- s, h, x, y and z for position
- t for time measured from a starting instant
- u and v for velocity
- a for acceleration.

The S.I. (Système International d'Unités) unit for *distance* is the metre (m), that for *time* is the second (s) and that for *mass* the kilogram (kg). Other units follow from these so speed is measured in metres per second, written ms^{-1}. S.I. units are used almost entirely in this book but occasional references are made to imperial and other units.

1.3 Speed and velocity

Speed is a scalar quantity and does not involve direction. *Velocity* is the vector related to speed; its magnitude is the speed but it also has a direction. When an object is moving in the negative direction, its velocity is negative.

Amy has to post a letter on her way to college. The post box is 500 m east of her house and the college is 2.5 km to the west. Amy cycles at a steady speed of 10 ms^{-1} and takes 10 s at the post box to find the letter and post it.

The diagram shows Amy's journey using east as the positive direction. The distance of 2.5 km has been changed to metres so that the units are consistent.

Figure 1.6

After she leaves the post box Amy is travelling west so her velocity is negative. It is $-10\,\text{ms}^{-1}$.

The distances and times for the three parts of Amy's journey are:

Home to post box	500 m	$\frac{500}{10} = 50$ s
At post box	0 m	10 s
Post box to college	3000 m	$\frac{3000}{10} = 300$ s

These can be used to draw the position–time position–time graph using home as the origin, as in figure 1.7.

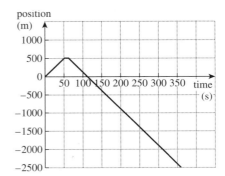

Figure 1.7

QUESTION 1.5   Calculate the gradient of the three portions of this graph. What does the gradient represent?

The velocity is the rate at which the position changes.

Velocity is represented by the gradient of the position–time graph.

Figure 1.8 is the velocity–time graph.

Note

By drawing the graphs below each other with the same horizontal scales, you can see how they correspond to each other.

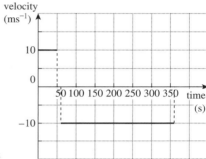

Figure 1.8

Distance–time graphs

Figure 1.9 is the distance–time graph of Amy's journey. It differs from the position–time graph because it shows how far she travels irrespective of her direction. There are no negative values.

The gradient of this graph represents Amy's speed rather than her velocity.

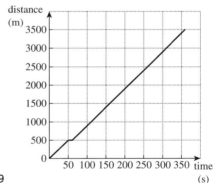

Figure 1.9

QUESTION 1.6   It has been assumed that Amy starts and stops instantaneously. What would more realistic graphs look like? Would it make a lot of difference to the answers if you tried to be more realistic?

Average speed and average velocity

You can find Amy's average speed on her way to college by using the definition

$$average \; speed = \frac{total \; distance \; travelled}{total \; time \; taken}$$

When the distance is in metres and the time in seconds, speed is found by dividing metres by seconds and is written as ms^{-1}. So Amy's average speed is

$$\frac{3500 \text{ m}}{360 \text{ s}} = 9.72 \text{ ms}^{-1}.$$

Amy's average velocity is different. Her displacement from start to finish is -2500 m so

$$\boxed{average\ velocity = \frac{displacement}{time\ taken}}$$

> The college is in the negative direction.

$$= \frac{-2500}{360} = -6.94 \text{ ms}^{-1}$$

If Amy had taken the same time to go straight from home to college at a steady speed, this steady speed would have been 6.94 ms^{-1}.

Velocity at an instant

The position–time graph for a marble thrown straight up into the air at 5 m^{-1} is curved because the velocity is continually changing.

The velocity is represented by the gradient of the position–time graph. When a position–time graph is curved like this you can find the velocity at an instant of time by drawing a tangent as in figure 1.10.

The velocity at P is approximately

$$\frac{0.6}{0.25} = 2.4 \text{ ms}^{-1}.$$

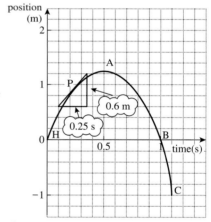

Figure 1.10

The velocity–time graph is shown in figure 1.11.

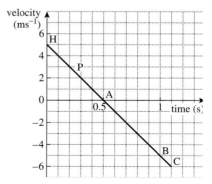

Figure 1.11

QUESTION 1.7 What is the velocity at H, A, B and C? The speed of the marble increases after it reaches the top. What happens to the velocity?

At the point A, the velocity and gradient of the position–time graph are zero. We say the marble is *instantaneously at rest*. The velocity at H is positive because the marble is moving in the positive direction (upwards). The velocity at B and at C is negative because the marble is moving in the negative direction (downwards).

1.4 Acceleration

In everyday language, the word 'accelerate' is usually used when an object speeds up and 'decelerate' when it slows down. The idea of deceleration is sometimes used in a similar way by mathematicians but in mathematics the word *acceleration* is used whenever there is a change in velocity, whether an object is speeding up, slowing down or changing direction. Acceleration is *the rate at which the velocity changes*.

Over a period of time

$$average\ acceleration = \frac{change\ in\ velocity}{time}$$

Acceleration is represented by the gradient of a velocity–time graph. It is a vector and can take different signs in a similar way to velocity. This is illustrated by Tom's cycle journey which is shown in figure 1.12.

Tom turns on to the main road at $4\ \text{ms}^{-1}$, accelerates uniformly, maintains a constant speed and then slows down uniformly to stop when he reaches home.

Between A and B, Tom's velocity increases by $(10 - 4) = 6\ \text{ms}^{-1}$ in 6 seconds, that is by 1 metre per second every second.

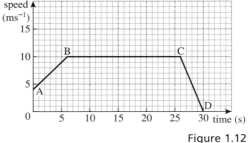

Figure 1.12

This acceleration is written as $1\ \text{ms}^{-2}$ (one metre per second squared) and is the gradient of AB.

From B to C   acceleration $= 0\ \text{ms}^{-2}$ ← ⎛There is no change in velocity.⎞

From C to D   acceleration $= \dfrac{(0 - 10)}{(30 - 26)} = -2.5\ \text{ms}^{-2}$

From C to D, Tom is slowing down while still moving in the positive direction towards home, so his acceleration, the gradient of the graph, is negative.

The sign of acceleration

Think again about the marble thrown up into the air with a speed of $5\,\mathrm{ms}^{-1}$.

Figure 1.13 represents the velocity when *upwards* is taken as the positive direction and shows that the velocity *decreases* from $+5\,\mathrm{ms}^{-1}$ to $-5\,\mathrm{ms}^{-1}$ in 1 second.

This means that the gradient, and hence the acceleration, is *negative*. It is $-10\,\mathrm{ms}^{-2}$. (You might recognise the number 10 as an approximation to g. See Chapter 1 page 3.)

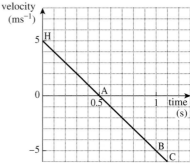

Figure 1.13

QUESTION 1.8 A car accelerates away from a set of traffic lights. It accelerates to a maximum speed and at that instant starts to slow down to stop at a second set of lights. Which of the graphs below could represent

i) the distance–time graph
ii) the velocity–time graph
iii) the acceleration–time graph of its motion?

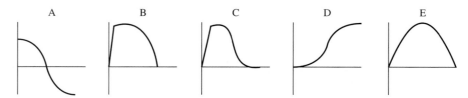

Figure 1.14

1.5 Using areas to find distances and displacements

These distance–time and speed–time graphs model the motion of a stone falling from rest.

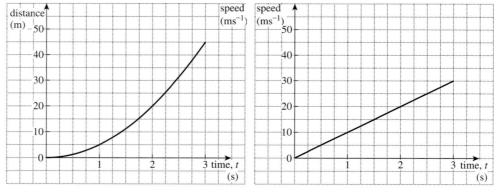

Figure 1.15   Figure 1.16

QUESTION 1.9   Calculate the area between the speed–time graph and the x axis from

i) $t = 0$ to 1   **ii)** $t = 0$ to 2   **iii)** $t = 0$ to 3.

Compare your answers with the distance that the stone has fallen, shown on the distance–time graph, at $t = 1$, 2 and 3. What conclusions do you reach?

> *The area between a speed–time graph and the x axis represents the distance travelled.*

There is further evidence for this if you consider the units on the graphs. Multiplying metres per second by seconds gives metres. A full justification relies on the calculus methods you will learn in Section 1.7.

Finding the area under speed–time graphs

Many of these graphs consist of straight-line sections. The area is easily found by splitting it up into triangles, rectangles or trapezia.

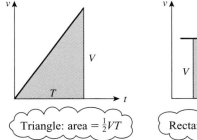

Triangle: area $= \frac{1}{2}VT$

Rectangle: area $= VT$

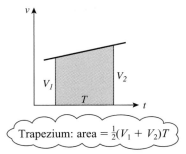

Trapezium: area $= \frac{1}{2}(V_1 + V_2)T$

Figure 1.17

EXAMPLE 1.1   The graph shows Tom's journey from the time he turns on to the main road until he arrives home. How far does Tom cycle?

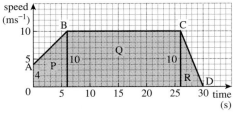

Figure 1.18

SOLUTION

The area under the speed–time graph is found by splitting it into three regions.

P	trapezium: area $= \frac{1}{2}(4 + 10) \times 6$	$=$	42 m
Q	rectangle: area $= 10 \times 20$	$=$	200 m
R	triangle: area $= \frac{1}{2} \times 10 \times 4$	$=$	20 m
	total area	$=$	262 m

Tom cycles 262 m.

QUESTION 1.10   What is the meaning of the area between a velocity–time graph and the x axis?

The area between a velocity–time graph and the *x* axis

EXAMPLE 1.2

David walks east for 6 s at $2\ \text{ms}^{-1}$ then west for 2s at $1\ \text{ms}^{-1}$. Draw

i) a diagram of the journey
ii) the speed–time graph
iii) the velocity–time graph.

Interpret the area under each graph.

SOLUTION

i) David's journey is illustrated below.

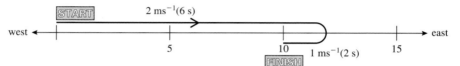

Figure 1.19

ii) Speed–time graph

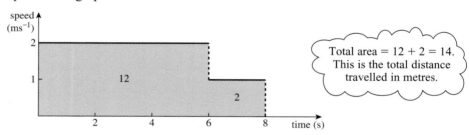

Total area $= 12 + 2 = 14$. This is the total distance travelled in metres.

Figure 1.20

iii) Velocity–time graph

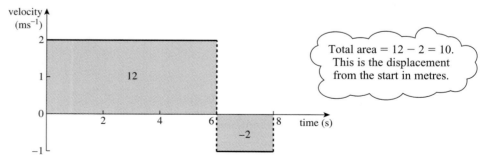

Total area $= 12 - 2 = 10$. This is the displacement from the start in metres.

Figure 1.21

> *The area between a velocity time graph and the x axis represents the change in position, that is the displacement.*

When the velocity is negative, the area is below the axis and represents a displacement in the negative direction, west in this case.

Estimating areas

Sometimes the velocity–time graph does not consist of straight lines so you have to make the best estimate you can by counting the squares underneath it or by replacing the curve by a number of straight lines as for the trapezium rule.

QUESTION 1.11   This speed–time graph shows the motion of a dog over a 60 s period.

Estimate how far the dog travelled during this time.

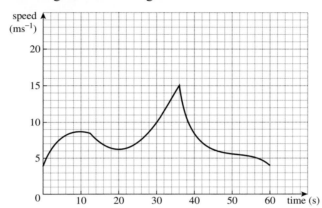

Figure 1.22

EXAMPLE 1.3   On the London underground, Oxford Circus and Piccadilly Circus are 0.8 km apart. A train accelerates uniformly to a maximum speed when leaving Oxford Circus and maintains this speed for 90 s before decelerating uniformly to stop at Piccadilly Circus. The whole journey takes 2 minutes. Find the maximum speed.

SOLUTION

The sketch of the speed–time graph of the journey shows the given information, with suitable units.
The maximum speed is $v\,\mathrm{ms}^{-1}$.

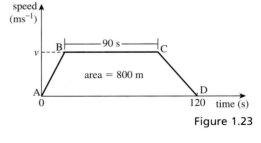

The area is $\frac{1}{2}(120 + 90) \times v = 800$

$$v = \frac{800}{105}$$

$$= 7.619$$

The maximum speed of the train is $7.6\,\mathrm{ms}^{-1}$.

Figure 1.23

QUESTION 1.12   Does it matter how long the train takes to speed up and slow down?

1.6 The constant acceleration formulae

The velocity–time graph shows part of the motion of a car on a fairground ride as it picks up speed. The graph is a straight line so the velocity increases at a constant rate and the car has a constant acceleration which is equal to the gradient of the graph.

The velocity increases from $4\,\text{ms}^{-1}$ to $24\,\text{ms}^{-1}$ in 10s so its acceleration is

$$\frac{24 - 4}{10} = 2\,\text{ms}^{-2}.$$

Figure 1.24

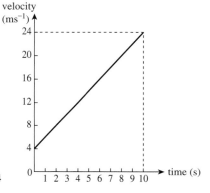

In general, when the initial velocity is $u\,\text{ms}^{-1}$ and the velocity a time t s later is $v\,\text{ms}^{-1}$, as in figure 1.25, the increase in velocity is $(v - u)\,\text{ms}^{-1}$ and the constant acceleration $a\,\text{ms}^{-2}$ is given by

$$\frac{v - u}{t} = a$$

so $v - u = at$

$$\boxed{v = u + at.}$$ ①

Figure 1.25

The area under the graph represents the distance travelled. For the fairground car, that is represented by a trapezium of area

$$\frac{(4 + 24)}{2} \times 10 = 140\,\text{m}.$$

In the general situation, the area represents the displacement s metres and is

$$\boxed{s = \frac{(u + v)}{2} \times t.}$$ ②

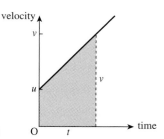

Figure 1.26

QUESTION 1.13 The two equations, ① and ②, can be used as formulae for solving problems when the acceleration is constant. Check that they work for the fairground ride.

There are other useful formulae as well. For example, you might want to find the displacement, s, without involving v in your calculations. This can be done by looking at the area under the velocity–time graph in a different way, using the rectangle R and the triangle T. In the diagram

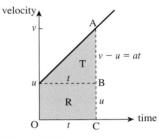

Figure 1.27

$$AC = v \text{ and } BC = u$$

so
$$AB = v - u$$

$$= at \qquad \text{from equation } ①$$

$$\text{total area} = \text{area of R} + \text{area of T}$$

so
$$s = ut + \tfrac{1}{2} \times t \times at$$

giving
$$s = ut + \tfrac{1}{2}at^2 \qquad ③$$

To find a formula which does not involve t, you need to eliminate t. One way to do this is first to rewrite equations ① and ② as

$$v - u = at \quad \text{and} \quad v + u = \frac{2s}{t}$$

and then multiplying them gives

$$(v - u)(v + u) = at \times \frac{2s}{t}$$

$$v^2 - u^2 = 2as$$

$$v^2 = u^2 + 2as. \qquad ④$$

You might have seen the equations ① to ④ before. They are sometimes called the *suvat* equations or formulae and they can be used whenever an object can be assumed to be moving with *constant acceleration*.

 When solving problems it is important to remember the requirement for constant acceleration and also to remember to specify positive and negative directions clearly.

EXAMPLE 1.4

A bus leaving a bus stop accelerates at 0.8 ms^{-2} for 5s and then travels at a constant speed for 2 minutes before slowing down uniformly at 4 ms^{-2} to come to rest at the next bus stop. Calculate

i) the constant speed

ii) the distance travelled while the bus is accelerating

iii) the total distance travelled.

SOLUTION

i) The diagram shows the information for the first part of the motion.

Let the constant speed be $v \text{ ms}^{-1}$.

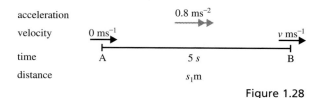

Figure 1.28

$u = 0, a = 0.8, t = 5$, so use $v = u + at$

$v = 0 + 0.8 \times 5$

$= 4$

> Want v
> know $u = 0, t = 5, a = 0.8$
> $v^2 = u^2 + 2as$ ✗ $v = u + at$ ✓

The constant speed is 4 ms^{-1}.

ii) Let the distance travelled be s_1 m.

> Use the suffix because there are three distances to be found in this question.

$u = 0, a = 0.8, t = 5$, so use $s = ut + \frac{1}{2}at^2$

$s_1 = 0 + \frac{1}{2} \times 0.8 \times 5^2$

$= 10$

> Want s
> know $u = 0, t = 5, a = 0.8$
> $s = \frac{1}{2}(u + v)t$ ✗
> $s = ut + \frac{1}{2}at^2$ ✓

The bus accelerates over 10 m.

iii) The diagram gives all the information for the rest of the journey.

> velocity decreases so acceleration is negative

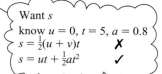

Figure 1.29

Between B and C the velocity is constant so the distance travelled is $4 \times 120 = 480$ m.

Let the distance between C and D be s_3 m.

$u = 4, a = -0.4, v = 0$, so use $v^2 = u^2 + 2as$

$0 = 16 + 2(-0.4)s_3$

$0.8s_3 = 16$

$s_3 = 20$

> Want s
> know $u = 4, a = -0.4, v = 0$
> $v = u + at$ ✗
> $s = ut + \frac{1}{2}at^2$ ✗
> $s = \frac{1}{2}(u + v)t$ ✗
> $v^2 = u^2 + 2as$ ✓

Distance taken to slow down = 20 m.

The total distance travelled is $(10 + 480 + 20) \text{ m} = 510$ m.

Units in the *suvat* equations

Constant acceleration usually takes place over short periods of time so it is best to use ms^{-2} for this. When you don't need to use a value for the acceleration you can, if you wish, use the *suvat* equations with other units provided they are consistent. This is shown in the next example.

EXAMPLE 1.5   When leaving a town, a car accelerates from 30 mph to 60 mph in 5 s. Assuming the acceleration is constant, find the distance travelled in this time.

SOLUTION

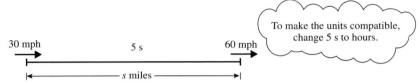

To make the units compatible, change 5 s to hours.

Figure 1.30

Let the distance travelled be s miles. You want s and are given $u = 30$, $v = 60$ and $t = 5 \div 3600$ so you need an equation involving u, v, t and s.

$$s = \frac{(u + v)}{2} \times t$$

$$s = \frac{(30 + 60)}{2} \times \frac{5}{3600}$$

$$= \frac{1}{16}$$

The distance travelled is $\frac{1}{16}$ mile or 110 yards. (One mile is 1760 yards.)

QUESTION 1.14   Write all the measurements in example 1.5 in terms of metres and seconds and then find the resulting distance. (110 yards is about 100 m.)

 In examples 1.4 and 1.5, the bus and the car are always travelling in the positive direction so it is safe to use s for distance. Remember that s is not the same as the distance travelled if the direction changes during the motion.

The acceleration due to gravity

When a model ignoring air resistance is used, all objects falling freely under gravity fall with the same constant acceleration, g ms^{-2}. This varies over the surface of the earth. In this book it is assumed that all the situations occur in a place where it is 9.8 ms^{-2} or sometimes 10 ms^{-2} as an approximation. Most answers are given correct to three significant figures so that you can check your working.

EXAMPLE 1.6

A coin is dropped from rest at the top of a building of height 12 m and travels in a straight line with constant acceleration 10 ms^{-2}.
Find the time it takes to reach the ground and the speed of impact.

SOLUTION

Suppose the time taken to reach the ground is t seconds. Using S.I. units, $u = 0$, $a = 10$ and $s = 12$ when the coin hits the ground, so you need to use a formula involving u, a, s and t.

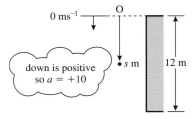

down is positive so $a = +10$

$$s = ut + \tfrac{1}{2}at^2$$
$$12 = 0 + \tfrac{1}{2} \times 10 \times t^2$$
$$t^2 = 2.4$$
$$t = 1.55$$

Figure 1.31

To find the velocity, v, a formula involving s, u, a and v is required.

$$v^2 = u^2 + 2as$$
$$v^2 = 0 + 2 \times 10 \times 12$$
$$v^2 = 240$$
$$v = 15.5$$

The coin takes 1.55 s to hit the ground and has speed 15.5 ms^{-1} on impact.

Summary of *suvat* formulae

The equations for motion with constant acceleration are

① $v = u + at$ ② $s = \dfrac{(u + v)}{2} \times t$

③ $s = ut + \tfrac{1}{2}at^2$ ④ $v^2 = u^2 + 2as$.

QUESTION 1.15   Derive equation ③ algebraically by substituting for v from equation ① into equation ②.

If you look at these equations you will see that each omits one variable. But there are five variables and only four equations; there isn't one without u. An equation omitting u is

⑤ $s = vt - \tfrac{1}{2}at^2$.

QUESTION 1.16   How can you derive this by referring to a graph or using substitution?

 When using these equations make sure that the units you use are consistent. For example, when the time is t seconds and the distance s metres, any speed involved is in ms^{-1}.

Further examples

The next two examples illustrate ways of dealing with more complex problems. In example 1.7, none of the possible equations has only one unknown and there are also two situations, so simultaneous equations are used.

EXAMPLE 1.7

James practises using the stopwatch facility on his new watch by measuring the time between lamp posts on a car journey. As the car speeds up, two consecutive times are 1.2 s and 1 s. Later he finds out that the lamp posts are 30 m apart.

i) Calculate the acceleration of the car (assumed constant) and its speed at the first lamp post.

ii) Assuming the same acceleration, find the time the car took to travel the 30 m before the first lamp post.

SOLUTION

i) The diagram shows all the information assuming the acceleration is a ms^{-2} and the velocity at A is u ms^{-1}.

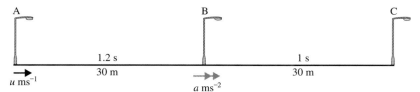

Figure 1.32

For AB, $s = 30$ and $t = 1.2$. You are using u and you want a so you use

$$s = ut + \tfrac{1}{2}at^2$$
$$30 = 1.2u + \tfrac{1}{2}a \times 1.2^2$$
$$30 = 1.2u + 0.72a \qquad\qquad ①$$

To use the same equation for the part BC you would need the velocity at B and this brings in another unknown. It is much better to go back to the beginning and consider the whole of AC with $s = 60$ and $t = 2.2$. Then again using $s = ut + \tfrac{1}{2}at^2$

$$60 = 2.2u + \tfrac{1}{2}a \times 2.2^2$$
$$60 = 2.2u + 2.42a \qquad\qquad ②$$

These two simultaneous equations in two unknowns can be solved more easily if they are simplified. First make the coefficients of u integers.

$$① \times 10 \div 12 \qquad 25 = u + 0.6a \qquad\qquad ③$$
$$② \times 5 \qquad 300 = 11u + 12.1a \qquad\qquad ④$$
then $\qquad ③ \times 11 \qquad 275 = 11u + 6.6a \qquad\qquad ⑤$

Subtracting gives

$$25 = 0 + 5.5a$$
$$a = 4.545$$

Now substitute 4.545 for a in ③ to find

$$u = 25 - 0.6 \times 4.545 = 22.273.$$

The acceleration of the car is 4.55 ms^{-2} and the initial speed is 22.3 ms^{-1} (correct to 3 s.f.).

ii)

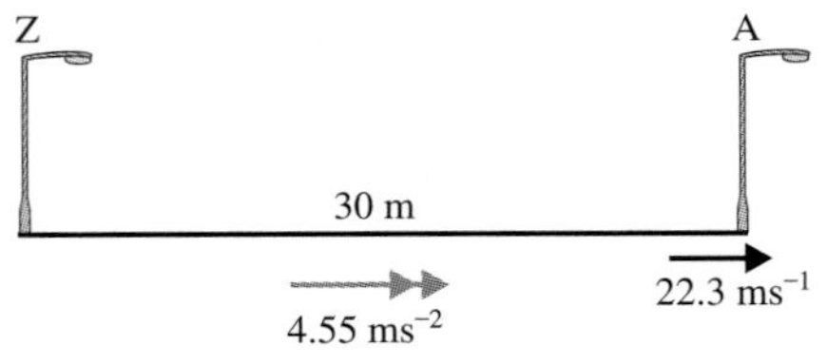

Figure 1.33

For this part, you know that $s = 30$, $v = 22.3$ and $a = 4.55$ and you want t so you use the fifth formula.

$$s = vt - \tfrac{1}{2}at^2$$

$$30 = 22.3 \times t - \tfrac{1}{2} \times 4.55 \times t^2$$

$$\Rightarrow \qquad 2.275t^2 - 22.3t + 30 = 0$$

Solving this using the quadratic formula gives $t = 1.61$ and $t = 8.19$.

The most sensible answer to this particular problem is 1.61 s.

QUESTION 1.17   Calculate u when $t = 8.19$, $v = 22.3$ and $a = 4.55$. Is $t = 8.19$ a possible answer?

Using a non-zero initial displacement

What, in the constant acceleration equations, are v and s when $t = 0$?

Putting $t = 0$ in the *suvat* equations gives the *initial values*, u for the velocity and $s = 0$ for the position.

Sometimes, however, it is convenient to use an origin which gives a non-zero value for s when $t = 0$. For example, when you model the motion of a rubber thrown vertically upwards you might decide to find its height above the ground rather than from the point from which it was thrown.

What is the effect on the various *suvat* equations if the initial position is s_0 rather than zero?

If the height of the rubber above the ground is s at time t and s_0 when $t = 0$, the displacement over time t is $s - s_0$. You then need to replace equation ③ with

$$s - s_0 = ut + \tfrac{1}{2}at^2$$

The next example avoids this in the first part but it is very useful in part **ii)**.

EXAMPLE 1.8

A juggler throws a ball up in the air with initial speed $5\ \mathrm{ms^{-1}}$ from a height of $1.2\ \mathrm{m}$. It has a constant acceleration of $10\ \mathrm{ms^{-2}}$ vertically downwards due to gravity.

i) Find the maximum height of the ball above the ground and the time it takes to reach it.

At the instant that the ball reaches its maximum height, the juggler throws up another ball with the same speed and from the same height.

ii) Where and when will the balls pass each other?

SOLUTION

i) In this example it is very important to draw a diagram and to be clear about the position of the origin. When O is $1.2\ \mathrm{m}$ above the ground and s is the height in metres above O after t s, the diagram looks like figure 1.34.

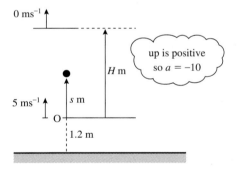

Figure 1.34

At the point of maximum height, let $s = H$ and $t = t_1$. ◄———— *Use the suffix because there are two times to be found in this question.*

The ball stops instantaneously before falling so at the top $v = 0$.

An equation involving u, v, a and s is required.

$$v^2 = u^2 + 2as$$

$$0 = 5^2 + 2 \times (-10) \times H$$ ◄———— *The acceleration given is constant, $a = -10$; $u = +5$; $v = 0$ and $s = H$.*

$$H = 1.25$$

The maximum height of the ball above the ground is $1.25 + 1.2 = 2.45\ \mathrm{m}$.

To find t_1, given $v = 0$, $a = -10$ and $u = +5$ requires a formula in v, u, a and t.

$$v = u + at$$
$$0 = 5 + (-10)t_1$$
$$t_1 = 0.5$$

The ball takes half a second to reach its maximum height.

ii) Now consider the motion from the instant the first ball reaches the top of its path and the second is thrown up.

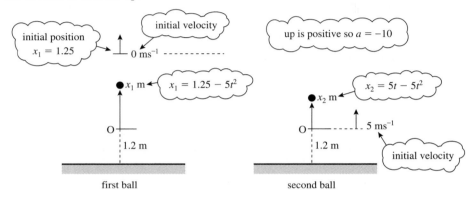

Figure 1.35

Suppose that the balls have displacements above the origin of x_1 m and x_2 m, as shown in the diagram, at a general time t s after the second ball is thrown up. The initial position of the second ball is zero, but the initial position of the first ball is $+1.25$ m.

For each ball you know u and a. You want to involve t and s so you use
$$s - s_0 = ut + \tfrac{1}{2}at^2$$

i.e. $s = s_0 + ut + \tfrac{1}{2}at^2$

For the first ball:

$$x_1 = 1.25 + 0 \times t + \tfrac{1}{2} \times (-10) \times t^2$$
$$x_1 = 1.25 - 5t^2 \quad \text{①}$$

> This makes $x_1 = 1.25$ when $t = 0$.

> x_1 decreases as t increases.

For the second ball:

$$x_2 = 0 + 5 \times t + \tfrac{1}{2} \times (-10) \times t^2$$
$$x_2 = 5t - 5t^2 \quad \text{②}$$

Suppose the balls pass after a time t s. This is when they are at the same height, so equate x_1 and x_2 from equations ① and ②.

$$1.25 - 5t^2 = 5t - 5t^2$$
$$1.25 = 5t$$
$$t = 0.25$$

Then substituting $t = 0.25$ in ① and ② gives

$$x_1 = 1.25 - 5 \times 0.25^2 = 0.9375$$

and

$$x_2 = 5 \times 0.25 - 5 \times 0.25^2 = 0.9375.$$

> These are the same, as expected.

The balls pass after 0.25 seconds at a height of $1.2 + 0.94\,\text{m} = 2.14\,\text{m}$ above the ground (correct to the nearest centimetre).

Note

The balls pass after half the time to reach the top, but not half-way up.

QUESTION 1.18   Why don't they travel half the distance in half the time?

1.7 Motion in one dimension (variable acceleration)

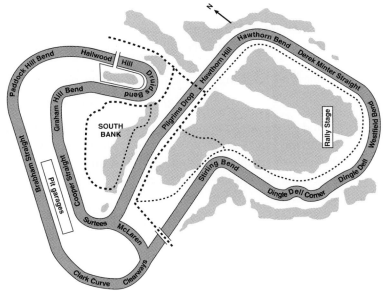

Figure 1.36

So far you have studied motion with constant acceleration in a straight line, but the motion of a car round the Brand's Hatch racing circuit shown in the diagram is much more complex.

The equations you have used for constant acceleration do not apply when the acceleration varies. You need to go back to first principles.

Consider how displacement, velocity and acceleration are related to each other. The velocity of an object is the rate at which its position changes with time. When the velocity is not constant the position–time graph is a curve.

The rate of change of the position is the gradient of the tangent to the curve. You can find this by differentiating.

$$v = \frac{\mathrm{d}s}{\mathrm{d}t} \qquad ①$$

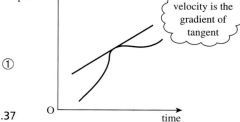

velocity is the gradient of tangent

Figure 1.37

Similarly, the acceleration is the rate at which the velocity changes, so

$$a = \frac{\mathrm{d}v}{\mathrm{d}t} = \frac{\mathrm{d}^2 s}{\mathrm{d}t^2}. \qquad ②$$

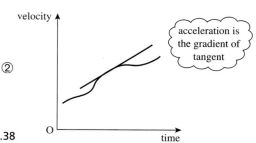

acceleration is the gradient of tangent

Figure 1.38

Using differentiation

When you are given the position of a moving object in terms of time, you can use equations ① and ② to solve problems even when the acceleration is not constant.

EXAMPLE 1.9

An object moves along a straight line so that its position at time t in seconds is given by

$$x = 2t^3 - 6t \text{ (in metres)} \quad (t \geqslant 0).$$

i) Find expressions for the velocity and acceleration of the object at time t.
ii) Find the values of x, v and a when $t = 0$, 1, 2 and 3.
iii) Sketch the graphs of x, v and a against time.
iv) Describe the motion of the object.

SOLUTION

i) Position $\qquad x = 2t^3 - 6t \qquad\qquad\qquad ①$

Velocity $\qquad v = \dfrac{\mathrm{d}x}{\mathrm{d}t} = 6t^2 - 6 \qquad\qquad ②$

Acceleration $\quad a = \dfrac{\mathrm{d}v}{\mathrm{d}t} = 12t \qquad\qquad ③$

You can now use these three equations to solve problems about the motion of the object.

ii) When

$t =$	0	1	2	3
From ① $x =$	0	−4	4	36
From ② $v =$	−6	0	18	48
From ③ $a =$	0	12	24	36

iii) The graphs are drawn under each other so that you can see how they relate.

iv) The object starts at the origin and moves towards the negative direction, gradually slowing down.

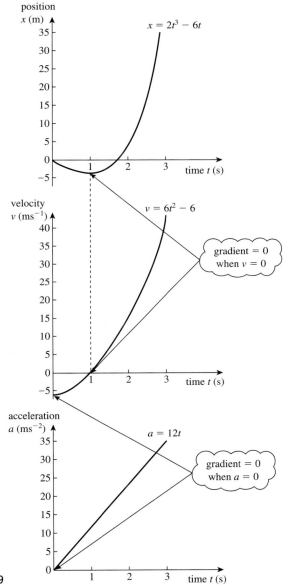

At $t = 1$ it stops instantaneously and changes direction, returning to its initial position at about $t = 1.7$.

It then continues moving in the positive direction with increasing speed.

The acceleration is increasing at a constant rate. This cannot go on for much longer or the speed will become excessive.

Figure 1.39

Finding displacement from velocity

How can you find an expression for the position of an object when you know its velocity in terms of time?

One way of thinking about this is to remember that $v = \dfrac{ds}{dt}$, so you need to do the opposite of differentiation, that is integrate, to find s.

$$s = \int v\, dt$$

The dt indicates that you must write v in terms of t before integrating.

EXAMPLE 1.10   The velocity (in ms^{-1}) of a model train which is moving along straight rails is

$$v = 0.3t^2 - 0.5$$

Find its displacement from its initial position

i) after time t
ii) after 3 seconds.

SOLUTION

i) The displacement at any time is $s = \int v \, dt$

$$= \int (0.3t^2 - 0.5) \, dt$$
$$= 0.1t^3 - 0.5t + c$$

To find the train's displacement from its initial position, put $s = 0$ when $t = 0$.

This gives $c = 0$ and so $s = 0.1t^3 - 0.5t$.

You can use this equation to find the displacement at any time before the motion changes.

ii) After 3 seconds, $t = 3$ and $s = 2.7 - 1.5$.

The train is 1.2 m from its initial position.

⚠️   When using integration don't forget the constant. This is very important in mechanics problems and you are usually given some extra information to help you find the value of the constant.

The area under a velocity–time graph

In Section 5 you saw that the area under a velocity–time graph represents a displacement. Both the area under the graph and the displacement are found by integrating. To find a particular displacement you calculate the area under the velocity–time graph by integration using suitable limits.

The distance travelled between the times T_1 and T_2 is shown by the shaded area on the graph.

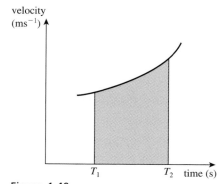
Figure 1.40

$$s = \text{area} = \int_{T_1}^{T_2} v \, dt$$

EXAMPLE 1.11   A car moves between two sets of traffic lights, stopping at both. Its speed v ms^{-1} at time t s is modelled by

$$v = \frac{1}{20} t (40 - t), \quad 0 \leqslant t \leqslant 40.$$

Find the times at which the car is stationary and the distance between the two sets of traffic lights.

SOLUTION

The car is stationary when $v = 0$. Substituting this into the expression for the speed gives

$$0 = \frac{1}{20} t (40 - t)$$

$$t = 0 \quad \text{or} \quad t = 40.$$

These are the times when the car starts to move away from the first set of traffic lights and stops at the second set.

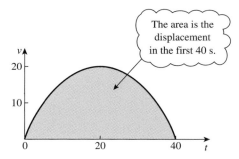

The area is the displacement in the first 40 s.

Figure 1.41

The distance between the two sets of lights is given by

$$\text{Distance} = \int_{0}^{40} \frac{1}{20} t (40 - t) \, dt$$

$$\text{Distance} = \frac{1}{20} \int_{0}^{40} 40t - t^2 \, dt$$

$$\text{Distance} = \frac{1}{20} \left[20t^2 - \frac{t^3}{3} \right]_{0}^{40}$$

$$= 533.\dot{3} \, \text{m}$$

Finding velocity from acceleration

You can also find the velocity from the acceleration by using integration.

$$a = \frac{dv}{dt}$$

$$\Rightarrow \quad v = \int a \, dt$$

The next example shows how you can obtain equations for motion using integration.

EXAMPLE 1.12

The acceleration of a particle (in ms^{-2}) at time t seconds is given by

$$a = 6 - t.$$

The particle is initially at the origin with velocity -2 ms^{-1}. Find an expression for

i) the velocity of the particle after t s
ii) the position of the particle after t s.

Hence find the velocity and position 6 s later.

SOLUTION

The information given may be summarised as follows:

at $t = 0$, $s = 0$ and $v = -2$;

at time t, $a = 6 - t$.   ①

i) $\dfrac{dv}{dt} = a = 6 - t$

Integrating gives

$$v = 6t - \tfrac{1}{2}t^2 + c.$$

When $t = 0$, $v = -2$

so $-2 = 0 - 0 + c$

$c = -2$

At time t

$$v = 6t - \tfrac{1}{2}t^2 - 2. \qquad ②$$

ii) $\dfrac{ds}{dt} = v = 6t - \dfrac{1}{2}t^2 - 2$

Integrating gives

$$s = 3t^2 - \tfrac{1}{6}t^3 - 2t + k.$$

When $t = 0$, $s = 0$

so $0 = 0 - 0 - 0 + k$

$k = 0.$

At time t

$$s = 3t^2 - \tfrac{1}{6}t^3 - 2t. \qquad ③$$

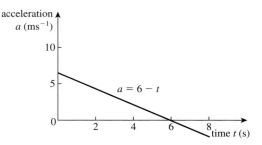

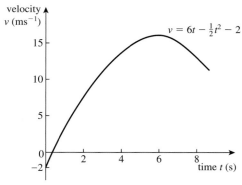

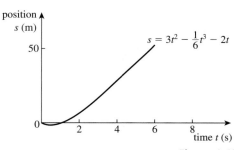

Figure 1.42

Notice that two different arbitrary constants (c and k) are necessary when you integrate twice. You could call them c_1 and c_2 if you wish.

The three numbered equations can now be used to give more information about the motion in a similar way to the *suvat* equations. (The *suvat* equations only apply when the acceleration is constant.)

When $t = 6$ $v = 36 - 18 - 2 = 16$ from ②

When $t = 6$ $s = 108 - 36 - 12 = 60$ from ③

The particle has a velocity of $+16\,\text{ms}^{-1}$ and is at $+60\,\text{m}$ after 6 s.

The constant acceleration equations revisited

You can use integration to prove the equations for constant acceleration. When a is constant (and only then)

$$v = \int a\,\mathrm{d}t = at + c_1.$$

u and a are both constant

When $t = 0$, $v = u$ $u = 0 + c_1$

$$\Rightarrow \quad v = u + at. \tag{①}$$

You can integrate this again to find $s = ut + \frac{1}{2}at^2 + c_2.$

If $s = s_0$ when $t = 0$, $c_2 = s_0$ and $s = ut + \frac{1}{2}at^2 + s_0.$ ②

QUESTION 1.19   How can you use these to derive the other equations for constant acceleration?

$s = \frac{1}{2}(u + v)\,t + s_0$ ③

$v^2 - u^2 = 2a\,(s - s_0)$ ④

$s = vt - \frac{1}{2}at^2 + s_0$ ⑤

Chapter 1 Exercises

1.1 A boy throws a ball vertically upwards so that its position y m at time t is as shown in the graph.
   i) Write down the position of the ball at times $t = 0$, 0.4, 0.8, 1.2, 1.6 and 2.
   ii) Calculate the displacement of the ball relative to its starting position at these times.
   iii) What is the total distance travelled
      a) during the first 0.8 s   **b)** during the 2 s of the motion?

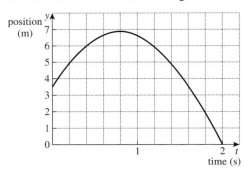

1.2 The position of a particle moving along a straight horizontal groove is given by $x = 2 + t(t - 3)$ ($0 \leqslant t \leqslant 5$) where x is measured in metres and t in seconds.
   i) What is the position of the particle at times $t = 0$, 1, 1.5, 2, 3, 4 and 5?
   ii) Draw a diagram to show the path of the particle, marking its position at these times.
   iii) Find the displacement of the particle relative to its initial position at $t = 5$.
   iv) Calculate the total distance travelled during the motion.

1.3 Draw a speed–time graph for Amy's journey on page 6.

1.4 The distance–time graph shows the relationship between distance travelled and time for a person who leaves home at 9.00 am, walks to a bus stop and catches a bus into town.

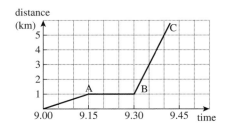

   i) Describe what is happening during the time from A to B.
   ii) The section BC is much steeper than OA; what does this tell you about the motion?
   iii) Draw the speed–time graph for the person.
   iv) What simplifications have been made in drawing these graphs?

1.5 **i)** Calculate the acceleration for each part of the following journey.

   ii) Use your results to sketch an acceleration–time graph.

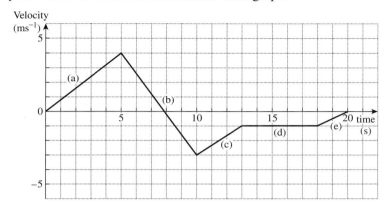

1.6 A particle moves so that its position x metres at time t seconds is $x = 2t^3 - 18t$.

   i) Calculate the position of the particle at times $t = 0, 1, 2, 3$ and 4.

   ii) Draw a diagram showing the position of the particle at these times.

   iii) Sketch a graph of the position against time.

   iv) State the times when the particle is at the origin and describe the direction in which it is moving at those times.

1.7 The graphs show the speeds of two cars travelling along a street.

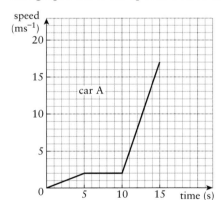
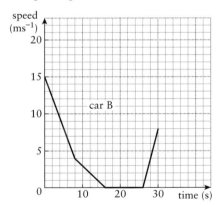

   For each car find

   i) the acceleration for each part of its motion

   ii) the total distance it travels in the given time

   iii) its average speed.

1.8 A car is moving at $20 \, \text{ms}^{-1}$ when it begins to increase speed. Every $10 \, \text{s}$ it gains $5 \, \text{ms}^{-1}$ until it reaches its maximum speed of $50 \, \text{ms}^{-1}$ which it retains.

   i) Draw the speed–time graph of the car.

   ii) When does the car reach its maximum speed of $50 \, \text{ms}^{-1}$?

   iii) Find the distance travelled by the car after $150 \, \text{s}$.

   iv) Write down expressions for the speed of the car t seconds after it begins to speed up.

1.9   Assuming no air resistance, a ball has an acceleration of $9.8 \, ms^{-2}$ when it is dropped from a window (so its initial speed, when $t = 0$, is zero). Calculate:
  i)   its speed after 1 s and after 10 s
  ii)   how far it has fallen after 1 s and after 10 s
  iii)   how long it takes to fall 19.6 m.

Which of these answers are likely to need adjusting to take account of air resistance? Would you expect your answer to be an over- or underestimate?

1.10 A top sprinter accelerates from rest to $9 \, ms^{-1}$ in 2 s. Calculate his acceleration, assumed constant, during this period and the distance travelled.

Use $g = 9.8 \, ms^{22}$ in exercises **1.11–1.13** *unless otherwise specified.*

1.11 Towards the end of a half-marathon Sabina is 100 m from the finish line and is running at a constant speed of $5 \, ms^{-1}$. Daniel, who is 140 m from the finish and is running at $4 \, ms^{-1}$, decides to accelerate to try to beat Sabina. If he accelerates uniformly at $0.25 \, ms^{-2}$ does he succeed?

1.12 Nathan hits a tennis ball straight up into the air from a height of 1.25 m above the ground. The ball hits the ground after 2.5 seconds. Assuming $g = 10 \, ms^{-2}$, find
  i)   the speed Nathan hits the ball
  ii)   the greatest height above the ground reached by the ball
  iii)   the speed the ball hits the ground
  iv)   how high the ball bounces if it loses 0.2 of its speed on hitting the ground.
  v)   Is your answer to part **i)** likely to be an over- or underestimate given that you have ignored air resistance?

1.13

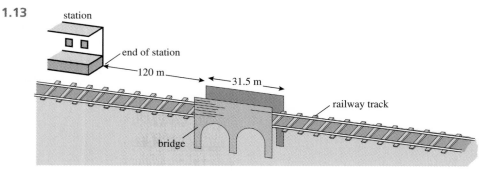

A train accelerates along a straight, horizontal section of track. The driver notes that he reaches a bridge 120 m from the station in 8 s and that he crosses the bridge, which is 31.5 m long, in a further 2 s.

The motion of the train is modelled by assuming constant acceleration. Take the speed of the train when leaving the station to be $u \, ms^{-1}$ and the acceleration to have the value $a \, ms^{-2}$.
  i)   By considering the part of the journey from the station to the bridge, show that $u + 4a = 15$.
  ii)   Find a second equation involving u and a.
  iii)   Solve the two equations for u and a to show that a is 0.15 and find the value of u.
  iv)   If the driver also notes that he travels 167 m in the 10 s after he crosses the bridge, have you any evidence to reject the modelling assumption that the acceleration is constant?

1.14 In each of the following cases
  i) $v = 4t + 3$
  ii) $v = 6t^2 - 2t + 1$
  iii) $v = 7t - 5$
    a) find expressions for the acceleration
    b) use your equations to write down the initial velocity and acceleration.

1.15 Until it stops moving, the speed of a bullet t s after entering water is modelled by
$v = 216 - t^3$ (in ms^{-1}).
  i)   When does the bullet stop moving?
  ii) How far has it travelled by this time?

1.16 Nick watches a golfer putting her ball 24 m from the edge of the green and into the hole and he decides to model the motion of the ball. Assuming that the ball is a particle travelling along a straight line he models its distance, s metres, from the golfer at time t seconds by

$$s = -\tfrac{3}{2}t^2 + 12t \quad 0 \leqslant t \leqslant 4.$$

  i)   Find the value of s when $t = 0, 1, 2, 3$ and 4.
  ii)   Explain the restriction $0 \leqslant t \leqslant 4$.
  iii) Find the velocity of the ball at time t seconds.
  iv)   With what speed does the ball enter the hole?
  v)   Find the acceleration of the ball at time t seconds.

Personal Tutor Exercises

1.1 A train was scheduled to travel at 50 ms^{-1} for 15 minutes on part of its journey. The velocity–time graph illustrates the actual progress of the train which was forced to stop because of signals.

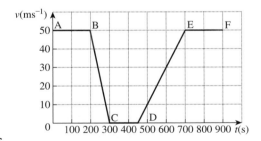

  i)   Without carrying out any calculations, describe what was happening to the train in each of the stages BC, CD and DE.
  ii) Find the deceleration of the train while it was slowing down and the distance travelled during this stage.
  iii) Find the acceleration of the train when it starts off again and the distance travelled during this stage.
  iv) Calculate by how long the stop will have delayed the train.
  v)   Sketch the distance–time graph for the journey between A and F, marking the points A, B, C, D, E and F.

1.2 A bird leaves its nest for a short horizontal flight along a straight line and then returns. Michelle models its displacement, s metres, from the nesting box at time t seconds by

$$s = 25t - \tfrac{5}{2}t^2, \quad 0 \leqslant t \leqslant 10. \tag{A}$$

i) Find the value of s when $t = 2$.

ii) Explain the restriction $0 \leqslant t \leqslant 10$.

iii) Find the velocity, $v\,\text{ms}^{-1}$, of the bird at time t seconds.

iv) What is the greatest distance of the bird from the nesting box?

v) Michelle's treacher tells her that a better model would be

$$s = 10t^2 - 2t^3 + \tfrac{1}{10}t^4. \tag{B}$$

Compare the two models.

1.1   Using a mathematical model

● Make simplifying assumptions by deciding what is most relevant. For example: a car is a *particle* with no dimensions
a road is a *straight line* with one dimension
acceleration is constant.

● Define variables and set up equations.

● Solve the equations.

● Check that the answer is sensible. If not, think again.

1.2   Vectors (with magnitude and direction)   **Scalars** (magnitude only)

Vectors (with magnitude and direction)	**Scalars** (magnitude only)
Displacement	Distance
Position – displacement from a fixed origin	
Velocity – rate of change of position	Speed – magnitude of velocity
Acceleration – rate of change of velocity	
	Time

● *Vertical* is towards the centre of the earth; *horizontal* is perpendicular to vertical.

1.3 Diagrams

● Motion along a line can be illustrated vertically or horizontally (as shown)

Positive direction ⟶

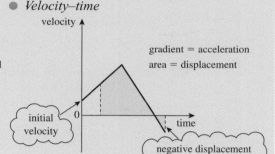

position

zero position

position − | position +

velocity

velocity +
⟶ ⟶

velocity −
⟵ ⟵

acceleration

	velocity $+$ ⟶	velocity $-$ ⟵
acceleration $+$ ⟹	speed increases	speed decreases
acceleration $-$ ⟸	speed decreases	speed increases

● *Average speed* $= \dfrac{\text{total distance travelled}}{\text{total time taken}}$

● *Average velocity* $= \dfrac{\text{displacement}}{\text{time taken}}$

1.4 Graphs

● *Position–time*

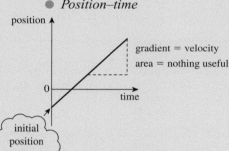

position

gradient = velocity
area = nothing useful

0 ⟶ time

initial position

● *Velocity–time*

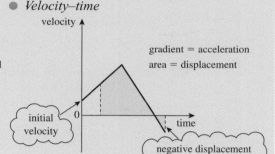

velocity

gradient = acceleration
area = displacement

0 ⟶ time

initial velocity

negative displacement

● *Distance–time*

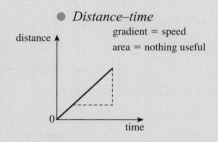

distance

gradient = speed
area = nothing useful

0 ⟶ time

● *Speed–time*

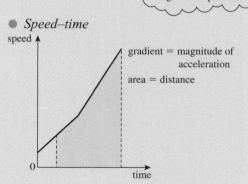

speed

gradient = magnitude of acceleration

area = distance

0 ⟶ time

1.5 The *suvat* equations

● The equations for motion with constant acceleration are

① $v = u + at$ ② $s = \dfrac{(u + v)}{2} \times t$ ③ $s = ut + \frac{1}{2}at^2$

④ $v^2 = u^2 + 2as$ ⑤ $s = vt - \frac{1}{2}at^2$

● a is the constant acceleration; s is the displacement from the starting position at time t; v is the velocity at time t; u is the velocity when $t = 0$. If $s = s_0$ when $t = 0$, replace s in each equation with $(s - s_0)$.

1.6 Vertical motion under gravity

- The acceleration due to gravity ($g\,\text{ms}^{-2}$) is $9.8\,\text{ms}^{-2}$ vertically downwards.
- Always draw a diagram and decide in advance where your origin is and which way is positive.
- Make sure that your units are compatible.

1.7 Relationships between the variables describing motion

	Position	$\rightarrow$	Velocity	$\rightarrow$	Acceleration
			differentiate	$\longrightarrow$	
In one dimension	s		$v = \dfrac{ds}{dt}$		$a = \dfrac{dv}{dt} = \dfrac{d^2s}{dt^2}$

	Acceleration	$\rightarrow$	Velocity	$\rightarrow$	Position
			integrate		
In one dimension	a		$v = \displaystyle\int a\,dt$		$s = \displaystyle\int v\,dt$

1.8 Acceleration may be due to change in direction or change in speed or both.

2 Forces and Newton's laws of motion

Nature to him was an open book. He stands before us, strong, certain and alone.

Einstein on Newton

2.1 Force diagrams

The picture shows a crate of medical supplies being dropped into a remote area by parachute. What forces are acting on the crate of supplies and the parachute?

One force which acts on every object near the earth's surface is its own *weight*. This is the force of gravity pulling it towards the centre of the earth. The weight of the crate acts on the crate and the weight of the parachute acts on the parachute.

The parachute is designed to make use of *air resistance*. A resistance force is present whenever a solid object moves through a liquid or gas. It acts in the opposite direction to the motion and depends on the speed of the object. The crate also experiences air resistance, but to a lesser extent than the parachute.

Other forces are the *tensions* in the guy lines attaching the crate to the parachute. These pull upwards on the crate and downwards on the parachute.

All these forces can be shown most clearly if you draw *force diagrams* for the crate and the parachute.

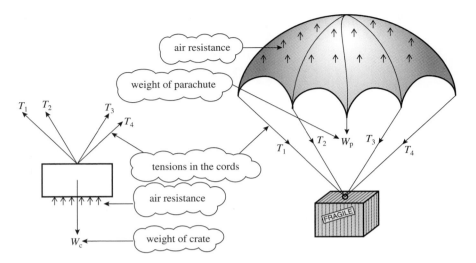

Figure 2.1: Forces acting on the crate   Figure 2.2: Forces acting on the parachute

Force diagrams are essential for the understanding of most mechanical situations. A force is a vector: it has a magnitude, or size, and a direction. It also has a *line of action*. This line often passes through a point of particular interest. Any force diagram should show clearly

- the direction of the force
- the magnitude of the force
- the line of action.

In figures 2.1 and 2.2 each force is shown by an arrow along its line of action. The air resistance has been depicted by a lot of separate arrows but this is not very satisfactory. It is much better if the combined effect can be shown by one arrow. When you have learned more about vectors, you will see how the tensions in the guy lines can also be combined into one force if you wish. The forces on the crate and parachute can then be simplified.

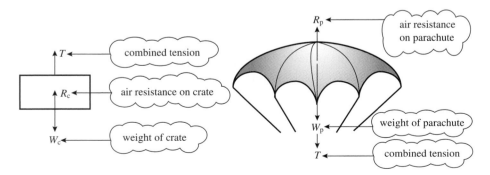

Figure 2.3: Forces acting on the crate   Figure 2.4: Forces acting on the parachute

Centre of mass and the particle model

When you combine forces you are finding their *resultant*. The weights of the crate and parachute are also found by combining forces; they are the resultant of the weights of all their separate parts. Each weight acts through a point called the *centre of mass* or centre of gravity.

Think about balancing a pen on your finger. The diagrams show the forces acting on the pen.

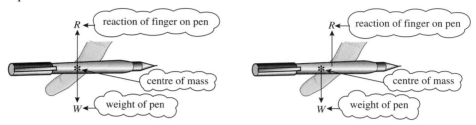

Figure 2.5   Figure 2.6

So long as you place your finger under the centre of mass of the pen, as in figure 2.5, it will balance. There is a force called a *reaction* between your finger and the pen which balances the weight of the pen. The forces on the pen are then said to be *in equilibrium*. If you place your finger under another point, as in figure 2.6, the pen will fall. The pen can only be in equilibrium if the two forces have the same line of action.

If you balance the pen on two fingers, there is a reaction between each finger and the pen at the point where it touches the pen. These reactions can be combined into one resultant vertical reaction acting through the centre of mass.

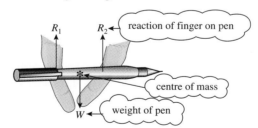

Figure 2.7

The behaviour of objects which are liable to rotate under the action of forces is covered later in this book. This chapter deals with situations where the resultant of the forces does not cause rotation. An object can then be modelled as a particle, that is a point mass, situated at its centre of mass.

Newton's third law of motion

Sir Isaac Newton (1642–1727) is famous for his work on gravity and the mechanics you learn in this book is often called Newtonian Mechanics because it is based entirely on Newton's three laws of motion. These laws provide us with an extremely powerful model of how objects, ranging in size from specks of dust to planets and stars, behave when they are influenced by forces.

We start with Newton's *third law* which says that

> *When one object exerts a force on another there is always a reaction of the same kind which is equal, and opposite in direction, to the acting force.*

You might have noticed that the combined tensions acting on the parachute and the crate in figures 2.3 and 2.4 are both marked with the same letter, *T*. The crate applies a force on the parachute through the supporting guy lines and the parachute applies an equal and opposite force on the crate. When you apply a force to a chair by sitting on it, it responds with an equal and opposite force on you. Figure 2.8 shows the forces acting when someone sits on a chair.

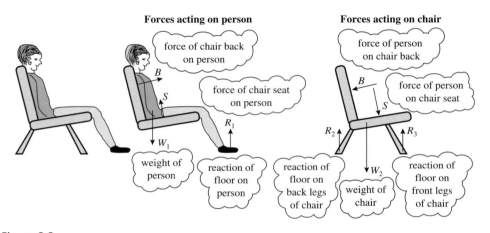

Figure 2.8

The reactions of the floor on the chair and on your feet act where there is contact with the floor. You can use R_1, R_2 and R_3 to show that they have different magnitudes. There are equal and opposite forces acting on the floor, but the forces on the floor are not being considered so do not appear here.

QUESTION 2.1 Why is the weight of the person not shown on the force diagram for the chair?

Gravitational forces obey Newton's third law just as other forces between bodies. According to Newton's universal law of gravitation, the earth pulls us towards its centre and we pull the earth in the opposite direction. However, in this book we are only concerned with the gravitational force on us and not the force we exert on the earth.

All the forces you meet in mechanics apart from the gravitational force are the result of physical contact. This might be between two solids or between a solid and a liquid or gas.

Friction and normal reaction

When you push your hand along a table, the table reacts in two ways.

● Firstly there are forces which stop your hand going through the table. Such forces are always present when there is any contact between your hand and the table. They are at right angles to the surface of the table and their resultant is called the *normal reaction* between your hand and the table.

● There is also another force which tends to prevent your hand from sliding. This is the *friction* and it acts in a direction which opposes the sliding.

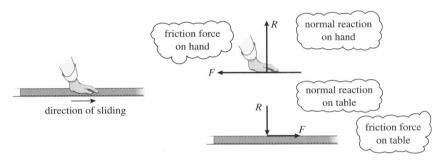

Figure 2.9

Figure 2.9 shows the reaction forces acting on your hand and on the table. By Newton's third law they are equal and opposite to each other. The frictional force is due to tiny bumps on the two surfaces (see electronmicrograph below). When you hold your hands together you will feel the normal reaction between them. When you slide them against each other you will feel the friction.

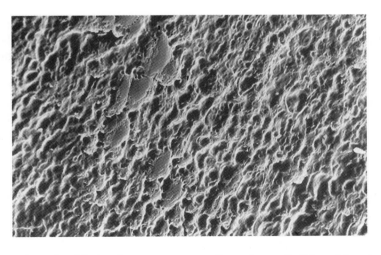

Etched glass magnified to high resolution, showing the tiny bumps.

When the friction between two surfaces is negligible, at least one of the surfaces is said to be *smooth*. This is a modelling assumption which you will meet frequently in this book. Oil can make surfaces smooth and ice is often modelled as a smooth surface.

> *When the contact between two surfaces is smooth, the only force between them is at right angles to any possible sliding and is just the normal reaction.*

QUESTION 2.2   What direction is the reaction between the sweeper's broom and the smooth ice?

EXAMPLE 2.1   A TV set is standing on a small table. Draw a diagram to show the forces acting on the TV and on the table as seen from the front.

SOLUTION

The diagram shows the forces acting on the TV and on the table. They are all vertical because the weights are vertical and there are no horizontal forces acting.

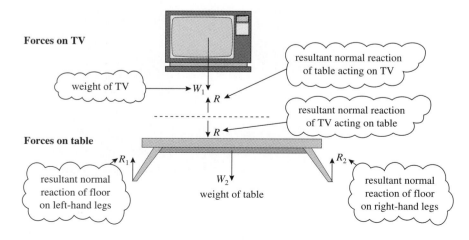

Figure 2.10

| EXAMPLE 2.2 | Draw diagrams to show the forces acting on a tennis ball which is hit downwards across the court: |

i) at the instant it is hit by the racket
ii) as it crosses the net
iii) at the instant it lands on the other side.

SOLUTION

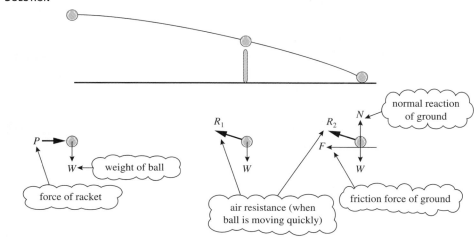

Figure 2.11

2.2 Force and motion

| QUESTION 2.3 | How are the rails and handles provided in buses and trains used by standing passengers? |

Newton's first law

Newton's *first law* can be stated as follows.

> *Every particle continues in a state of rest or uniform motion in a straight line unless acted on by a resultant external force.*

Newton's first law provides a reason for the handles on trains and buses. When you are on a train which is stationary or moving at constant speed in a straight line you can easily stand without support. But when the velocity of the train changes a force is required to change your velocity to match. This happens when the train slows down or speeds up. It also happens when the train goes round a bend even if the speed does not change. The velocity changes because the direction changes.

QUESTION 2.4   Why is Josh's car in the pond?

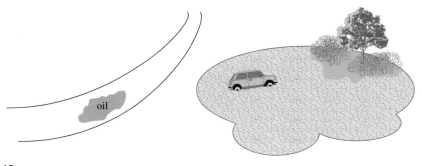

oil

Figure 2.12

EXAMPLE 2.3   A coin is balanced on your finger and then you move it upwards.

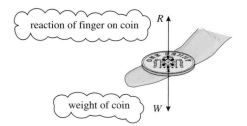

reaction of finger on coin R

weight of coin W

Figure 2.13

By considering Newton's first law, what can you say about W and R in these situations?

i)   The coin is stationary.
ii) The coin is moving upwards with a constant velocity.
iii) The speed of the coin is increasing as it moves upwards.
iv) The speed of the coin is decreasing as it moves upwards.

Solution

i) When the coin is stationary the velocity does not change. The forces are in equilibrium and $R = W$.

ii) When the coin is moving upwards with a constant velocity the velocity does not change. The forces are in equilibrium and $R = W$.

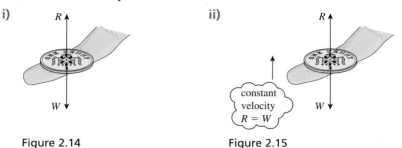

Figure 2.14   Figure 2.15

iii) When the speed of the coin is increasing as it moves upwards there must be a net upwards force to make the velocity increase in the upwards direction so $R > W$. The net force is $R - W$.

iv) When the speed of the coin is decreasing as it moves upwards there must be a net downwards force to make the velocity decrease and slow the coin down as it moves upwards. In this case $W > R$ and the net force is $W - R$.

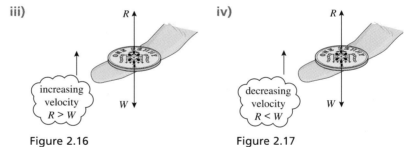

Figure 2.16   Figure 2.17

Driving forces and resistances to the motion of vehicles

In problems about such things as cycles, cars and trains, all the forces acting along the line of motion will usually be reduced to two or three, the *driving force* forwards, the *resistance* to motion (air resistance, etc.) and possibly a *braking force* backwards.

Resistances due to air or water always act in a direction opposite to the velocity of a vehicle or boat and are usually more significant for fast-moving objects.

Tension and thrust

The lines joining the crate of supplies to the parachute described at the beginning of this chapter are in tension. They pull upwards on the crate and downwards on the parachute. You are familiar with tensions in ropes and strings, but rigid objects can also be in tension.

When you hold the ends of a pencil, one with each hand, and pull your hands apart, you are pulling on the pencil. What is the pencil doing to each of your hands? Draw the forces acting on your hands and on the pencil.

Now draw the forces acting on your hands and on the pencil when you push the pencil inwards.

Your first diagram might look like figure 2.18. The pencil is in tension so there is an inward *tension* force on each hand.

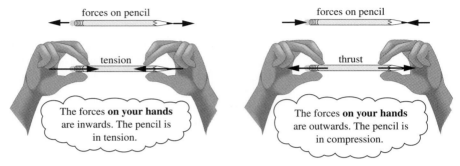

Figure 2.18   Figure 2.19

When you push the pencil inwards the forces on your hands are outwards as in figure 2.19. The pencil is said to be *in compression* and the outward force on each hand is called a *thrust*.

If each hand applies a force of 2 units on the pencil, the tension or thrust acting on each hand is also 2 units because each hand is in equilibrium.

QUESTION 2.5   Which of the above diagrams is still possible if the pencil is replaced by a piece of string?

Resultant forces and equilibrium

You have already met the idea that a single force can have the same effect as several forces acting together. Imagine that several people are pushing a car. A single rope pulled by another car can have the same effect. The force of the rope is equivalent to the resultant of the forces of the people pushing the car. When there is no resultant force, the forces are in equilibrium and there is no change in motion.

EXAMPLE 2.4   A car is using a tow bar to pull a trailer along a straight, level road. There are resisting forces R acting on the car and S acting on the trailer. The driving force of the car is D and its braking force is B.

Draw diagrams showing the horizontal forces acting on the car and the trailer

i)   when the car is moving at constant speed
ii) when the speed of the car is increasing
iii) when the car brakes and slows down rapidly.

In each case write down the resultant force acting on the car and on the trailer.

SOLUTION

i) When the car moves at constant speed, the forces are as shown in figure 2.20. The tow bar is in tension and the effect is a forward force on the trailer and an equal and opposite backwards force on the car.

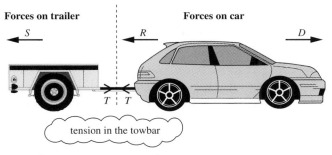

Figure 2.20: Car travelling at constant speed

There is no resultant force on either the car or the trailer when the speed is constant; the forces on each are in equilibrium.

For the trailer: $T - S = 0$

For the car: $D - R - T = 0$

ii) When the car speeds up, the same diagram will do, but now the magnitudes of the forces are different. There is a resultant *forwards* force on both the car and the trailer.

For the trailer: resultant $= T - S$

For the car: resultant $= D - R - T$

iii) When the car brakes a resultant *backwards* force is required to slow down the trailer. When the resistance S is not sufficiently large to do this, a thrust in the tow bar comes into play as shown in the figure 2.21.

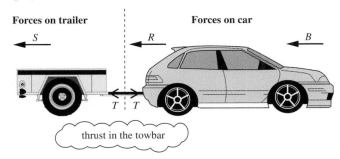

Figure 2.21: Car braking

For the trailer: resultant $= T + S$

For the car: resultant $= B + R - T$

Newton's second law

Newton's *second law* gives us more information about the relationship between the magnitude of the resultant force and the change in motion. Newton said that

The change in motion is proportional to the force.

For objects with constant mass, this can be interpreted as *the force is proportional to the acceleration*.

$$\text{Resultant force} = \text{a constant} \times \text{acceleration} \qquad \text{①}$$

The constant in this equation is proportional to the mass of the object; a more massive object needs a larger force to produce the same acceleration. For example, you and your friends would be able to give a car a greater acceleration than you would be able to give a lorry.

Newton's second law is so important that a special unit of force, the *newton* (*N*), has been defined so that the constant in equation ① is actually equal to the mass. A force of 1 newton will give a mass of 1 kilogram an acceleration of $1\ \text{ms}^{-2}$. The equation then becomes:

$$\text{Resultant force} = \text{mass} \times \text{acceleration} \qquad \text{②}$$

This is written: $F = ma.$

The resultant force and the acceleration are always in the same direction.

Relating mass and weight

The *mass* of an object is related to the amount of matter in the object. It is a *scalar*. The *weight* of an object is a force. It has magnitude and direction and so is a *vector*.

The mass of an astronaut on the moon is the same as his mass on the earth but his weight is only about one-sixth of his weight on the earth. This is why he can bounce around more easily on the moon. The gravitational force on the moon is less because the mass of the moon is less than that of the earth.

When Buzz Aldrin made the first landing on the moon in 1969 with Neil Armstrong, one of the first things he did was to drop a feather and a hammer to demonstrate that they fell at the same rate. Their accelerations due to the gravitational force of the moon were equal, even though they had very different masses. The same is true on earth. If other forces were negligible all objects would fall with an acceleration g.

When the weight is the only force acting on an object, Newton's second law means that

$$\text{Weight in newtons} = \text{mass in kg} \times g \text{ in ms}^{-2}.$$

Using standard letters: $W = mg.$

Even when there are other forces acting, the weight can still be written as *mg*.

A good way to visualise a force of 1 N is to think of the weight of an apple. 1 kg of apples weighs (1×9.8) N that is approximately 10 N. There are about 10 small to medium-sized apples in 1 kg, so each apple weighs about 1 N.

Note

Anyone who says 1 kg of apples *weighs* 1 kg is not strictly correct. The terms weight and mass are often confused in everyday language but it is very important for your study of mechanics that you should understand the difference.

EXAMPLE 2.5

What is the weight of

i)   a baby of mass 3 kg

ii) a golf ball of mass 46 g?

SOLUTION

i)   The baby's weight is $3 \times 9.8 = 29.4$ N.

ii) Mass of golf ball $= 46$ g

$$= 0.046 \text{ kg}$$
$$\text{Weight} = 0.046 \times 9.8 \text{ N}$$
$$= 0.45 \text{ N (to 2 s.f.)}.$$

QUESTION 2.6

Most weighing machines have springs or some other means to measure force even though they are calibrated to show mass. Would something appear to weigh the same on the moon if you used one of these machines? What could you use to find the mass of an object irrespective of where you measure it?

Pulleys

In the remainder of this section weight will be represented by mg. You will learn to apply Newton's second law more generally in the next section.

A pulley can be used to change the direction of a force; for example it is much easier to pull down on a rope than to lift a heavy weight. When a pulley is well designed it takes a relatively small force to make it turn and such a pulley is modelled as being *smooth and light*. Whatever the direction of the string passing over this pulley, its tension is the same on both sides.

Figure 2.22 shows the forces acting when a pulley is used to lift a heavy parcel.

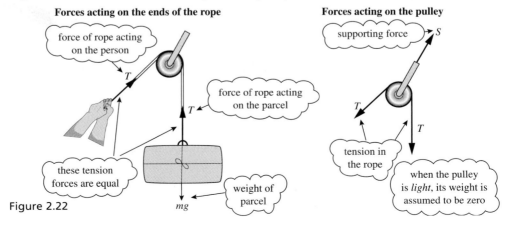

Forces acting on the ends of the rope

force of rope acting on the person

T

these tension forces are equal

force of rope acting on the parcel

T

weight of parcel

mg

Forces acting on the pulley

supporting force

S

T

T

tension in the rope

when the pulley is *light*, its weight is assumed to be zero

Figure 2.22

Note

The rope is in tension. It is not possible for a rope to exert a thrust force.

EXAMPLE 2.6

In this diagram the pulley is smooth and light and the 2 kg block, A, is on a rough surface.

i) Draw diagrams to show the forces acting on each of A and B.

ii) If the block A does not slip, find the tension in the string and calculate the magnitude of the friction force on the block.

iii) Write down the resultant force acting on each of A and B if the block slips and accelerates.

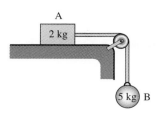

Figure 2.23

SOLUTION

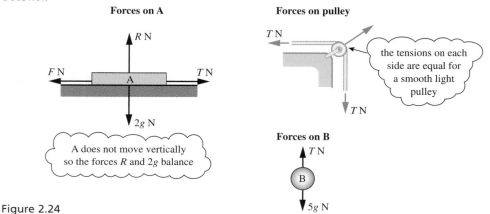

Figure 2.24

Note

The masses of 2 kg and 5 kg are not shown in the force diagram. The weights $2g$ N and $5g$ N are more appropriate.

ii) When the block does not slip, the forces on B are in equilibrium so

$$5g - T = 0$$
$$T = 5g.$$

The tension throughout the string is $5g$ N.

For A, the resultant horizontal force is zero so

$$T - F = 0$$
$$F = T = 5g.$$

The friction force is $5g$ N towards the left.

iii) When the block slips, the forces are not in equilibrium and T and F have different magnitudes.

The resultant horizontal force on A is $(T - F)$N towards the right.
The resultant force on B is $(5g - T)$ N vertically downwards.

2.3 Applying Newton's second law along a line

QUESTION 2.7   What would you observe if you stood on some bathroom scales in a moving lift?

Hold a heavy book on your hand and move it up and down.

What force do you feel on your hand?

Equation of motion

Suppose you make the book accelerate upwards at a ms^{-2}. Figure 2.25 shows the forces acting on the book and the acceleration.

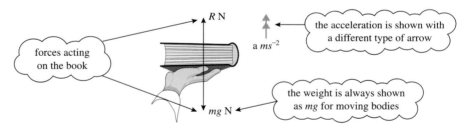

Figure 2.25

By Newton's first law, a resultant force is required to produce an acceleration. In this case the resultant upwards force is $R - mg$ newtons.

You were introduced to Newton's second law earlier in this chapter. When the forces are in newtons, the mass in kilograms and the acceleration in metres per second squared, this law is:

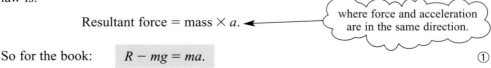

Resultant force = mass $\times a$.

So for the book: $R - mg = ma.$ ①

When Newton's second law is applied, the resulting equation is called *the equation of motion*.

When you give a book of mass 0.8 kg an acceleration of 0.5 ms^{-2} equation ① becomes

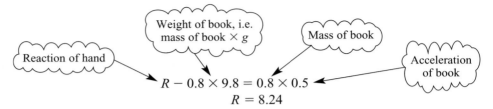

$$R - 0.8 \times 9.8 = 0.8 \times 0.5$$
$$R = 8.24$$

When the book is accelerating upwards the reaction force of your hand on the book is 8.24 N. This is equal and opposite to the force experienced by you so the book feels heavier than its actual weight, mg, which is $0.8 \times 9.8 = 7.84$ N.

EXAMPLE 2.7

A lift and its passengers have a total mass of 400 kg. Find the tension in the cable supporting the lift when

i) the lift is at rest
ii) the lift is moving at constant speed
iii) the lift is accelerating upwards at 0.8 ms^{-2}
iv) the lift is accelerating downwards at 0.6 ms^{-2}.

SOLUTION

Before starting the calculations you must define a direction as positive. In this example the upward direction is chosen to be positive.

i) **At rest**
As the lift is at rest the forces must be in equilibrium. The equation of motion is

$$T - mg = 0$$
$$T - 400 \times 9.8 = 0$$
$$T = 3920.$$

The tension in the cable is 3920 N.

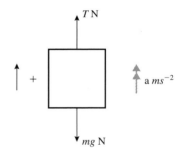

Figure 2.26

ii) **Moving at constant speed**
Again, the forces on the lift must be in equilibrium because it is moving at a constant speed, so the tension is 3920 N.

iii) **Accelerating upwards**
The resultant upward force on the lift is $T - mg$ so the equation of motion is

$$T - mg = ma$$

which in this case gives

$$T - 400 \times 9.8 = 400 \times 0.8$$
$$T - 3920 = 320$$
$$T = 4240.$$

The tension in the cable is 4240 N.

iv) **Accelerating downwards**

The equation of motion is

$$T - mg = ma.$$

In this case, a is negative so

$$T - 400 \times 9.8 = 400 \times (-0.6)$$
$$T - 3920 = -240$$
$$T = 3680.$$

> A downward acceleration of $0.6\,\text{ms}^{-2}$ is an upward acceleration of $-0.6\,\text{ms}^{-2}$

QUESTION 2.8   How is it possible for the tension to be $3680\,\text{N}$ upwards but the lift to accelerate downwards?

EXAMPLE 2.8   This example shows how the suvat equations for motion with constant acceleration, which you met in Chapter 1, can be used with Newton's second law.

A supertanker of mass 500000 tonnes is travelling at a speed of $10\,\text{ms}^{-1}$ when its engines fail. It then takes half an hour for the supertanker to stop.

i) Find the force of resistance, assuming it to be constant, acting on the supertanker.

When the engines have been repaired it takes the supertanker 10 minutes to return to its full speed of $10\,\text{ms}^{-1}$.

ii) Find the driving force produced by the engines, assuming this also to be constant.

SOLUTION

Use the direction of motion as positive.

i) First find the acceleration of the supertanker, which is constant for constant forces. Figure 2.27 shows the velocities and acceleration.

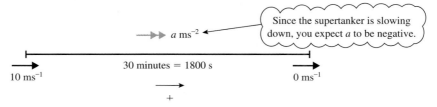

> Since the supertanker is slowing down, you expect a to be negative.

Figure 2.27

You know $u = 10$, $v = 0$, $t = 1800$ and you want a, so use $v = u + at$.

$$0 = 10 + 1800a$$

$$a = -\tfrac{1}{180}$$

> The acceleration is negative because the supertanker is slowing down.

Now we can use Newton's second law (Newton II) to write down the equation of motion. Figure 3.28 shows the horizontal forces and the acceleration.

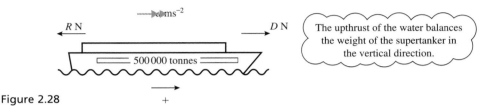

R N a ms^{-2} D N

500 000 tonnes

> The upthrust of the water balances the weight of the supertanker in the vertical direction.

Figure 2.28 $+$

The resultant forwards force is $D - R$ newtons. When there is no driving force $D = 0$ so Newton II gives

$$0 - R = 500\,000\,000 \times a$$

> the mass must be in kg

so when $a = -\tfrac{1}{180}$, $-R = 500\,000\,000 \times (-\tfrac{1}{180})$

The resistance to motion is 2.78×10^6 N or 2780 kN (correct to 3 s.f.).

 You have to be very careful with signs here: the resultant force and acceleration are both positive towards the right.

---

ii) Now $u = 0$, $v = 10$ and $t = 600$, and you want a, so use $v = u + at$ again.

$$10 = 0 + a \times 600$$

$$a = \tfrac{1}{60}$$

Using Newton's second law again

$$D - R = 500\,000\,000 \times a$$
$$D - 2.78 \times 10^6 = 500\,000\,000 \times \tfrac{1}{60}$$
$$D = 2.78 \times 10^6 + 8.33 \times 10^6$$

The driving force is 11.11×10^6 N or 11100 kN (correct to 3 s.f.).

Tackling mechanics problems

When you tackle mechanics problems such as these you will find them easier if you:

● always draw a clear diagram
● clearly indicate the positive direction
● label each object (A, B, etc. or whatever is appropriate)
● show all the forces acting on each object
● make it clear which object you are referring to when writing an equation of motion.

2.4 Newton's second law applied to connected objects

This section is about using Newton's second law for more than one object. It is important to be very clear which forces act on which object in these cases.

A stationary helicopter is raising two people of masses 90 kg and 70 kg as shown in the diagram.

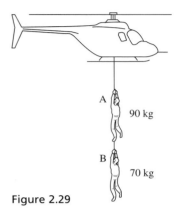

A   90 kg

B   70 kg

Figure 2.29

QUESTION 2.9   Imagine that you are each person in turn. Your eyes are shut so you cannot see the helicopter or the other person. What forces act on you?

Remember that all the forces acting, apart from your weight, are due to contact between you and something else.

Which forces acting on A and B are equal in magnitude? What can you say about their accelerations?

EXAMPLE 2.9

i)   Draw a diagram to show the forces acting on the two people being raised by the helicopter in figure 4.5 and their acceleration.

ii) Write down the equation of motion for each person.

iii) When the force applied to the first person, A, by the helicopter is $180g$ N, calculate
a) the acceleration of the two people being raised
b) the tension in the ropes.
Use 10 ms^{-2} for g.

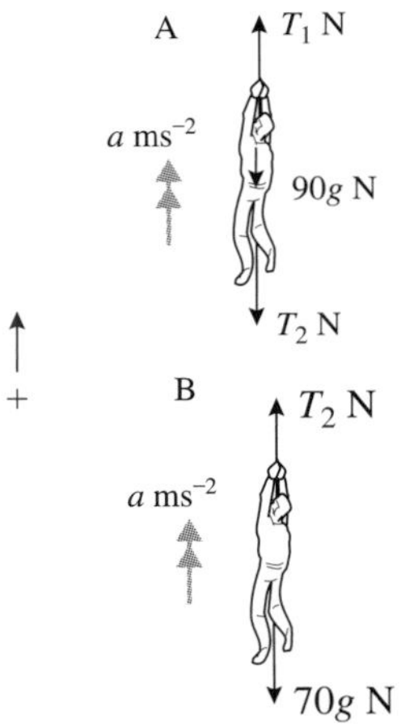

A T_1 N

$a \text{ ms}^{-2}$

$90g$ N

T_2 N

B T_2 N

$a \text{ ms}^{-2}$

$70g$ N

SOLUTION

i)   Figure 2.30 shows the acceleration and forces acting on the two people.

Figure 2.30

ii) When the helicopter applies a force T_1 N to A, the resultant upward forces are

A $(T_1 - 90g - T_2)$ N   B $T_2 - 70g$

Their equations of motion are

A (↑) $T_1 - 90g - T_2 = 90a$ ①   B (↑) $T_2 - 70g = 70a$ ②

iii) You can eliminate T_2 from equations ① and ② by adding:

$$T_1 - 90g - T_2 + T_2 - 70g = 160a$$
$$T_1 - 160g = 160a \qquad\qquad ③$$

When the force applied by the helicopter is $T_1 = 180g$

$$20g = 160a$$
$$a = 1.25$$

Substituting for a in equation ② gives $\quad T_2 = 70 \times 1.25 + 70g$
$$= 787.5$$

The acceleration is 1.25 ms^{-2} and the tensions in the ropes are 1800 N and 787.5 N.

QUESTION 2.10 The force pulling downwards on A is 787.5 N. This is not equal to B's weight (700 N). Why are they different?

Treating the system as a whole

When two objects are moving in the same direction with the same velocity at all times they can be treated as one. In example 2.9 the two people can be treated as one object and then the equal and opposite forces T_2 cancel out. They are *internal forces* similar to the forces between your head and your body.

The resultant upward force on both people is $T_1 - 90g - 70g$ and the total mass is 160 kg so the equation of motion is:

$$T_1 - 90g - 70g = 160a \quad \longleftarrow \quad \text{as equation ③ above}$$

So you can find *a* directly

$$\text{when } T_1 = 180g$$
$$20g = 160a$$
$$a = 1.25 \text{ when } g = 10$$

Treating the system as a whole finds *a*, but not the internal force T_2.

You need to consider the motion of B separately to obtain equation ②.

$$T_2 - 70g = 70a \qquad\qquad ②$$
$$T_2 = 787.5 \quad \longleftarrow \quad \text{as before}$$

Using this method, equation ① can be used to check your answers. Alternatively, you could use equation ① to find T_2 and equation ② to check your answers.

When several objects are joined there are always more equations possible than are necessary to solve a problem and they are not all independent. In the above example, only two of the equations were necessary to solve the problem. The trick is to choose the most relevant ones.

A further note on mathematical modelling

Several modelling assumptions have been made in the solution to example 2.9. It is assumed that:

- the only forces acting on the people are their weights and the tensions in the ropes. Forces due to the wind or air turbulence are ignored.
- the motion is vertical and nobody swings from side to side.
- the ropes do not stretch (i.e. they are inextensible) so the accelerations of the two people are equal.
- the people are rigid bodies which do not change shape and can be treated as particles.

All these modelling assumptions make the problem simpler. In reality, if you were trying to solve such a problem you might work through it first using these assumptions. You would then go back and decide which ones needed to be modified to produce a more realistic solution.

In the next example one person is moving vertically and the other horizontally. You might find it easier to decide on which forces are acting if you imagine you are Alvin or Bernard and you can't see the other person.

EXAMPLE 2.10   Alvin is using a snowmobile to pull Bernard out of a crevasse. His rope passes over a smooth block of ice at the top of the crevasse as shown in figure 2.31 and Bernard hangs freely away from the side. Alvin and his snowmobile together have a mass of 300 kg and Bernard's mass is 75 kg. Ignore any resistance to motion.

Figure 2.31

i) Draw diagrams showing the forces on the snowmobile (including Alvin) and on Bernard.
ii) Calculate the driving force required for the snowmobile to give Bernard an upward acceleration of 0.5 ms^{-2} and the tension in the rope for this acceleration.
iii) How long will it take for Bernard's speed to reach 5 ms^{-1} starting from rest and how far will he have been raised in this time?

SOLUTION

i) The diagram shows the essential features of the problem.

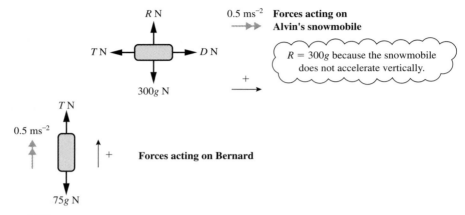

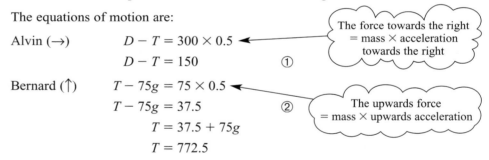

Figure 2.32

ii) Alvin and Bernard have the same acceleration providing the rope does not stretch. The tension in the rope is T newtons and Alvin's driving force is D newtons.

The equations of motion are:

Alvin ($\rightarrow$) $D - T = 300 \times 0.5$ [The force towards the right = mass × acceleration towards the right]

$D - T = 150$ ①

Bernard ($\uparrow$) $T - 75g = 75 \times 0.5$ [The upwards force = mass × upwards acceleration]

$T - 75g = 37.5$ ②

$T = 37.5 + 75g$

$T = 772.5$

Substituting in equation ①

$D - 772.5 = 150$

$D = 922.5$

The driving force required is 922.5 N and the tension in the rope is 772.5 N.

iii) When $u = 0$, $v = 5$, $a = 0.5$ and t is required

$$v = u + at$$
$$5 = 0 + 0.5 \times t$$
$$t = 10$$

The time taken is 10 seconds.

Then using $s = ut + \frac{1}{2}at^2$ to find s gives   [$v^2 = u^2 + as$ would also give s]

$$s = 0 + \frac{1}{2}at^2$$
$$s = \frac{1}{2} \times 0.5 \times 100$$
$$s = 25$$

QUESTION 2.11   Alvin thinks the rope will not stand a tension of more than 1.2kN. What is the maximum safe acceleration in this case? Under the circumstances, is Alvin likely to use this acceleration?

Make a list of the modelling assumptions made in this example and suggest what effect a change in each of these assumptions might have on the solution.

EXAMPLE 2.11   A woman of mass 60 kg is standing in a lift.

i) Draw a diagram showing the forces acting on the woman.

Find the normal reaction of the floor of the lift on the woman in the following cases.

ii) The lift is moving upwards at a constant speed of 3 ms^{-1}.
iii) The lift is moving upwards with an acceleration of 2 ms^{-2} upwards.
iv) The lift is moving downwards with an acceleration of 2 ms^{-2} downwards.
v) The lift is moving downwards and slowing down with a deceleration of 2 ms^{-2}.

In order to calculate the maximum number of occupants that can safely be carried in the lift, the following assumptions are made:

The lift has mass 300 kg, all resistances to motion may be neglected, the mass of each occupant is 75 kg and the tension in the supporting cable should not exceed 12 000 N.

vi) What is the greatest number of occupants that can be carried safely if the magnitude of the acceleration does not exceed 3 ms^{-2}?

SOLUTION

i) The diagram shows the forces acting on the woman and her acceleration.

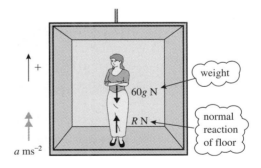

Figure 2.33

In general, when positive is upwards, her equation of motion is

($\uparrow$) $R - 60g = 60a.$

> This equation contains all the mathematics in the situation. It can be used to solve parts ii) to iv).

ii) When the speed is constant $a = 0$ so $R = 60g = 588$.
The normal reaction is 588 N.

iii) When $a = 2$

$$R - 60g = 60 \times 2$$
$$R = 120 + 588$$
$$= 708.$$

The normal reaction is 708 N.

iv) When the acceleration is downwards, $a = -2$ so

$$R - 60g = 60 \times (-2)$$
$$R = 468.$$

The normal reaction is 468 N.

v) When the lift is moving downwards and slowing down, the acceleration is negative downwards, so it is positive upwards, and $a = +2$. Then $R = 708$ as in part **iii)**.

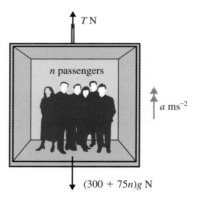

Figure 2.34

vi) When there are n passengers in the lift, the combined mass of these and the lift is $(300 + 75n)$ kg and their weight is $(300 + 75n)g$ N.

The equation of motion for the lift and passengers together is

$$T - (300 + 75n)g = (300 + 75n)\, a$$

So when $a = 3$ and $g = 9.8$, $\qquad T = (300 + 75n) \times 3 + (300 + 75n) \times 9.8$
$$= 12.8\,(300 + 75n).$$

For a maximum tension of 12 000 N

$$12\,000 = 12.8\,(300 + 75n)$$
$$937.5 = 300 + 75n$$
$$637.5 = 75n$$
$$n = 8.5$$

The lift cannot carry more than 8 passengers.

2.5 Reviewing a mathematical model: air resistance

QUESTION 2.12 Why does a leaf or a feather or a piece of paper fall more slowly than other objects?

The model you have used so far for falling objects has assumed no air resistance and this is clearly unrealistic in many circumstances. There are several possible models for air resistance but it is usually better when modelling to try simple models first. Having rejected the first model you could try a second one as follows.

Model 2: Air resistance is constant and the same for all objects.

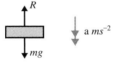

Figure 2.35

Assume an object of mass m falls vertically through the air.

The equation of motion is

$$mg - R = ma$$

$$a = g - \frac{R}{m}$$

The model predicts that a heavy object will have a greater acceleration than a lighter one because $\frac{R}{m}$ is smaller for larger m.

Steps in the modelling procedure

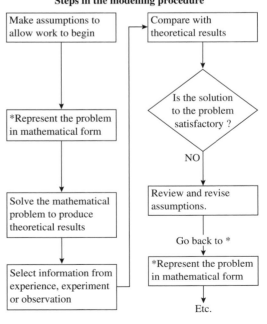

This seems to agree with our experience of dropping a piece of paper and a book, for example. The heavier book has a greater acceleration.

QUESTION 2.13   However, think again about air resistance. Is there a property of the object other than its mass which might affect its motion as it falls? How do people and animals maximise or minimise the force of the air?

Try dropping two identical sheets of paper from a horizontal position, but fold one of them. The folded one lands first even though they have the same mass.

This contradicts the prediction of model 2. A large surface at right angles to the motion seems to increase the resistance.

Model 3: Air resistance is proportional to the area perpendicular to the motion.

Assume the air resistance is kA where k is constant and A is the area of the surface perpendicular to the motion.

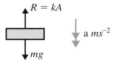

Figure 2.36

The equation of motion is now $mg - kA = ma$

$$a = g - \frac{kA}{m}$$

According to this model, the acceleration depends on the ratio of the area to the mass.

QUESTION 2.14   All the above models ignore one important aspect of air resistance. What is that?

Historical note

Isaac Newton was born in Lincolnshire in 1642. He was not an outstanding scholar either as a schoolboy or as a university student, yet later in life he made remarkable contributions in dynamics, optics, astronomy, chemistry, music theory and theology. He became Member of Parliament for Cambridge University and later Warden of the Royal Mint. His tomb in Westminster Abbey reads 'Let mortals rejoice That there existed such and so great an Ornament to the Human Race'.

Chapter 2 Exercises

2.1 Draw a clear diagram to show the forces acting on a *cricket ball* which follows the path shown on the right. Draw diagrams for each of the three positions A, B and C (include air resistance). Clarity is more important than realism when drawing the diagram.

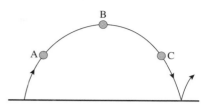

2.2 Draw a clear diagram to show the forces acting on a *paving stone* leaning against a wall. Clarity is more important than realism when drawing the diagram.

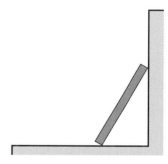

2.3 A book is resting on an otherwise empty table.
  i) Draw diagrams showing the forces acting on
    a) the book   b) the table as seen from the side.
  ii) Write down equations connecting the forces acting on the book and on the table.

Data: On the earth g = 9.8 ms⁻². On the moon g = 1.6 ms⁻².
1000 newtons (N) = 1 kilonewton (kN).

Data: On the earth $g = 9.8\,ms^{-2}$. On the moon $g = 1.6\,ms^{-2}$.
1000 newtons (N) = *1 kilonewton* (kN).

2.4 Calculate the magnitude of the force of gravity on the following objects on the earth.
  i) A suitcase of mass 15 kg.
  ii) A car of mass 1.2 tonnes. (1 tonne = 1000 kg)
  iii) A letter of mass 50 g.

2.5 Find the mass of each of these objects on the earth.
  i) A girl of weight 600 N.   ii) A lorry of weight 11 kN.

In exercise **2.6** you are asked to draw force diagrams using the various types of force you have met in this chapter. Remember that all the forces you need, other than weight, occur when objects are in contact or joined together in some way. Where motion is involved, indicate its direction clearly.

2.6 Ten boxes each of mass 5 kg are stacked on top of each other on the floor.
  i) What forces act on the top box?
  ii) What forces act on the bottom box?

2.7 What is the reaction between a book of mass 0.8 kg and your hand when it is
  i) accelerating downwards at 0.3 ms⁻²
  ii) moving upwards at constant speed.

2.8   A top sprinter of mass 65 kg starting from rest reaches a speed of 10 ms^{-1} in 2 s.
  i)   Calculate the force required to produce this acceleration, assuming it is uniform.
  ii) Compare this to the force exerted by a weight lifter holding a mass of 180 kg above the ground.

2.9   An ice skater of mass 65 kg is initially moving with speed 2 ms^{-1} and glides to a halt over a distance of 10 m. Assuming that the force of resistance is constant, find
  i)   the size of the resistance force
  ii) the distance he would travel gliding to rest from an initial speed of 6 ms^{-1}
  iii) the force he would need to apply to maintain a steady speed of 10 ms^{-1}.

Remember: Always make it clear which object each equation of motion refers to.

2.10 In this question you should take g to be 10 ms^{-2}. The diagram shows a block of mass 5 kg lying on a smooth table. It is attached to blocks of mass 2 kg and 3 kg by strings which pass over smooth pulleys. The tensions in the strings are T_1 and T_2, as shown, and the blocks have acceleration $a \text{ ms}^{-2}$.

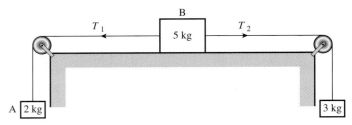

  i)   Draw a diagram for each block showing all the forces acting on it and its acceleration.
  ii) Write down the equation of motion for each of the blocks.
  iii) Use your equations to find the values of a, T_1 and T_2.

In practice, the table is not truly smooth and a is found to be 0.5 ms^{-2}.

  iv) Repeat parts i) and ii) including a frictional force on B and use your new equations to find the frictional force that would produce this result.

2.11 The diagram shows a goods train consisting of an engine of mass 40 tonnes and two trucks of 20 tonnes each. The engine is producing a driving force of $5 \times 10^4 \text{ N}$, causing the train to accelerate. The ground is level and resistance forces may be neglected.

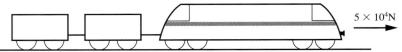

  i)   By considering the motion of the whole train, find its acceleration.
  ii) Draw a diagram to show the forces acting on the engine and use this to help you to find the tension in the first coupling.
  iii) Find the tension in the second coupling.

The brakes on the first truck are faulty and suddenly engage, causing a resistance of 10^4 N.

  iv) What effect does this have on the tension in the coupling to the last truck?

2.12 A man of mass 70 kg is standing in a lift which has an upward acceleration a ms^{-2}.

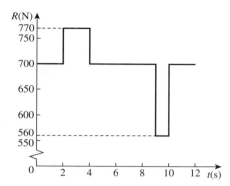

i) Draw a diagram showing the man's weight, the force, R N, that the lift floor exerts on him and the direction of his acceleration.

ii) Taking g to be 10 ms^{-2} find the value of a when $R = 770$ N.

The graph shows the value of R from the time ($t = 0$) when the man steps into the lift to the time ($t = 12$) when he steps out.

iii) Explain what is happening in each section of the journey.

iv) Draw the corresponding speed–time graph.

v) To what height does the man ascend?

Personal Tutor Exercise

2.1 Pat and Nicholas are controlling the movement of a canal barge by means of long ropes attached to each end. The tension in the ropes may be assumed to be horizontal and parallel to the line and direction of motion of the barge, as shown in the diagrams.

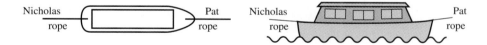

The mass of the barge is 12 tonnes and the total resistance to forward motion may be taken to be 250 N at all times. Initially Pat pulls the barge forwards from rest with a force of 400 N and Nicholas leaves his rope slack.

i) Write down the equation of motion for the barge and hence calculate its acceleration.

Pat continues to pull with the same force until the barge has moved 10 m.

ii) What is the speed of the barge at this time and for what length of time did Pat pull?

Pat now lets her rope go slack and Nicholas brings the barge to rest by pulling with a constant force of 150 N.

iii) Calculate
   a) how long it takes the barge to come to rest
   b) the total distance travelled by the barge from when it first moved
   c) the total time taken for the motion.

2.1 Newton's laws of motion

I Every object continues in a state of rest or uniform motion in a straight line unless it is acted on by a resultant external force.

II Resultant force = mass × acceleration or $F = ma$.

III When one object exerts a force on another there is always a reaction which is equal, and opposite in direction, to the acting force.

- *Force* is a vector; *mass* is a scalar.
- The weight of an object is the force of gravity pulling it towards the centre of the earth. Weight = mg vertically downwards.

2.2 S.I. units

- length: metre (m)
- time: second (s)
- velocity: ms^{-1}
- acceleration: ms^{-2}
- mass: kilogram (kg)

2.3 Force

1 newton (N) is the force required to give a mass of 1 kg an acceleration of $1\ ms^{-2}$.

A force of 1000 newtons (N) = 1 kilonewton (kN).

2.4 Commonly used modelling terms

- inextensible   does not vary in length
- light   negligible mass
- negligible   small enough to ignore
- particle   negligible dimensions
- smooth   negligible friction
- uniform   the same throughout

2.5 Types of force

- Forces due to contact between surfaces
- Forces in a joining rod or string

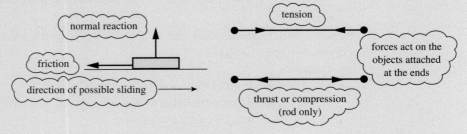

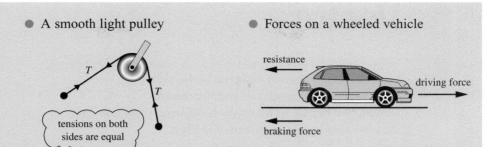

● A smooth light pulley

tensions on both sides are equal

● Forces on a wheeled vehicle

resistance
driving force
braking force

2.6 The equation of motion

Newton's second law gives the equation of motion for an object.

$$\text{The resultant force} = \text{mass} \times \text{acceleration} \quad \text{or} \quad F = ma$$

The acceleration is always in the same direction as the resultant force.

2.7 Connected objects

● Reaction forces between two objects (such as tension forces in joining rods or strings) are equal and opposite.
● When connected objects are moving along a line, the equations of motion can be obtained for each one separately or for a system containing more than one object. The number of independent equations is equal to the number of separate objects.

2.8 Reviewing a model

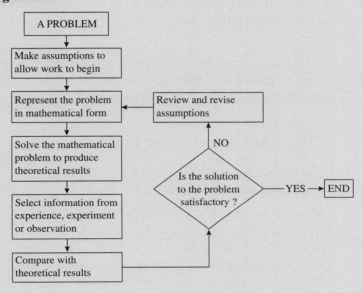

3 Vectors

But the principal failing occurred in the sailing
And the bellman, perplexed and distressed,
Said he had hoped, at least when the wind blew due East
That the ship would not travel due West.

Lewis Carroll

3.1 Adding vectors

QUESTION 3.1 If you walk 12 m east and then 5 m north, how far and in what direction will you be from your starting point?

A bird is caught in a wind blowing east at 12 ms^{-1} and flies so that its speed would be 5 ms^{-1} north in still air. What is its actual velocity?

A sledge is being pulled with forces of 12 N east and 5 N north. What single force would have the same effect?

All these questions involve vectors – displacement, velocity and force. When you are concerned only with the magnitude and direction of these vectors, the three problems can be reduced to one. They can all be solved using the same vector techniques.

Displacement vectors

The instruction 'walk 12 m east and then 5 m north' can be modelled mathematically using a scale diagram, as in figure 3.1. The arrowed lines AB and BC are examples of vectors.

We write the vectors as $\overrightarrow{AB}$ and $\overrightarrow{BC}$. The arrow above the letters is very important as it indicates the direction of the vector. $\overrightarrow{AB}$ means from A to B. $\overrightarrow{AB}$ and $\overrightarrow{BC}$ are examples of *displacement vectors*. Their lengths represent the magnitude of the displacements.

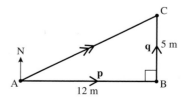

Figure 3.1

It is often more convenient to use a single letter to denote a vector. For example in textbooks and exam papers you might see the displacement vectors $\overrightarrow{AB}$ and $\overrightarrow{BC}$ written as **p** and **q** (i.e. in bold print). When writing these vectors yourself, you should underline your letters, e.g. p̱ and q̱.

The magnitudes of **p** and **q** are then shown as $|\mathbf{p}|$ and $|\mathbf{q}|$ or p and q (in italics). These are scalar quantities.

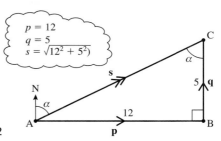

Figure 3.2

The combined effect of the two displacements $\overrightarrow{AB}$ (=**p**) and $\overrightarrow{BC}$ (=**q**) is $\overrightarrow{AC}$ and this is called the *resultant vector*. It is marked with two arrows to distinguish it from **p** and **q**. The process of combining vectors in this way is called *vector addition*. We write $\overrightarrow{AB} + \overrightarrow{BC} = \overrightarrow{AC}$ or $\mathbf{p} + \mathbf{q} = \mathbf{s}$.

You can calculate the resultant using Pythagoras' theorem and trigonometry.

In triangle ABC $\qquad\qquad$ $AC = \sqrt{12^2 + 5^2} = 13$

and $\qquad\qquad\qquad\qquad$ $\tan \alpha = \frac{12}{5}$

$\qquad\qquad\qquad\qquad\qquad$ $\alpha = 67°$ (to the nearest degree)

The distance from the starting point is 13 m and the direction is 067°.

Velocity and force

The other two problems that begin this chapter are illustrated in these diagrams.

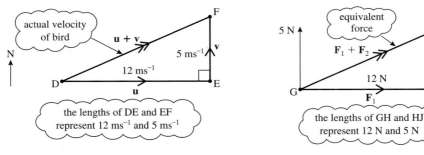

Figure 3.3 $\qquad\qquad\qquad\qquad\qquad\qquad\qquad$ Figure 3.4

When $\overrightarrow{DE}$ represents the velocity (**u**) of the wind and $\overrightarrow{EF}$ represents the velocity (**v**) of the bird in still air, the vector $\overrightarrow{DF}$ represents the resultant velocity, **u** + **v**.

QUESTION 3.2 $\quad$ Why does the bird move in the direction DF? Think what happens in very small intervals of time.

In figure 3.4, the vector $\overrightarrow{GJ}$ represents the equivalent (resultant) force. You know that it acts at the same point as the children's forces, but its magnitude and direction can be found using the triangle GHJ which is similar to the two triangles, ABC and DEF.

The same diagram does for all; you just have to supply the units. The bird travels at 13 ms^{-1} in the direction of 067° and one child would have the same effect as the others by pulling with a force of 13 N in the direction 067°. In most of this chapter vectors are treated in the abstract. You can then apply what you learn to different real situations.

Free vectors

Free vectors have magnitude and direction only. All vectors which are in the same direction and have the same magnitude are equal.

If the vector **p** represents a velocity of 3 km h^{-1} north-east, what do $-$**p** and 2**p** represent?

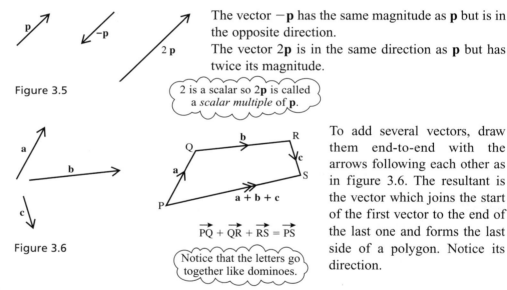

Figure 3.5

The vector $-$**p** has the same magnitude as **p** but is in the opposite direction.

The vector 2**p** is in the same direction as **p** but has twice its magnitude.

2 is a scalar so 2**p** is called a *scalar multiple* of **p**.

Figure 3.6

$$\overrightarrow{PQ} + \overrightarrow{QR} + \overrightarrow{RS} = \overrightarrow{PS}$$

Notice that the letters go together like dominoes.

To add several vectors, draw them end-to-end with the arrows following each other as in figure 3.6. The resultant is the vector which joins the start of the first vector to the end of the last one and forms the last side of a polygon. Notice its direction.

EXAMPLE 3.1   The diagram shows several vectors.

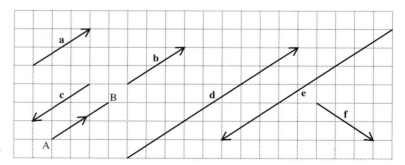

Figure 3.7

i)   Write each of the other vectors in terms of the vector **a**.

ii) Draw scale diagrams to show

a) **a** + **f**   b) **a** − **f**   c) 2**c** + **f**   d) **a** + **f** + **c**

SOLUTION

i) $\mathbf{b} = \mathbf{a}$, $\overrightarrow{AB} = \mathbf{a}$, $\mathbf{c} = -\mathbf{a}$, $\mathbf{d} = 3\mathbf{a}$, $\mathbf{e} = -3\mathbf{a}$. The vector $\mathbf{f}$ cannot be written in terms of $\mathbf{a}$ but $|\mathbf{f}| = |\mathbf{a}|$.

ii) Using vector addition the solutions are as shown below.

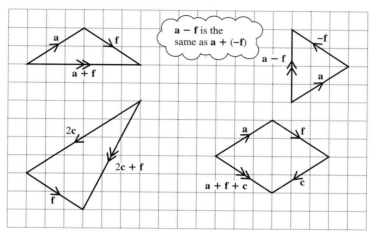

Figure 3.8

Adding parallel vectors

You can add parallel vectors by thinking of them as positive and negative, or by drawing diagrams as in Example 3.2.

EXAMPLE 3.2

The vector $\mathbf{p}$ is 10 units north-west and $\mathbf{q}$ is 6 units north-west.
i) Describe the vector $\mathbf{p} - \mathbf{q}$ and write the answer in terms of $\mathbf{p}$.
ii) Write $\mathbf{p} + \mathbf{q}$ and $\mathbf{q} - \mathbf{p}$ in terms of $\mathbf{p}$.
iii) The vector $\mathbf{s}$ is 5 units south-east. What is $\mathbf{p} + 2\mathbf{s}$?

SOLUTION

i) The diagram shows the vectors $\mathbf{p}$, $\mathbf{q}$ and $\mathbf{p} - \mathbf{q}$.

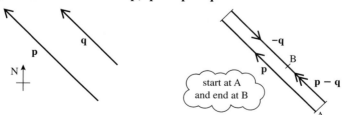

Figure 3.9

   $\mathbf{p} - \mathbf{q}$ is in the direction of $\mathbf{p}$ and of magnitude $10 - 6 = 4$ units.
   $\mathbf{p} - \mathbf{q} = 0.4\mathbf{p}$
ii) $\mathbf{p} + \mathbf{q}$ is $(10 + 6) = 16$ units NW so $\mathbf{p} + \mathbf{q} = 1.6\mathbf{p}$
   $\mathbf{q} - \mathbf{p}$ is $(6 - 10) = -4$ units NW or 4 units SE, so $\mathbf{q} - \mathbf{p} = -(\mathbf{p} - \mathbf{q}) = -0.4\mathbf{p}$
iii) $\mathbf{s} = -0.5\mathbf{p}$ so $\mathbf{p} + 2\mathbf{s} = 0\mathbf{p} = \mathbf{0}$

Note

We use **0** (in bold) and not 0 (in ordinary type) on the right-hand side of this expression to show that the quantity is still a vector.

3.2 Components of a vector

So far you have added two vectors to make one resultant vector. Alternatively, it is often convenient to write one vector in term of two others called *components*.

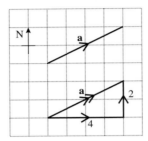

Figure 3.10

The vector **a** in the diagram can be split into two components in an infinite number of ways. All you need to do is to make **a** one side of a triangle. It is most sensible, however, to split vectors into components in convenient directions and these directions are usually perpendicular.

Using the given grid, **a** is 4 units east combined with 2 units north.

Unit vectors i and j

You can write **a** in figure 3.10 as $4\mathbf{i} + 2\mathbf{j}$ where **i** represents a vector of one unit to the east and **j** a vector of one unit to the north. **i** and **j** are called unit vectors.

Alternatively **a** can be written as $\begin{pmatrix} 4 \\ 2 \end{pmatrix}$. This is called a *column vector*.

When using the standard Cartesian co-ordinate system, **i** is a vector of one unit along the x axis and **j** is a vector of one unit along the y axis. Any other vector drawn in the xy plane can then be written in terms of **i** and **j**.

You may define the unit vectors **i** and **j** to be in any two perpendicular directions if it is convenient to do so.

QUESTION 3.3   What is **a** in terms of **i** and **j** if **i** is north-east and **j** is south-east?

Note

You have already worked with vectors in components. The total reaction between two surfaces is often split into two components. One (friction) is opposite to the direction of possible sliding and the other (normal reaction) is perpendicular to it.

EXAMPLE 3.3

The four vectors **a**, **b**, **c** and **d** are shown in the diagram.
i)   Write them in component form.
ii) Draw a diagram to show 2**c** and −**d** and write them in component form.

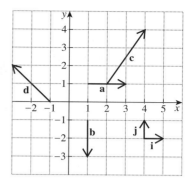

Figure 3.11

SOLUTION

i) $\mathbf{a} = 2\mathbf{i} = \begin{pmatrix} 2 \\ 0 \end{pmatrix}$ $\mathbf{b} = -2\mathbf{j} = \begin{pmatrix} 0 \\ -2 \end{pmatrix}$

$\mathbf{c} = 2\mathbf{i} + 3\mathbf{j} = \begin{pmatrix} 2 \\ 3 \end{pmatrix}$ $\mathbf{d} = -2\mathbf{i} + 2\mathbf{j} = \begin{pmatrix} -2 \\ 2 \end{pmatrix}$

ii) $2\mathbf{c} = 2(2\mathbf{i} + 3\mathbf{j})$ or $2\begin{pmatrix} 2 \\ 3 \end{pmatrix}$

$= 4\mathbf{i} + 6\mathbf{j}$ or $\begin{pmatrix} 4 \\ 6 \end{pmatrix}$

$-\mathbf{d} = -(-2\mathbf{i} + 2\mathbf{j})$ or $-\begin{pmatrix} -2 \\ 2 \end{pmatrix}$

$= 2\mathbf{i} - 2\mathbf{j}$ or $\begin{pmatrix} 2 \\ -2 \end{pmatrix}$

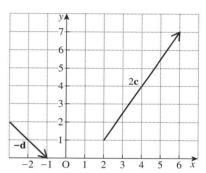

Figure 3.12

Equal vectors and parallel vectors

When two vectors, **p** and **q**, are equal then they must be equal in both magnitude and direction. If they are written in component form their components must be equal.

So if $\mathbf{p} = a_1\mathbf{i} + b_1\mathbf{j}$

and $\mathbf{q} = a_2\mathbf{i} + b_2\mathbf{j}$

then $a_1 = a_2$ and $b_1 = b_2$.

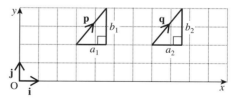

Figure 3.13

Thus in two dimensions, the statement **p** = **q** is the equivalent of two equations (and in three dimensions, three equations).

If **p** and **q** are *parallel but not equal*, they make the same angle with the *x* axis.

then $\dfrac{b_1}{a_1} = \dfrac{b_2}{a_2}$ or $\dfrac{a_1}{a_2} = \dfrac{b_1}{b_2}$

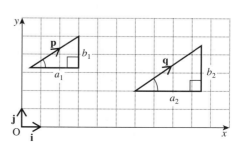

Figure 3.14

QUESTION 3.4   If $\begin{pmatrix} 4 \\ 3 \end{pmatrix}$ is parallel to $\begin{pmatrix} -8 \\ y \end{pmatrix}$ what is y?

Position vectors

When an object is modelled as a particle or a point moving in space its *position* is its *displacement relative to a fixed origin*.

When O is the fixed origin, the vector $\overrightarrow{OA}$ (or **a**) is called the *position vector* of A.

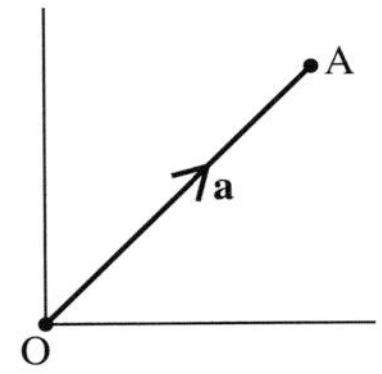

If a point P has coordinates (3, 2) and **i** and **j** are in the direction of the x and y axes, the position of the point P is

$$\overrightarrow{OP} = 3\mathbf{i} + 2\mathbf{j} \text{ or } \begin{pmatrix} 3 \\ 2 \end{pmatrix}.$$

Figure 3.15

Adding vectors in component form

In component form, addition and subtraction of vectors is simply carried out by adding or subtracting the components of the vectors.

EXAMPLE 3.4   Two vectors **a** and **b** are given by $\mathbf{a} = 2\mathbf{i} + 3\mathbf{j}$ and $\mathbf{b} = -\mathbf{i} + 4\mathbf{j}$.
i)   Find the vectors $\mathbf{a} + \mathbf{b}$ and $\mathbf{a} - \mathbf{b}$.
ii) Verify that your results are the same if you use a scale drawing.

SOLUTION

i)   Using **i** and **j**:

$$\mathbf{a} + \mathbf{b} = (2\mathbf{i} + 3\mathbf{j}) + (-\mathbf{i} + 4\mathbf{j})$$
$$= 2\mathbf{i} - \mathbf{i} + 3\mathbf{j} + 4\mathbf{j}$$
$$= \mathbf{i} + 7\mathbf{j}$$

$$\mathbf{a} - \mathbf{b} = (2\mathbf{i} + 3\mathbf{j}) - (-\mathbf{i} + 4\mathbf{j})$$
$$= 2\mathbf{i} + \mathbf{i} + 3\mathbf{j} - 4\mathbf{j}$$
$$= 3\mathbf{i} - \mathbf{j}$$

Using column vectors:

$$\mathbf{a} + \mathbf{b} = \begin{pmatrix} 2 \\ 3 \end{pmatrix} + \begin{pmatrix} -1 \\ 4 \end{pmatrix}$$
$$= \begin{pmatrix} 1 \\ 7 \end{pmatrix}$$

$$\mathbf{a} - \mathbf{b} = \begin{pmatrix} 2 \\ 3 \end{pmatrix} - \begin{pmatrix} -1 \\ 4 \end{pmatrix}$$
$$= \begin{pmatrix} 3 \\ -1 \end{pmatrix}$$

ii)

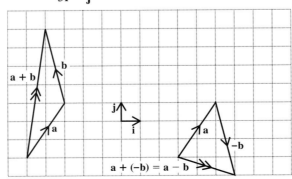

Figure 3.16

From the diagram you can see that $\mathbf{a} + \mathbf{b} = \mathbf{i} + 7\mathbf{j}$ or $\begin{pmatrix} 1 \\ 7 \end{pmatrix}$

and $\mathbf{a} - \mathbf{b} = 3\mathbf{i} - \mathbf{j}$ or $\begin{pmatrix} 3 \\ -1 \end{pmatrix}$.

These vectors are the same as those obtained in part **i)**.

QUESTION 3.5   **a** and **b** are the position vectors of points A and B as shown in the diagram.

How can you write the displacement vector $\overrightarrow{AB}$ in terms of **a** and **b**?

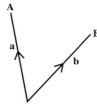

Figure 3.17   O

3.3 The magnitude and direction of vectors written in component form

At the beginning of this chapter the magnitude of a vector was found by using Pythagoras' theorem. The direction was given using bearings, measured clockwise from the north.

When the vectors are in an xy plane, a mathematical convention is used for direction. Starting from the x axis, angles measured anticlockwise are positive and angles in a clockwise direction are negative as in figure 3.18.

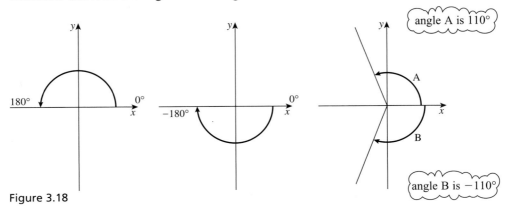

Figure 3.18

Using the notation in figure 3.19, the magnitude and direction can be written in general form.

Magnitude of the vector $|a_1\mathbf{i} + a_2\mathbf{j}| = \sqrt{a_1^2 + a_2^2}$

Direction $\tan \theta = \dfrac{a_2}{a_1}$

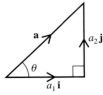

Figure 3.19

EXAMPLE 3.5   Find the magnitude and direction of the vectors $4\mathbf{i} + 3\mathbf{j}$, $4\mathbf{i} - 3\mathbf{j}$, $-4\mathbf{i} + 3\mathbf{j}$ and $-4\mathbf{i} - 3\mathbf{j}$ in the xy plane.

SOLUTION

First draw diagrams so that you can see which lengths and acute angles to find.

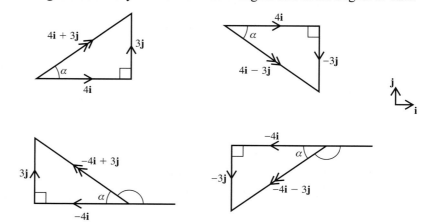

Figure 3.20

The vectors in each of the diagrams have the same magnitude and using Pythagoras' theorem, the resultants all have magnitude $\sqrt{4^2 + 3^2} = 5$.

The angles α are also the same size in each diagram and can be found using

$$\tan \alpha = \tfrac{3}{4}$$
$$\alpha = 37°.$$

The angles the vectors make with the $\mathbf{i}$ direction specify their directions:

$4\mathbf{i} + 3\mathbf{j}$	$37°$
$4\mathbf{i} - 3\mathbf{j}$	$-37°$
$-4\mathbf{i} + 3\mathbf{j}$	$180° - 37° = 143°$
$-4\mathbf{i} - 3\mathbf{j}$	$-143°$

Unit vectors

Sometimes you need to write down a unit vector (i.e. a vector of magnitude 1) in the direction of a given vector. You have seen that the vector $a_1\,\mathbf{i} + a_2\,\mathbf{j}$ has magnitude $\sqrt{a_1^2 + a_2^2}$ and so it follows that the vector

$$\frac{a_1\,\mathbf{i} + a_2\,\mathbf{j}}{\sqrt{a_1^2 + a_2^2}} = \frac{a_1}{\sqrt{a_1^2 + a_2^2}}\,\mathbf{i} + \frac{a_2}{\sqrt{a_1^2 + a_2^2}}\,\mathbf{j} \text{ has magnitude } 1.$$

EXAMPLE 3.6

i)   Find a unit vector in the direction of $-4\mathbf{i} + 3\mathbf{j}$.

ii) A force $\mathbf{F}$ has magnitude 8 N and is in the direction $-4\mathbf{i} + 3\mathbf{j}$. Write $\mathbf{F}$ in terms of $\mathbf{i}$ and $\mathbf{j}$.

SOLUTION

i) $\left|-4\mathbf{i} + 3\mathbf{j}\right| = \sqrt{4^2 + 3^2} = 5$

So a unit vector in this direction is

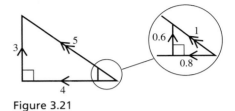

$$\frac{-4\mathbf{i} + 3\mathbf{j}}{5} = -0.8\mathbf{i} + 0.6\mathbf{j}.$$

Notice that $\sqrt{0.8^2 + 0.6^2} = 1$, so this is a unit vector.

Figure 3.21

Also $\dfrac{0.6}{-0.8} = -\dfrac{3}{4} = \tan 143°$ so it is in the same direction as $-4\mathbf{i} + 3\mathbf{j}$.

ii) $\mathbf{F}$ is 8 times the unit vector in the same direction, so

$\mathbf{F} = 8(-0.8\mathbf{i} + 0.6\mathbf{j}) = -6.4\mathbf{i} + 4.8\mathbf{j}$ N.

Vectors in three dimensions

In three dimensions there is a third, z, axis perpendicular to both the x and y axes through the origin O of the xy plane. When the x and y axes are drawn on a horizontal sheet, the z axis is drawn vertically upwards as shown in figure 3.22. Imagine turning the sheet so that the positive region faces you. The point Q $(2, 3, 6)$ would then be plotted as shown in figure 3.23.

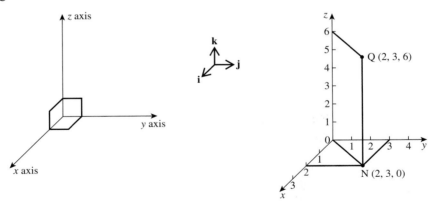

Figure 3.22

Figure 3.23

A third unit vector $\mathbf{k}$ is introduced to represent the positive z direction. For example, the position vector of the point Q $(2, 3, 6)$ is written as

$$\overrightarrow{OQ} = 2\mathbf{i} + 3\mathbf{j} + 6\mathbf{k} = \begin{pmatrix} 2 \\ 3 \\ 6 \end{pmatrix}.$$

In general the position vector of point P (x, y, z) is written as $\mathbf{r} = x\mathbf{i} + y\mathbf{j} + z\mathbf{k}$.

You can use Pythagoras' theorem to find the magnitude of a vector in space.

In figure 3.23
$$OQ^2 = ON^2 + NQ^2$$
$$= (2^2 + 3^2) + 6^2$$
$$OQ = \sqrt{2^2 + 3^2 + 6^2} = 7.$$

A unit vector in the direction of OQ is $\frac{2}{7}\mathbf{i} + \frac{3}{7}\mathbf{j} + \frac{6}{7}\mathbf{k} = 0.29\mathbf{i} + 0.43\mathbf{j} + 0.86\mathbf{k}$.

In general, the vector $\mathbf{a} = a_1\,\mathbf{i} + a_2\,\mathbf{j} + a_3\,\mathbf{k}$ or $\begin{pmatrix} a_1 \\ a_2 \\ a_3 \end{pmatrix}$

has a magnitude of $\sqrt{a_1{}^2 + a_2{}^2 + a_3{}^2}$ and a unit vector in the direction of $\mathbf{a}$ is

$$\frac{a_1}{\sqrt{a_1{}^2 + a_2{}^2 + a_3{}^2}}\,\mathbf{i} + \frac{a_2}{\sqrt{a_1{}^2 + a_2{}^2 + a_3{}^2}}\,\mathbf{j} + \frac{a_3}{\sqrt{a_1{}^2 + a_2{}^2 + a_3{}^2}}\,\mathbf{k}.$$

3.4 Resolving vectors

A vector has magnitude 10 units and it makes an angle of 60° with the $\mathbf{i}$ direction. How can it be represented in component form?

In the diagram:

$$\frac{AC}{AB} = \cos 60° \quad \text{and} \quad \frac{BC}{AB} = \sin 60°$$

$$AC = AB \cos 60° \qquad BC = AB \sin 60°$$
$$= 10 \cos 60° \qquad\quad = 10 \sin 60°$$

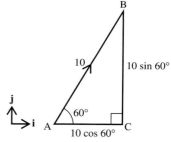

Figure 3.24

The vector can then be written as $10 \cos 60° \,\mathbf{i} + 10 \sin 60° \,\mathbf{j} = 5\mathbf{i} + 8.66\,\mathbf{j}$ (to 3 s.f.).

In a similar way, any vector $\mathbf{a}$ with magnitude a which makes an angle α with the $\mathbf{i}$ direction can be written in component form as

$\mathbf{a} = a \cos \alpha\,\mathbf{i} + a \sin \alpha\,\mathbf{j}$.

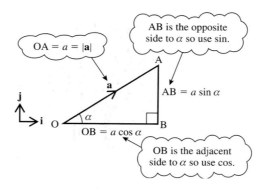

Figure 3.25

When α is an obtuse angle, this expression is still true. For example, when $\alpha = 120°$ and $a = 10$,

$$\mathbf{a} = a \cos \alpha \, \mathbf{i} + a \sin \alpha \, \mathbf{j}$$
$$= 10 \cos 120° \, \mathbf{i} + 10 \sin 120° \, \mathbf{j}$$
$$= -5\mathbf{i} + 8.66\mathbf{j}$$

However, it is usually easier to write

$$\mathbf{a} = -10 \cos 60° \, \mathbf{i} + 10 \sin 60° \, \mathbf{j}$$

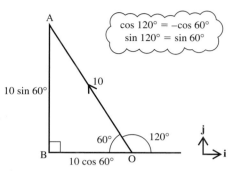

$$\cos 120° = -\cos 60°$$
$$\sin 120° = \sin 60°$$

Figure 3.26

EXAMPLE 3.7

Two forces **P** and **Q** have magnitudes 4 and 5 in the directions shown in the diagram.

Find the magnitude and direction of the resultant force **P** + **Q**.

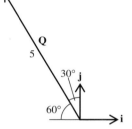

Figure 3.27

SOLUTION

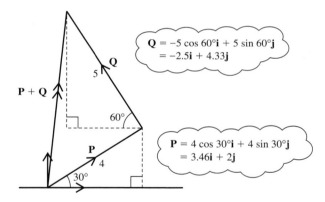

$$\mathbf{Q} = -5 \cos 60°\mathbf{i} + 5 \sin 60°\mathbf{j}$$
$$= -2.5\mathbf{i} + 4.33\mathbf{j}$$

$$\mathbf{P} = 4 \cos 30°\mathbf{i} + 4 \sin 30°\mathbf{j}$$
$$= 3.46\mathbf{i} + 2\mathbf{j}$$

Figure 3.28

$$\mathbf{P} + \mathbf{Q} = (3.46\mathbf{i} + 2\mathbf{j}) + (-2.5\mathbf{i} + 4.33\mathbf{j})$$
$$= 0.96\mathbf{i} + 6.33\mathbf{j}$$

This resultant is shown in the diagram on the right.

Magnitude
$$|\mathbf{P} + \mathbf{Q}| = \sqrt{0.96^2 + 6.33^2}$$
$$= \sqrt{40.99}$$
$$= 6.4$$

Direction
$$\tan \theta = \frac{6.33}{0.96}$$
$$= 6.59$$
$$\theta = 81.4°$$

The force **P** + **Q** has magnitude 6.4 and direction 81.4° relative to the positive x direction.

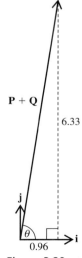

Figure 3.29

Note

If you choose to use the angle that **P** makes with the y axis, 60°, as in figure 3.30, the components are 4 sin 60° **i** + 4 cos 60° **j**.

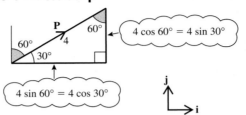

Figure 3.30

This is the same as before because sin 60° = cos 30° and cos 60° = sin 30°.

3.5 Velocity triangles

When a boat moves in a current or a plane in a wind, the *velocity relative to the water or the air* is the same as the velocity in still water or in still air. It is the velocity the boat or plane would have if there were no current or wind. The resultant vector is found by adding the intended velocity and the velocity of the current or wind.

EXAMPLE 3.8

A swimmer is attempting to cross a river which has a current of 5 km h^{-1} parallel to its banks. She aims at a point directly opposite her starting point and can swim at 4 km h^{-1} in still water. Find her resultant velocity.

SOLUTION

The swimmer's velocity in still water and the velocity of the current are shown in the velocity diagram.

By Pythagoras' theorem in triangle ABC:

Actual speed $= \sqrt{4^2 + 5^2}$

$= 6.4$ km h^{-1}

Direction $\tan \alpha = \dfrac{5}{4}$

$\alpha = 51°$

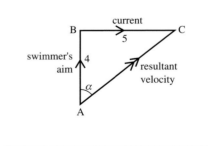

Figure 3.31

The swimmer has a velocity of 6.4 km h^{-1} at an angle of 90° − 51° = 39° to the bank.

EXAMPLE 3.9

A small motorboat moving at 8 km h^{-1} relative to the water travels directly between two lighthouses which are 10 km apart, the bearing of the second lighthouse from the first being 135°. The current has a constant speed of 4 km h^{-1} from the east. Find

i)   the course that the boat must set

ii)   the time for the journey.

Solution

First draw a clear diagram showing the triangle of velocities. The boat is required to achieve a course of 135° and this must be the direction of the resultant of the boat's own velocity and that of the current.

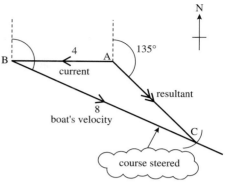

Figure 3.32

Draw a line representing the resultant direction and mark on it a point A. Then draw a line AB to represent the current (4 km h^{-1} from the east). Finally, draw a line BC to represent the boat's velocity relative to the water (8 km h^{-1}); this must be to the same scale as AB and C must lie on the resultant. (If you were to solve the problem by scale drawing you would need to use a compass to find point C.)

⚠ Notice that the 10 km *displacement* is not part of the *velocity* diagram.

---

i)   To calculate the course, use the sine rule in triangle ABC.

$$\angle BAC = 135°, AB = 4, BC = 8.$$

$$\frac{\sin C}{4} = \frac{\sin 135°}{8}$$

$$\angle ACB = 20.7°$$

$$\angle ABC = 180° - 135° - 20.7° = 24.3°$$

Therefore the course steered is $90° + 24° = 114°$ (to the nearest degree).

ii)   The resultant speed is represented by AC and is also found by the sine rule.

$$\frac{AC}{\sin 24.3°} = \frac{8}{\sin 135°}$$

$$AC = 4.656 \,(\text{in km h}^{-1})$$

Now you can use the resultant speed to find the journey time.

$$\text{Time} = \frac{10}{4.656} = 2.15 \text{ hours.}$$

3.6 Motion in two and three dimensions

In your work on projectile motion you have met the idea that the position of an object can be represented by a vector

$$\mathbf{r} = \begin{pmatrix} x \\ y \end{pmatrix} \quad \text{or} \quad x\mathbf{i} + y\mathbf{j}.$$

When a ball is thrown into the air, for example, its position might be given by

$$\mathbf{r} = \begin{pmatrix} 5t \\ 12t - 5t^2 \end{pmatrix}$$

so in this case $x = 5t$ and $y = 12t - 5t^2$. You can plot the path of the ball by finding the cartesian equation as in Chapter 4 (page 95) or by finding the values of $\mathbf{r}$ and hence x and y for several values of t.

t	0	0.5	1	1.5	2	2.4
$\mathbf{r}$	$\begin{pmatrix} 0 \\ 0 \end{pmatrix}$	$\begin{pmatrix} 2.5 \\ 4.75 \end{pmatrix}$	$\begin{pmatrix} 5 \\ 7 \end{pmatrix}$	$\begin{pmatrix} 7.5 \\ 6.75 \end{pmatrix}$	$\begin{pmatrix} 10 \\ 4 \end{pmatrix}$	$\begin{pmatrix} 12 \\ 0 \end{pmatrix}$

QUESTION 3.6   Why is 2.4 chosen as the last value for t?

Figure 3.33 shows the path of the ball and also its position $\mathbf{r}$ when $t = 2$.

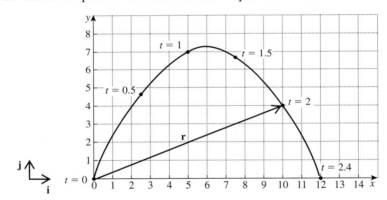

Figure 3.33

Finding the velocity and acceleration

The equation for $\mathbf{r}$ can be differentiated to give the velocity and acceleration.

When

$$\mathbf{r} = \begin{pmatrix} 5t \\ 12t - 5t^2 \end{pmatrix}$$

$$\mathbf{v} = \begin{pmatrix} 5 \\ 12 - 10t \end{pmatrix}$$

$$\mathbf{a} = \begin{pmatrix} 0 \\ -10 \end{pmatrix}$$

Note

The direction of the acceleration is not at all obvious when you look at the diagram of the path of the ball.

Newton's notation

When you write derivatives in column vectors, the notation becomes very cumbersome so many people use Newton's notation when differentiating with respect to time. In this notation a dot is placed over the variable for each differentiation. For example $\dot{x} = \dfrac{dx}{dt}$ and $\ddot{x} = \dfrac{d^2x}{dt^2}$.

$$\mathbf{v} = \dot{\mathbf{r}} = \dot{x}\mathbf{i} + \dot{y}\mathbf{j} \quad \text{or} \quad \begin{pmatrix} \dot{x} \\ \dot{y} \end{pmatrix}$$

Speed
$$|\mathbf{v}| = \sqrt{\dot{x}^2 + \dot{y}^2}$$
Direction of motion
$$\tan \alpha = \frac{\dot{y}}{\dot{x}}$$

$$\mathbf{a} = \dot{\mathbf{v}} = \ddot{\mathbf{r}}$$

$$\mathbf{a} = \ddot{x}\mathbf{i} + \ddot{y}\mathbf{j} \quad \text{or} \quad \begin{pmatrix} \ddot{x} \\ \ddot{y} \end{pmatrix}$$

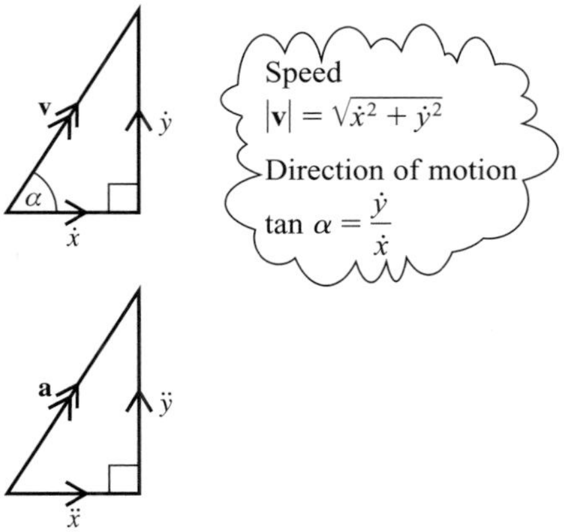

Figure 3.34

The next example shows how you can use these results.

EXAMPLE 3.10

Relative to an origin on a long, straight beach, the position of a speedboat is modelled by the vector

$$\mathbf{r} = (2t + 2)\mathbf{i} + (12 - t^2)\mathbf{j}$$

where $\mathbf{i}$ and $\mathbf{j}$ are unit vectors perpendicular and parallel to the beach. Distances are in metres and the time t is in seconds.

i)   Calculate the distance of the boat from the origin, O, when the boat is 6 m from the beach.

ii) Sketch the path of the speedboat for $0 \leqslant t \leqslant 3$.

iii) Find expressions for the velocity and acceleration of the speedboat at time t. Is the boat ever at rest? Explain your answer.

iv) For $t = 3$, calculate the speed of the boat and the angle its direction of motion makes to the line of the beach.

v)   Suggest why this model for the motion of the speedboat is unrealistic for large t.

SOLUTION

i) $\mathbf{r} = (2t + 2)\mathbf{i} + (12 - t^2)\mathbf{j}$ so the boat is 6 m from the beach when
$x = 2t + 2 = 6$

Then $t = 2$ and $y = 12 - t^2 = 8$

The distance from O is $\sqrt{6^2 + 8^2} = 10$ m.

ii) The table shows the position at different times and the path of the boat is shown on the graph in figure 3.35.

t	0	1	2	3
r	$2\mathbf{i} + 12\mathbf{j}$	$4\mathbf{i} + 11\mathbf{j}$	$6\mathbf{i} + 8\mathbf{j}$	$8\mathbf{i} + 3\mathbf{j}$

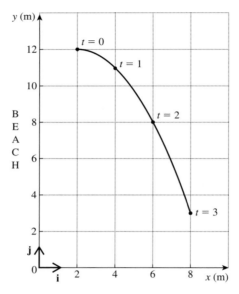

Figure 3.35

iii)
$$\mathbf{r} = (2t + 2)\,\mathbf{i} + (12 - t^2)\,\mathbf{j}$$
$$\Rightarrow \quad \mathbf{v} = \dot{\mathbf{r}} = 2\,\mathbf{i} - 2t\,\mathbf{j}$$
$$\text{and} \quad \mathbf{a} = \dot{\mathbf{v}} = -2\mathbf{j}$$

The boat is at rest if both components of velocity ($\dot{x}$ and $\dot{y}$) are zero at the same time. But $\dot{x}$ is always 2, so the velocity can never be zero.

iv) When $t = 3$ $\mathbf{v} = 2\mathbf{i} - 6\mathbf{j}$

The angle $\mathbf{v}$ makes with the beach is α as shown where

$$\tan \alpha = \frac{2}{6}$$

$$\alpha = 18.4°$$

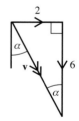

Figure 3.36

v) According to this model, the speed after time t is

$$|\mathbf{v}| = |2\mathbf{i} - 2t\,\mathbf{j}| = \sqrt{2^2 + (-2t)^2} = \sqrt{4 + 4t^2}$$

As t increases, the speed increases at an increasing rate so there must come a time when the boat is incapable of going at the predicted speed and the model cannot then apply.

 Notice that the direction of motion is found using the velocity and not the position.

Using integration

When you are given the velocity or acceleration and wish to work backwards to the displacement, you need to integrate. The next two examples show how you can do this with vectors.

EXAMPLE 3.11

An aircraft is dropping a crate of supplies on to level ground. Relative to an observer on the ground, the crate is released at the point with position vector $\begin{pmatrix} 650 \\ 576 \end{pmatrix}$ m and with initial velocity $\begin{pmatrix} -100 \\ 0 \end{pmatrix}$ ms^{-1}, where the directions are horizontal and vertical. Its acceleration is modelled by

$$\mathbf{a} = \begin{pmatrix} -t + 12 \\ \frac{1}{2}t - 10 \end{pmatrix} \quad \text{for} \quad t \leqslant 12 \text{ s.}$$

i)   Find an expression for the velocity vector of the crate at time t.
ii) Find an expression for the position vector of the crate at time t.
iii) Verify that the crate hits the ground 12 s after its release and find how far from the observer this happens.

SOLUTION

i)
$$\mathbf{a} = \frac{d\mathbf{v}}{dt} = \begin{pmatrix} -t + 12 \\ \frac{1}{2}t - 10 \end{pmatrix} \qquad ①$$

> ① You can treat horizontal and vertical motion separately if you wish.

Integrating gives $\mathbf{v} = \begin{pmatrix} -\frac{1}{2}t^2 + 12t + c_1 \\ \frac{1}{4}t^2 - 10t + c_2 \end{pmatrix}$

At $t = 0$ $\mathbf{v} = \begin{pmatrix} -100 \\ 0 \end{pmatrix} \Rightarrow \begin{array}{l} 0 + 0 + c_1 = -100 \\ 0 - 0 + c_2 = 0 \end{array}$

Velocity $\mathbf{v} = \begin{pmatrix} -\frac{1}{2}t^2 + 12t - 100 \\ \frac{1}{4}t^2 - 10t \end{pmatrix} \qquad ②$

ii)
$$\mathbf{v} = \frac{d\mathbf{r}}{dt}$$

Integrating again gives $\mathbf{r} = \begin{pmatrix} -\frac{1}{6}t^3 + 6t^2 - 100t + k_1 \\ \frac{1}{12}t^3 - 5t^2 + k_2 \end{pmatrix}$

At $t = 0$ $\mathbf{r} = \begin{pmatrix} 650 \\ 576 \end{pmatrix} \Rightarrow \begin{array}{l} k_1 = 650 \\ k_2 = 576 \end{array}$

Position vector $\mathbf{r} = \begin{pmatrix} -\frac{1}{6}t^3 + 6t^2 - 100t + 650 \\ \frac{1}{12}t^3 - 5t^2 + 576 \end{pmatrix} \qquad ③$

iii) At $t = 12$ $\mathbf{r} = \begin{pmatrix} -\frac{1}{6} \times 12^3 + 6 \times 12^2 - 100 \times 12 + 650 \\ \frac{1}{12} \times 12^3 - 5 \times 12^2 + 576 \end{pmatrix}$

$$\mathbf{r} = \begin{pmatrix} 26 \\ 0 \end{pmatrix}$$

Since $y = 0$, the crate hits the ground after 12 s and it is then $x = 26$ m in front of the observer.

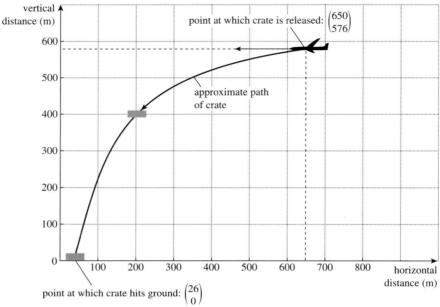

Figure 3.37

Note

When you integrate a vector in two dimensions you need a constant of integration for each direction, for example c_1 and c_2 as above.

It is also a good idea to number your equations for **a**, **v** and **r** so that you can find them easily if you want to use them later.

Force as a function of time

When the force acting on an object is given as a function of t you can use Newton's second law to find out about its motion. You can now write this as $\mathbf{F} = m\,\mathbf{a}$ because force and acceleration are both vectors.

EXAMPLE 3.12   A force of $(12\mathbf{i} + 3t\,\mathbf{j})$ N, where t is the time in seconds, acts on a particle of mass 6 kg. The directions of **i** and **j** correspond to east and north respectively.

i)   Show that the acceleration is $(2\mathbf{i} + 0.5t\,\mathbf{j})\,\text{ms}^{-2}$ at time t.

ii)   Find the acceleration and magnitude of the acceleration when $t = 12$.

iii)   At what time is the acceleration directed north-east (i.e. a bearing of $045°$)?

iv)   If the particle starts with a velocity of $(2\mathbf{i} - 3\mathbf{j})\,\text{ms}^{-1}$ when $t = 0$, what will its velocity be when $t = 3$?

v)   When $t = 3$ a second **constant** force begins to act. Given that the acceleration of the particle at that time due to both forces is $4\,\text{ms}^{-2}$ due south, find the second force.

SOLUTION

i) By Newton's second law the force $=$ mass $\times$ acceleration

$$(12\mathbf{i} + 3t\,\mathbf{j}) = 6\mathbf{a}$$
$$\mathbf{a} = \tfrac{1}{6}(12\mathbf{i} + 3t\,\mathbf{j})$$
$$\mathbf{a} = 2\mathbf{i} + 0.5t\,\mathbf{j} \qquad \textcircled{1}$$

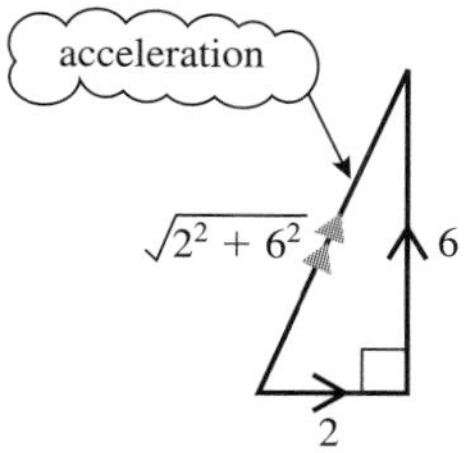

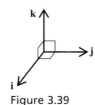

acceleration

$\sqrt{2^2 + 6^2}$ 6

2

Figure 3.38

ii) When $t = 12$ $\mathbf{a} = 2\mathbf{i} + 6\mathbf{j}$

magnitude of a $|\mathbf{a}| = \sqrt{2^2 + 6^2}$

The acceleration is $2\mathbf{i} + 6\mathbf{j}\ \text{ms}^{-2}$
with magnitude $6.32\ \text{ms}^{-2}$.

iii) The acceleration is north-east when its northerly component is equal to its easterly component. From $\textcircled{1}$, this is when $2 = 0.5t$ i.e. when $t = 4$.

iv) The velocity at time t is $\int \mathbf{a}\,\mathrm{d}t = \int (2\mathbf{i} + 0.5t\,\mathbf{j})\,\mathrm{d}t$

$$\Rightarrow \quad \mathbf{v} = 2t\,\mathbf{i} + 0.25t^2\,\mathbf{j} + \mathbf{c}$$

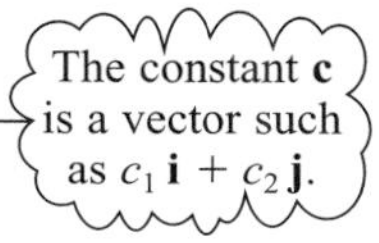

The constant $\mathbf{c}$ is a vector such as $c_1\,\mathbf{i} + c_2\,\mathbf{j}$.

When $t = 0$, $\mathbf{v} = 2\mathbf{i} - 3\mathbf{j}$

so $2\mathbf{i} - 3\mathbf{j} = 0\mathbf{i} + 0\mathbf{j} + \mathbf{c} \quad \Rightarrow \quad \mathbf{c} = 2\mathbf{i} - 3\mathbf{j}$

$$\mathbf{v} = 2t\,\mathbf{i} + 0.25t^2\,\mathbf{j} + 2\mathbf{i} - 3\mathbf{j}$$
$$\mathbf{v} = (2t + 2)\,\mathbf{i} + (0.25t^2 - 3)\,\mathbf{j}. \qquad \textcircled{2}$$

When $t = 3$ $\mathbf{v} = 8\mathbf{i} - 0.75\,\mathbf{j}$.

v) Let the second force be $\mathbf{F}$ so the total force when $t = 3$ is $(12\mathbf{i} + 3 \times 3\mathbf{j}) + \mathbf{F}$.
The acceleration is $-4\mathbf{j}$, so by Newton II

$$(12\mathbf{i} + 9\mathbf{j}) + \mathbf{F} = 6 \times -4\mathbf{j}$$
$$\mathbf{F} = -24\mathbf{j} - 12\mathbf{i} - 9\mathbf{j}$$
$$\mathbf{F} = -12\mathbf{i} - 33\mathbf{j}.$$

The second force is $-12\mathbf{i} - 33\mathbf{j}$.

Note

You can use the same methods in three dimensions just by including a third direction $\mathbf{k}$, for example

$\mathbf{r} = x\mathbf{i} + y\mathbf{j} + z\mathbf{k}$, $\mathbf{v} = \dot{x}\mathbf{i} + \dot{y}\mathbf{j} + \dot{z}\mathbf{k}$ and $\mathbf{a} = \ddot{x}\mathbf{i} + \ddot{y}\mathbf{j} + \ddot{z}\mathbf{k}$

and similarly in column vectors.

$\mathbf{k}$

$\mathbf{j}$

$\mathbf{i}$

Figure 3.39

Historical Note

Newton's work on motion required more mathematical tools than were generally used at the time. He had to invent his own ways of thinking about continuous change and in about 1666 he produced a theory of 'fluxions' in which he imagined a quantity 'flowing' from one magnitude to another. This was the beginning of calculus. He did not publish his methods, however, and when Leibniz published his version in 1684 there was an enormous amount of controversy amongst their supporters about who was first to discover calculus. The sharing of ideas between mathematicians in Britain and the rest of Europe was hindered for a century. The contributions of both men are remembered today by their notation. Leibniz's $\dfrac{\mathrm{d}x}{\mathrm{d}t}$ is common and Newton's $\dot{x}$ is widely used in mechanics.

Chapter 3 Exercises

Remember to underline all your vectors (these are printed in bold here).

3.1   A child climbs up the ladder attached to a slide and then slides down. What three vectors model the displacement of the child during this activity?

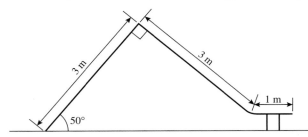

3.2   A crane moves a crate from the ground 10m vertically upward, then 6 m horizontally and 2 m vertically downward. Draw a scale diagram of the path of the crate. What single translation would move the crate to its final position from its initial position on the ground?

3.3   In the parallelogram, $\overrightarrow{OA} = \mathbf{a}$, $\overrightarrow{OC} = \mathbf{c}$, and M is the mid-point of AB. Express the displacements below in terms of $\mathbf{a}$ and $\mathbf{c}$.

i) $\overrightarrow{CB}$ ii) $\overrightarrow{OB}$
iii) $\overrightarrow{AC}$ iv) $\overrightarrow{CA}$
v) $\overrightarrow{BO}$ vi) $\overrightarrow{AM}$
vii) $\overrightarrow{OM}$ viii) $\overrightarrow{MC}$

3.4   A, B and C are the points $(0, -3)$, $(2, 5)$ and $(3, 9)$.

i)   Write down in terms of $\mathbf{i}$ and $\mathbf{j}$ the position vectors of these three points.

ii)   Find the displacements $\overrightarrow{AB}$ and $\overrightarrow{BC}$.

iii)   Show that the three points all lie on a straight line.

3.5   A, B, C and D are the points $(4, 2)$, $(1, 3)$, $(0, 10)$ and $(3, d)$.

i)   Find the value of d so that DC is parallel to AB.

ii)   Find a relationship between $\overrightarrow{BC}$ and $\overrightarrow{AD}$. What is ABCD?

*Make use of sketches to help you in exercises **3.6** and **3.7**.*

3.6   i)   Find a unit vector in the direction of $5\mathbf{i} - 12\mathbf{j}$.
A force F acts in the direction $5\mathbf{i} - 12\mathbf{j}$ and has magnitude 39 N.

ii)   Use your answer to part i) to write **F** in component form.

3.7   The position vectors of the points A, B and C are $\mathbf{a} = \mathbf{i} + \mathbf{j} - 2\mathbf{k}$, $\mathbf{b} = 6\mathbf{i} - 3\mathbf{j} + \mathbf{k}$ and $\mathbf{c} = -2\mathbf{i} + 2\mathbf{j}$.

i)   Find the vectors $\overrightarrow{AC}$, $\overrightarrow{AB}$ and $\overrightarrow{BC}$.

ii)   Find $|\mathbf{a}|$, $|\mathbf{b}|$ and $|\mathbf{c}|$.

iii)   Show that $|\mathbf{a} + \mathbf{b} - \mathbf{c}|$ is *not* equal to $|\mathbf{a}| + |\mathbf{b}| - |\mathbf{c}|$.

3.8 The diagram shows the big wheel ride at
a fairground. The radius of the wheel is 5 m
and the length of the arms that support each
carriage is 1 m.
Express the position vector of the carriages
A, B, C and D in terms of **i** and **j**.

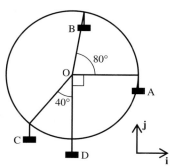

3.9 **a)** Write down each of the following vectors in terms of **i** and **j**.
  b) Find the resultant of each set of vectors in terms of **i** and **j**.

  i)   **ii)**

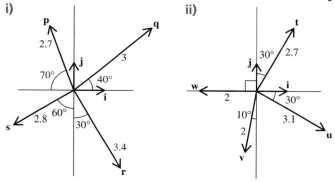

3.10 i) Find the distance and bearing of Sean relative to his starting point if he goes
for a walk with the following three stages.
Stage 1: 600 m on a bearing 030°
Stage 2: 1 km on a bearing 100°
Stage 3: 700 m on a bearing 340°
  ii) Shona sets off from the same place at the same time as Sean. She walks at the
same speed but takes the stages in the order 3–1–2.
How far apart are Sean and Shona at the end of their walks?

Draw diagrams to help you answer exercises **3.11** *and* **3.12**.

3.11 A plane leaves Heathrow airport at noon travelling due east towards Prague,
1200 km away. The speed of the plane in still air is 400 km h^{-1}.
  i) Initially there is no wind. Estimate the time of arrival of the plane.

After flying for one hour the plane runs into a storm with strong winds of
75 km h^{-1} from the south-west.
  ii) If the pilot fails to adjust the heading of the plane how many kilometres from
Prague will the plane be after the estimated journey time?
  iii) What new course should the pilot set?

3.12 Rain falling vertically hits the windows of an intercity train which is travelling at
50 ms^{-1}. A passenger watches the rain streaks on the window and estimates them
to be at an angle of 20° to the horizontal.
  i) Calculate the speed the rain would run down a stationary window.
  ii) The train starts to slow down with a uniform deceleration and 1.5 minutes later
the streaks are at 45° to the horizontal. What is the deceleration of the train?

3.13 A rocket moves with a velocity (in ms^{-1}) modelled by

$$\mathbf{v} = \tfrac{1}{10}t\,\mathbf{i} + \tfrac{1}{10}t^2\,\mathbf{j}$$

where $\mathbf{i}$ and $\mathbf{j}$ are horizontal and vertical unit vectors respectively and t is in seconds. Find

i) an expression for its position vector relative to its starting position at time t

ii) the displacement of the rocket after 10 s of its flight.

3.14 A speedboat is initially moving at $5\ \text{ms}^{-1}$ on a bearing of $135°$.

i) Express the initial velocity as a vector in terms of $\mathbf{i}$ and $\mathbf{j}$, which are unit vectors east and north respectively.

The boat then begins to accelerate with an acceleration modelled by

$$\mathbf{a} = 0.1t\,\mathbf{i} + 0.3t\,\mathbf{j} \quad \text{in ms}^{-2}.$$

ii) Find the velocity of the boat 10 s after it begins to accelerate and its displacement over the 10 s period.

3.15 Ship A is 5 km due west of ship B and is travelling on a course $035°$ at a constant but unknown speed $v\ \text{km h}^{-1}$. Ship B is travelling at a constant $10\ \text{km h}^{-1}$ on a course $300°$.

i) Write the velocity of each ship in terms of unit vectors $\mathbf{i}$ and $\mathbf{j}$ with directions east and north.

ii) Find the position vector of each ship at time t hours, relative to the starting position of ship A.

The ships are on a collision course.

iii) Find the speed of ship A.

iv) How much time elapses before the collision occurs?

Personal Tutor Exercise

3.1 A small, delicate microchip which is initially at rest is to be moved by a robot arm so that it is placed **gently** on to a horizontal assembly bench. Two mathematical models have been proposed for the motion which will be programmed into the robot. In each model the unit of length is the centimetre and time is measured in seconds. The unit vectors $\mathbf{i}$ and $\mathbf{j}$ have directions which are horizontal and vertical respectively and the origin is the point O on the surface of the bench, as shown in the diagram.

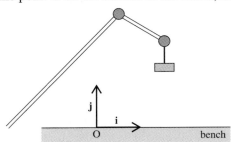

Model A for the position vector of the microchip at time t is

$$\mathbf{r}_A = 5t^2\,\mathbf{i} + (16 - 4t^2)\,\mathbf{j}, \quad (t \geqslant 0).$$

i)   How far above the table is the microchip initially (i.e. when $t = 0$)?

ii) Show that this model predicts that the microchip reaches the table after 2 s and state the horizontal distance moved in this time.

iii) Calculate the predicted horizontal and vertical components of velocity when $t = 0$ and $t = 2$.

Model B for the position vector at time t of the microchip is

$$\mathbf{r}_B = (15t^2 - 5t^3)\,\mathbf{i} + (16 - 24t^2 + 16t^3 - 3t^4)\,\mathbf{j}, \quad (t \geqslant 0).$$

iv) Show that model B predicts the same positions for the microchip at $t = 0$ and $t = 2$ as model A.

v)   Calculate the predicted horizontal and vertical components of velocity for the microchip at $t = 0$ and $t = 2$ from model B and comment, with brief reasons, on which model you think describes the more suitable motion for the microchip.

3.1 A vector has both magnitude and direction.

3.2 Vectors may be represented in either magnitude–direction form or in component form.

Magnitude–direction form   **Component form**

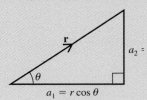   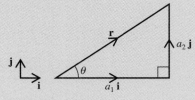

Magnitude, r; direction, θ $a_1\,\mathbf{i} + a_2\,\mathbf{j}$ or $\begin{pmatrix} a_1 \\ a_2 \end{pmatrix}$

Where $r = \sqrt{a_1{}^2 + a_2{}^2}$ $a_1 = r\cos\theta$

and $\tan\theta = \dfrac{a_2}{a_1}$ $a_2 = r\sin\theta$

$\qquad\qquad\qquad\qquad = r\cos(90 - \theta)$

3.3 When two or more vectors are added, the *resultant* is obtained. Vector addition may be done graphically or algebraically.

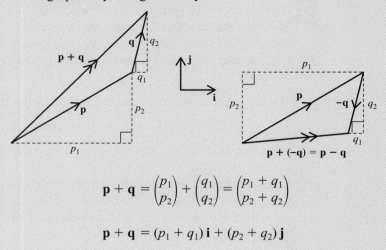

$$\mathbf{p} + \mathbf{q} = \begin{pmatrix} p_1 \\ p_2 \end{pmatrix} + \begin{pmatrix} q_1 \\ q_2 \end{pmatrix} = \begin{pmatrix} p_1 + q_1 \\ p_2 + q_2 \end{pmatrix}$$

$$\mathbf{p} + \mathbf{q} = (p_1 + q_1)\,\mathbf{i} + (p_2 + q_2)\,\mathbf{j}$$

3.4 Multiplication by a scalar

$$n(a_1\,\mathbf{i} + a_2\,\mathbf{j}) = na_1\,\mathbf{i} + na_2\,\mathbf{j} \qquad n\begin{pmatrix} a_1 \\ a_2 \end{pmatrix} = \begin{pmatrix} na_1 \\ na_2 \end{pmatrix}$$

3.5 The *position vector* of a point P is $\overrightarrow{OP}$, its displacement from a fixed origin.

3.6 When A and B have position vectors $\mathbf{a}$ and $\mathbf{b}$, $\overrightarrow{AB} = \mathbf{b} - \mathbf{a}$.

3.7 Equal vectors have equal magnitude and are in the same direction.

$$p_1\,\mathbf{i} + p_2\,\mathbf{j} = q_1\,\mathbf{i} + q_2\,\mathbf{j} \Rightarrow p_1 = q_1 \quad \text{and} \quad p_2 = q_2$$

3.8 When $p_1\,\mathbf{i} + p_2\,\mathbf{j}$ and $q_1\,\mathbf{i} + q_2\,\mathbf{j}$ are parallel, $\dfrac{p_1}{q_1} = \dfrac{p_2}{q_2}$.

3.9 The unit vector in the direction of $a_1\,\mathbf{i} + a_2\,\mathbf{j}$ is $\dfrac{a_1}{\sqrt{a_1{}^2 + a_2{}^2}}\,\mathbf{i} + \dfrac{a_2}{\sqrt{a_2{}^2 + a_2{}^2}}\,\mathbf{j}$.

3.10 Relationships between the variables describing motion

	Position	$\rightarrow$	Velocity	$\rightarrow$	Acceleration
			differentiate		

In one dimension $\quad s \qquad\qquad v = \dfrac{\mathrm{d}s}{\mathrm{d}t} \qquad\qquad a = \dfrac{\mathrm{d}v}{\mathrm{d}t} = \dfrac{\mathrm{d}^2 s}{\mathrm{d}t^2}+$

In two dimensions $\quad \mathbf{r} = x\,\mathbf{i} + y\,\mathbf{j} \qquad \mathbf{v} = \dfrac{\mathrm{d}\mathbf{r}}{\mathrm{d}t} = \dot{x}\,\mathbf{i} + \dot{y}\,\mathbf{j} \quad \mathbf{a} = \dfrac{\mathrm{d}\mathbf{v}}{\mathrm{d}t} = \ddot{x}\,\mathbf{i} + \ddot{y}\,\mathbf{j}$

$$= \begin{pmatrix} x \\ y \end{pmatrix} \qquad\qquad = \begin{pmatrix} \dot{x} \\ \dot{y} \end{pmatrix} \qquad\qquad = \begin{pmatrix} \ddot{x} \\ \ddot{y} \end{pmatrix}$$

	Acceleration	$\rightarrow$	Velocity	$\rightarrow$	Position
			integrate		

In one dimension $\quad a \qquad\qquad v = \displaystyle\int a\,\mathrm{d}t \qquad s = \displaystyle\int v\,\mathrm{d}t$

In two dimensions $\quad \mathbf{a} \qquad\qquad \mathbf{v} = \displaystyle\int \mathbf{a}\,\mathrm{d}t \qquad \mathbf{r} = \displaystyle\int \mathbf{v}\,\mathrm{d}t$

3.11 Acceleration may be due to change in direction or change in speed or both.

3.12 If the acceleration is *constant*:

$$\mathbf{v} = \mathbf{u} + \mathbf{a}t \quad \mathbf{r} = \mathbf{r}_0 + \mathbf{u}\,t + \tfrac{1}{2}\mathbf{a}t^2 \quad \mathbf{r} = \mathbf{r}_0 + \tfrac{1}{2}(\mathbf{u} + \mathbf{v})t.$$

3.13 Using vectors, Newton's second law is $\mathbf{F} = m\mathbf{a}$.

4 Projectiles

Swift of foot was Hiawatha;
He could shoot an arrow from him,
And run forward with such fleetness,
That the arrow fell behind him!
Strong of arm was Hiawatha;
He could shoot ten arrows upwards,
Shoot them with such strength and swiftness,
That the last had left the bowstring,
Ere the first to earth had fallen!

The Song of Hiawatha, Longfellow

Look at the water jet in the picture. Every drop of water in a water jet follows its own path which is called its *trajectory*. You can see the same sort of trajectory if you throw a small object across a room. Its path is a parabola. Objects moving through the air like this are called projectiles.

4.1 Modelling assumptions for projectile motion

The path of a cricket ball looks parabolic, but what about a boomerang? There are modelling assumptions which must be satisfied for the motion to be parabolic. These are

- a projectile is a particle
- it is not powered
- the air has no effect on its motion.

Equations for projectile motion

A projectile moves in two dimensions under the action of only one force, the force of gravity, which is constant and acts vertically downwards. This means that the acceleration of the projectile is g ms^{-2} vertically downwards and there is no horizontal acceleration. You can treat the horizontal and vertical motion separately using the equations for constant acceleration.

To illustrate the ideas involved, think of a ball being projected with a speed of 20 ms^{-1} at 60° to the ground as illustrated in figure 4.1. This could be a first model for a football, a chip shot from the rough at golf or a lofted shot at cricket.

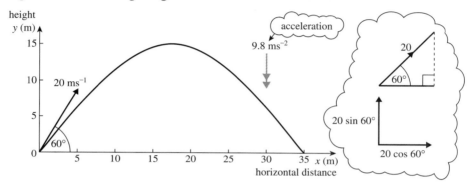

Figure 4.1

Using axes as shown, the components are:

	Horizontal	Vertical
Initial position	0	0
Acceleration	$a_x = 0$	$a_y = -9.8$

> This is negative because the positive y axis is upwards.

> as vectors $\mathbf{a} = \begin{pmatrix} 0 \\ -9.8 \end{pmatrix}$

Initial velocity	$u_x = 20 \cos 60°$	$u_y = 20 \sin 60°$
	$= 10$	$= 17.32$

> $\mathbf{u} = \begin{pmatrix} 20 \cos 60° \\ 20 \sin 60° \end{pmatrix}$

Using $v = u + at$ in the two directions gives the components of velocity.

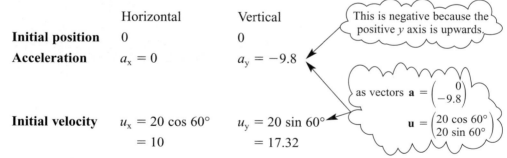

Velocity   Horizontal   Vertical

$a_x = 0 \Rightarrow v_x$ is constant

$v_x = 20 \cos 60°$ $v_y = 20 \sin 60° - 9.8t$

$v_x = 10$ ① $v_y = 17.32 - 9.8t$ ②

> $\mathbf{v} = \begin{pmatrix} 10 \\ 17.32 - 9.8t \end{pmatrix}$

Using $s = ut + \frac{1}{2}at^2$ in both directions gives the components of position.

Position   Horizontal   Vertical

$x = (20 \cos 60°)t$ $y = (20 \sin 60°)t - 4.9t^2$

$x = 10t$ ③ $y = 17.32t - 4.9t^2$ ④

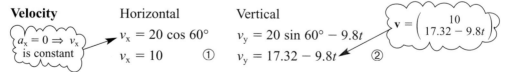

> $\mathbf{v} = \begin{pmatrix} 10 \\ 17.32 - 9.8t \end{pmatrix}$.

You can summarise these results in a table.

	Horizontal motion		Vertical motion	
initial position	0		0	
a	0		−9.8	
u	$u_x = 20 \cos 60° = 10$		$u_y = 20 \sin 60° = 17.32$	
v	$v_x = 10$	①	$v_y = 17.32 - 9.8t$	②
r	$x = 10t$	③	$y = 17.32t - 4.9t^2$	④

The four equations ①, ②, ③ and ④ for velocity and position can be used to find several things about the motion of the ball.

QUESTION 4.1   What is true at

i)   the top-most point of the path of the ball?

ii)   the point where it is just about to hit the ground?

When you have decided the answer to these questions you have sufficient information to find the greatest height reached by the ball, the time of flight and the total distance travelled horizontally before it hits the ground. This is called the range of the ball.

The maximum height

When the ball is at its maximum height, H m, the *vertical* component of its velocity is zero. It still has a horizontal component of 10 ms^{-1} which is constant.

Equation ② gives the vertical component as

$$v_y = 17.32 - 9.8t$$

At the top: $0 = 17.32 - 9.8t$

$$t = \frac{17.32}{9.8}$$

$$= 1.767$$

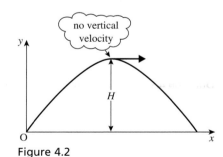

Figure 4.2

To find the maximum height, you now need to find y at this time. Substituting for t in equation ④ ,

$$y = 17.32t - 4.9t^2$$

$$y = 17.32 \times 1.767 - 4.9 \times 1.767^2$$

$$= 15.3.$$

The maximum height is 15.3 m.

The time of flight

The flight ends when the ball returns to the ground, that is when $y = 0$. Substituting $y = 0$ in equation ④,

$$y = 17.32t - 4.9t^2$$

$$17.32t - 4.9t^2 = 0$$

$$t(4.9t - 17.3) = 0$$

$$t = 0 \text{ or } t = 3.53$$

Clearly $t = 0$ is the time when the ball is thrown, so $t = 3.53$ is the time when it lands and the flight time is 3.53 s.

The range

The range, R m, of the ball is the horizontal distance it travels before landing.

R is the value of x when $y = 0$.

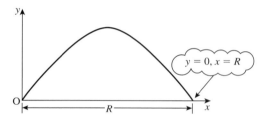

Figure 4.3

R can be found by substituting $t = 3.53$ in equation ③: $x = 10t$. The range is $10 \times 3.53 = 35.3$ m.

QUESTION 4.2 Notice in this example that the time to maximum height is half the flight time. Is this always the case?

QUESTION 4.3 Decide which of the following could be modelled as projectiles.

a balloon	a bird	a bullet shot from a gun	a glider
a golf ball	a parachutist	a rocket	a tennis ball

What special conditions would have to apply in particular cases?

Projectile problems

When doing projectile problems, you can treat each direction separately or you can write them both together as vectors. Example 4.1 shows both methods.

EXAMPLE 4.1

A ball is thrown horizontally at 5 ms^{-1} out of a window 4 m above the ground.
i)   How long does it take to reach the ground?
ii) How far from the building does it land?
iii) What is its speed just before it lands and at what angle to the ground is it moving?

SOLUTION

The diagram shows the path of the ball. It is important to decide at the outset where the origin and axes are. You may choose any axes that are suitable, but you must specify them carefully to avoid making mistakes. Here the origin is taken to be at ground level below the point of projection of the ball and upwards is positive. With these axes, the acceleration is $-g \text{ ms}^{-2}$.

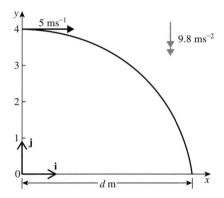

Figure 4.4

Method 1: Resolving into components

Position: Using axes as shown and $s = s_0 + ut + \frac{1}{2}at^2$ in the two directions

Horizontally $x_0 = 0$, $u_x = 5$, $a_x = 0$
$$x = 5t \qquad \qquad \text{①}$$

Vertically $y_0 = 4$, $u_y = 0$, $a_y = -9.8$
$$y = 4 - 4.9t^2 \qquad \qquad \text{②}$$

i)   The ball reaches the ground when $y = 0$. Substituting in equation ② gives

$$0 = 4 - 4.9t^2$$
$$t^2 = \frac{4}{4.9}$$
$$t = 0.904$$

The ball hits the ground after 0.9s.

ii) When the ball lands $x = d$ so, from equation ①, $d = 5t = 5 \times 0.904 = 4.52$. The ball lands 4.52 m from the building.

Velocity: Using $v = u + at$ in the two directions

Horizontally $v_x = 5 + 0$

Vertically $v_y = 0 - 9.8t$

iii) To find the speed and direction just before it lands: The ball lands when $t = 0.904$ so $v_x = 5$ and $v_y = -8.86$.

The components of velocity are shown in the diagram.

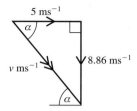

Figure 4.5

The speed of the ball is $\sqrt{5^2 + 8.86^2} = 10.17\,\text{ms}^{-1}$. It hits the ground moving downwards at an angle α to the horizontal where

$$\tan \alpha = \frac{8.86}{5}$$
$$\alpha = 60.6°$$

Method 2: Using vectors

Using the unit vectors shown, the initial position is $\mathbf{r}_0 = 4\mathbf{j}$ and the ball hits the ground when $\mathbf{r} = d\,\mathbf{i}$. The initial velocity, $\mathbf{u} = 5\mathbf{i}$ and the acceleration $\mathbf{a} = -9.8\mathbf{j}$. In column vectors these are

$$\mathbf{r}_0 = \begin{pmatrix} 0 \\ 4 \end{pmatrix} \quad \mathbf{r} = \begin{pmatrix} d \\ 0 \end{pmatrix} \quad \mathbf{u} = \begin{pmatrix} 5 \\ 0 \end{pmatrix} \quad \mathbf{a} = \begin{pmatrix} 0 \\ -9.8 \end{pmatrix}$$

Using

$$\mathbf{r} = \mathbf{r}_0 + \mathbf{u}t + \tfrac{1}{2}\mathbf{a}t^2$$

$$\begin{pmatrix} d \\ 0 \end{pmatrix} = \begin{pmatrix} 0 \\ 4 \end{pmatrix} + \begin{pmatrix} 5 \\ 0 \end{pmatrix}t + \tfrac{1}{2}\begin{pmatrix} 0 \\ -9.8 \end{pmatrix}t^2$$

$$d = 5t \qquad\qquad\qquad ①$$

and

$$0 = 4 - 4.9t^2 \qquad\qquad ②$$

i) Equation ② gives $t = 0.904$ and substituting this into ① gives **ii)** $d = 4.52$.

iii) The speed and direction of motion are the magnitude and direction of the velocity of the ball. Using

$$\mathbf{v} = \mathbf{u} + \mathbf{a}t$$

$$\begin{pmatrix} v_x \\ v_y \end{pmatrix} = \begin{pmatrix} 5 \\ 0 \end{pmatrix} + \begin{pmatrix} 0 \\ -9.8 \end{pmatrix}t$$

So when $t = 0.904$, $\begin{pmatrix} v_x \\ v_y \end{pmatrix} = \begin{pmatrix} 5 \\ -8.86 \end{pmatrix}$.

You can find the speed and angle as before.

Notice that in both methods the time forms a link between the motions in the two directions. You can often find the time from one equation and then substitute it in another to find out more information.

Representing projectile motion by vectors

The diagram shows a possible path for a marble which is thrown across a room from the moment it leaves the hand until just before it hits the floor.

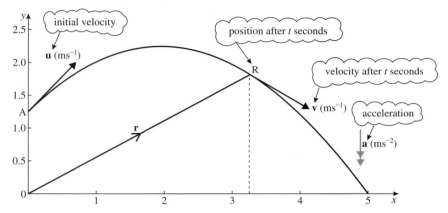

Figure 4.6

The vector $\mathbf{r} = \overrightarrow{OR}$ is the position vector of the marble after a time t seconds and the vector $\mathbf{v}$ represents its velocity in ms^{-1} at that instant of time (to a different scale).

⚠ Notice that the graph shows the trajectory of the marble. It is its path through space, not a position–time graph.

You can use equations for constant acceleration in vector form to describe the motion as in Example 4.1, Method 2.

velocity $\qquad\qquad\qquad \mathbf{v} = \mathbf{u} + \mathbf{a}t$

displacement $\qquad\qquad \mathbf{r} - \mathbf{r}_0 = \mathbf{u}t + \tfrac{1}{2}\mathbf{a}t \;$ so $\; \mathbf{r} = \mathbf{r}_0 + \mathbf{u}t + \tfrac{1}{2}\mathbf{a}t^2$

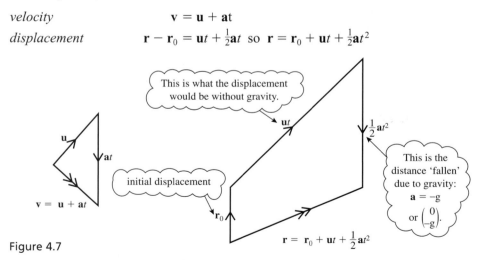

Figure 4.7

⚠ Always check whether or not the projectile starts at the origin. The change in position is the vector $\mathbf{r} - \mathbf{r}_0$. This is the equivalent of $s - s_0$ in one dimension.

Further examples

EXAMPLE 4.2

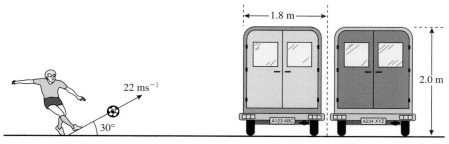

Figure 4.8

In this question use $10\,ms^{-2}$ for g and neglect air resistance.

In an attempt to raise money for a charity, participants are sponsored to kick a ball over some vans. The vans are each 2 m high and 1.8 m wide and stand on horizontal ground. One participant kicks the ball at an initial speed of $22\,ms^{-1}$ inclined at 30° to the horizontal.

i)   What are the initial values of the vertical and horizontal components of velocity?

ii) Show that while in flight the vertical height y metres at time t seconds satisfies the equation $y = 11t - 5t^2$.

Calculate at what times the ball is at least 2 m above the ground.

The ball should pass over as many vans as possible.

iii) Deduce that the ball should be placed about 3.8 m from the first van and find how many vans the ball will clear.

iv) What is the greatest vertical distance between the ball and the top of the vans?

SOLUTION

i)   Initial velocity
horizontally: $22 \cos 30° = 19.05\,ms^{-1}$
vertically: $22 \sin 30° = 11\,ms^{-1}$

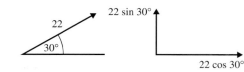

Figure 4.9

ii) *When the ball is above 2 m*
Using axes as shown and
$s = ut + \frac{1}{2}at^2$ vertically

$\Rightarrow \qquad y = 11t - 5t^2$

The ball is 2 m above the ground
when $y = 2$, then

$$2 = 11t - 5t^2$$
$$5t^2 - 11t + 2 = 0$$
$$(5t - 1)(t - 2) = 0$$
$$t = 0.2 \ \text{ or } \ 2$$

The ball is at least 2 m above the ground when $0.2 \leqslant t \leqslant 2$

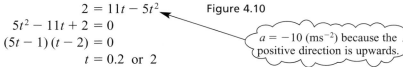

Figure 4.10

$a = -10\,(ms^{-2})$ because the positive direction is upwards.

iii) *How many vans?*

Horizontally, $s = ut + \frac{1}{2}at^2$ with $a = 0$

$$\Rightarrow \qquad x = 19.05t$$

When $t = 0.2$, $x = 3.81$ (at A)
when $t = 2$, $x = 38.1$ (at B)

To clear as many vans as possible,
the ball should be placed about
3.8 m in front of the first van.

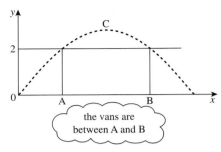

the vans are
between A and B

Figure 4.11

$$AB = 38.1 - 3.81 \text{ m} = 34.29 \text{ m}$$

$$\frac{34.29}{1.8} = 19.05.$$

The maximum possible number of vans is 19.

iv) *Maximum height*

At the top (C), vertical velocity $= 0$, so using $v = u + at$ vertically

$$\Rightarrow \qquad 0 = 11 - 10t$$
$$t = 1.1.$$

Substituting in $y = 11t - 5t^2$, maximum height is $11 \times 1.1 - 5 \times 1.1^2 = 6.05 \text{ m}$

The ball clears the tops of the vans by about 4 m.

EXAMPLE 4.3   Sharon is diving into a swimming pool. During her flight she may be modelled as a particle. Her initial velocity is 1.8 ms^{-1} at angle 30° above the horizontal and initial position 3.1 m above the water. Air resistance may be neglected.

i)   Find the greatest height above the water that Sharon reaches during her dive.

ii) Show that the time t, in seconds, that it takes Sharon to reach the water is given by $4.9t^2 - 0.9t - 3.1 = 0$ and solve this equation to find t.
Explain the significance of the other root to the equation.

Just as Sharon is diving a small boy jumps into the swimming pool. He hits the water at a point in line with the diving board and 1.5 m from its end.

iii) Is there an accident?

SOLUTION

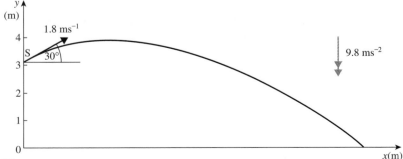

Figure 4.12

Referring to the axes shown

	Horizontal motion	Vertical motion
initial position	0	3.1
a	0	-9.8
u	$u_x = 1.8 \cos 30° = 1.56$	$u_y = 1.8 \sin 30° = 0.9$
v	$v_x = 1.56$ ①	$v_y = 0.9 - 9.8t$ ②
r	$x = 1.56t$ ③	$y = 3.1 + 0.9t - 4.9t^2$ ④

i)   At the top $v_y = 0$ $0 = 0.9 - 9.8t$ from ②
$$t = 0.092$$

When $t = 0.092$ $y = 3.1 + 0.9 \times 0.092 - 4.9 \times 0.092^2$ from ④
$$= 3.14$$

Sharon's greatest height above the water is 3.14 m.

ii)   Sharon reaches the water when $y = 0$
$$0 = 3.1 + 0.9t - 4.9t^2 \qquad \text{from ④}$$
$$4.9t^2 - 0.9t - 3.1 = 0$$
$$t = \frac{0.9 \pm \sqrt{0.9^2 + 4 \times 4.9 \times 3.1}}{9.8}$$
$$t = -0.71 \text{ or } 0.89$$

Sharon hits the water after 0.89 s. The negative value of t gives the point on the parabola at water level to the left of the point (S) where Sharon dives.

iii)   At time t the horizontal distance from diving board,
$$x = 1.56t \qquad \text{from ③}$$
When Sharon hits the water $x = 1.56 \times 0.89 = 1.39$

Assuming that the particles representing Sharon and the boy are located at their centres of mass, the difference of 11 cm between 1.39 m and 1.5 m is not sufficient to prevent an accident.

Note

When the point S is taken as the origin in the above example, the initial position is (0, 0) and $y = 0.9t - 4.9t^2$. In this case, Sharon hits the water when $y = -3.1$. This gives the same equation for t.

EXAMPLE 4.4   A boy kicks a small ball from the floor of a gymnasium with an initial velocity of 12 ms^{-1} inclined at an angle α to the horizontal. Air resistance may be neglected.

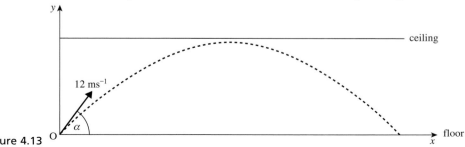

Figure 4.13

i)   Write down expressions in terms of α for the vertical speed of the ball and vertical height of the ball after t seconds.

The ball just fails to touch the ceiling which is 4 m high. The highest point of the motion of the ball is reached after T seconds.

ii)   Use one of your expressions to show that $6 \sin \alpha = 5T$ and the other to form a second equation involving $\sin\alpha$ and T.

iii)   Eliminate $\sin\alpha$ from your two equations to show that t has a value of about 0.89.

iv)   Find the horizontal range of the ball when kicked at 12 ms⁻¹ from the floor of the gymnasium so that it just misses the ceiling.

Use 10 ms⁻² for g in this question.

SOLUTION

i)   *Vertical components*

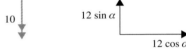

acceleration (ms⁻²)   initial velocity (ms⁻¹)

speed $v_y = 12 \sin \alpha - 10t$ ①

height $y = (12 \sin \alpha)t - 5t^2$ ②

Figure 4.14

ii)   *Time to highest point*
At the top $v_y = 0$ and $t = T$, so equation ① gives

$$12 \sin \alpha - 10T = 0$$
$$12 \sin \alpha = 10T$$
$$6 \sin \alpha = 5T \qquad ③$$

When $t = T$, $y = 4$ so from ②

$$4 = 12 \sin \alpha \, T - 5T^2 \qquad ④$$

iii)   Substituting for $6 \sin \alpha$ from ③ into ④ gives

$$4 = 2 \times 5T \times T - 5T^2$$
$$4 = 5T^2$$
$$T = \sqrt{0.8} = 0.89$$

iv)   *Range*
The path is symmetrical so the time of flight is $2T$ seconds. Horizontally $a = 0$ and $u_x = 12 \cos \alpha$

$$\Rightarrow \qquad x = (12 \cos \alpha) \, t$$

The range is $12 \cos \alpha \times 2T = 21.36 \cos \alpha$ m.

From ③ $6 \sin \alpha = 5T = 4.45$
$$\alpha = 47.87°$$

The range is $21.47 \cos 47.87° = 14.4$ m.

QUESTION 4.4   Two marbles start simultaneously from the same height. One (P) is dropped and the other (Q) is projected horizontally. Which reaches the ground first?

4.2 The path of a projectile

Look at the equations

$$x = 20t$$
$$y = 6 + 30t - 5t^2$$

They represent the path of a projectile.

QUESTION 4.5 What is the initial velocity of the projectile? What is its initial position? What value of g is assumed?

These equations give x and y in terms of a third variable t. (They are called *parametric equations* and t is the *parameter*.)

You can find the *cartesian equation* connecting x and y directly by eliminating t as follows

$$x = 20t \implies t = \frac{x}{20}$$

so

$$y = 6 + 30t - 5t^2$$

can be written as

$$y = 6 + 30 \times \frac{x}{20} - 5 \times \left(\frac{x}{20}\right)^2$$

$$y = 6 + 1.5x - \frac{x^2}{80} \longleftarrow \text{This is the cartesian equation.}$$

Accessible points

The equation of the path of a projectile can be used to decide whether certain points can be reached by the projectile. The next two examples illustrate how this can be done.

EXAMPLE 4.5 A projectile is launched from the origin with an initial velocity $20\ \text{ms}^{-1}$ at an angle of $30°$ to the horizontal.

i)   Write down the position of the projectile after time t.
ii) Show that the equation of the path is the parabola $y = 0.578x - 0.016x^2$.
iii) Find y when $x = 3$.
iv) Decide whether the projectile can hit a point 6 m above the ground.

SOLUTION

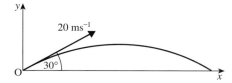

Figure 4.15

i) Using horizontal and vertical components for the position:

Horizontally: $x = (20 \cos 30°)t$

Vertically: $y = (20 \sin 30°)t - 4.9t^2$

$\Rightarrow \qquad x = 17.3t$ ①

and $\qquad y = 10t - 4.9t^2$ ②

ii) From equation ① $\qquad t = \dfrac{x}{17.3}$

Substituting this into equation ② for y gives

$$y = 10 \times \frac{x}{17.3} - 4.9 \times \frac{x^2}{17.3^2}$$

$$\Rightarrow \qquad y = 0.578x - 0.016x^2$$

iii) When $x = 3$ $\qquad y = 0.578 \times 3 - 0.016 \times 9 = 1.59$

iv) When $y = 6$ $\qquad 6 = 0.578x - 0.016x^2$

$$\Rightarrow \qquad 0.016x^2 - 0.578x + 6 = 0$$

In this quadratic equation the discriminant
$b^2 - 4ac = 0.578^2 - 4 \times 0.016 \times 6 = -0.0499$

You cannot find the square root of this negative number, so the equation cannot be solved for x. The projectile cannot hit a point 6 m above the ground. .

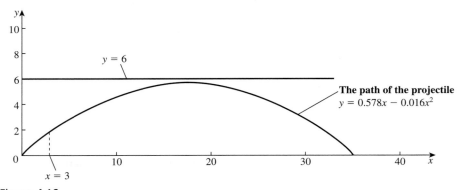

Figure 4.16

QUESTION 4.6 How high does the projectile reach?

Looking at the sign of the discriminant of a quadratic equation is a good way of deciding whether points are within the range of a projectile. The example above is a quadratic in x. Example 4.6 involves a quadratic equation in $\tan \alpha$ but the same idea is used in part **iii)** to determine whether certain points are within range.

EXAMPLE 4.6

A ball is hit from the origin with a speed of 14 ms^{-1} at an angle α to the horizontal.
i)   Find the equation of the path of the ball in terms of tan α.
ii) Find the two values of α which ensure that the ball passes through the point (5, 2.5).
iii) Decide whether the ball can pass through the points
    a) (10, 7.5)   **b)** (8, 9).

SOLUTION

i)   The components of the initial velocity are 14 cos α and 14 sin α, so the path of the ball is given by the equations

$$x = (14 \cos \alpha)t \quad \text{and} \quad y = (14 \sin \alpha)t - 4.9t^2$$

$$\Rightarrow \qquad t = \frac{x}{14 \cos \alpha} \quad \text{and} \quad y = 14 \sin \alpha \times \frac{x}{14 \cos \alpha} - 4.9 \times \frac{x^2}{(14 \cos \alpha)^2}$$

$$\Rightarrow \qquad\qquad\qquad y = x \tan \alpha - \frac{x^2}{40 \cos^2 \alpha}.$$

You can then express this in a more useful form by using two trigonometrical identities

$$\frac{1}{\cos \alpha} = \sec \alpha \quad \text{and} \quad \sec^2 \alpha = 1 + \tan^2\alpha$$

So the equation of the path is $y = x \tan \alpha - \frac{1}{40}x^2(1 + \tan^2\alpha)$ ①

ii) When $x = 5$ and $y = 2.5$ $2.5 = 5 \tan \alpha - \frac{25}{40}(1 + \tan^2\alpha)$

(multiply by $\frac{40}{25}$)   ➤ $4 = 8 \tan \alpha - (1 + \tan^2 \alpha)$

$$\tan^2 \alpha - 8 \tan \alpha + 5 = 0$$

This is a quadratic in tan α which has solution $\tan \alpha = \dfrac{8 \pm \sqrt{64 - 20}}{2}$.

So tan $\alpha = 7.32$ or 0.68 giving possible angles of projection of 82° and 34°.

iii) **a)** For the point (10, 7.5), equation ① gives

$$7.5 = 10 \tan \alpha - \frac{10^2}{40} (1 + \tan^2 \alpha)$$
$$7.5 = 10 \tan \alpha - 2.5 (1 + \tan^2 \alpha)$$
$$\tan^2\alpha - 4 \tan \alpha + 4 = 0 \quad\longleftarrow\quad \text{(divide by 2.5)}$$

The discriminant $b^2 - 4ac = 16 - 16 = 0$

This zero discriminant means that the equation has two equal roots tan $\alpha = 2$. The point (10, 7.5) lies on the path of the projectile but there is only one possible angle of projection.

b) For the point (8, 9), equation ① gives $9 = 8 \tan \alpha - \frac{8^2}{40} (1 + \tan^2 \alpha)$

$$1.6 \tan^2 \alpha - 8 \tan \alpha + 10.6 = 0$$

The discriminant is $64 - 4 \times 1.6 \times 10.6 = -3.84$ and this has no real square root.

This equation cannot be solved to find tan α so the projectile cannot pass through the point (8, 9) if it has an initial speed of 14 ms^{-1}.

The diagram shows all the results.

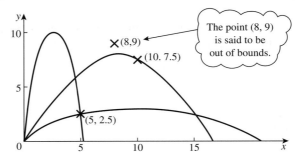

Figure 4.17

4.3 General equations

The work done in this chapter can now be repeated for the general case using algebra. Assume a particle is projected from the origin with speed u at an angle α to the horizontal and that the only force acting on the particle is the force due to gravity. The x and y axes are horizontal and vertical through the origin, O, in the plane of motion of the particle.

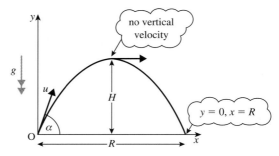

Figure 4.18

The components of velocity and position

	Horizontal motion		Vertical motion	
Initial position	0		0	
a	0		$-g$	
u	$u_x = u \cos \alpha$		$u_y = u \sin \alpha$	
v	$v_x = u \cos \alpha$	①	$v_y = u \sin \alpha - gt$	②
r	$x = ut \cos \alpha$	③	$y = ut \sin \alpha - \tfrac{1}{2}gt^2$	④

$ut \cos \alpha$ is preferable to $u \cos \alpha t$ because this could mean $u \cos (\alpha t)$ which is incorrect.

The maximum height

At its greatest height, the vertical component of velocity is zero.
From equation ②

$$u \sin \alpha - gt = 0$$

$$t = \frac{u \sin \alpha}{g}.$$

Substitute in equation ④ to obtain the height of the projectile:

$$y = u \times \frac{u \sin \alpha}{g} \times \sin \alpha - \tfrac{1}{2}g \times \frac{(u \sin \alpha)^2}{g^2}$$

$$= \frac{u^2 \sin^2\alpha}{g} - \frac{u^2 \sin^2\alpha}{2g}$$

The greatest height is $H = \dfrac{u^2 \sin^2\alpha}{2g}$

The time of flight

When the projectile hits the ground, $y = 0$.

From equation ④: $y = ut \sin \alpha - \tfrac{1}{2}gt^2$

$$0 = ut \sin \alpha - \tfrac{1}{2}gt^2$$

$$0 = t(u \sin \alpha - \tfrac{1}{2}gt)$$

The time of flight is $t = \dfrac{2u \sin \alpha}{g}.$

The range

The range of the projectile is the value of x when $t = \dfrac{2u \sin \alpha}{g}$

From equation ③: $x = ut \cos \alpha$

$$\Rightarrow \quad R = u \times \frac{2u \sin \alpha}{g} \times \cos \alpha$$

$$R = \frac{2u^2 \sin \alpha \cos \alpha}{g}$$

It can be shown that $2 \sin \alpha \cos \alpha = \sin 2\alpha$, so the range can be expressed as

$$R = \frac{u^2 \sin 2\alpha}{g}$$

The range is a maximum when $\sin 2\alpha = 1$, that is when $2\alpha = 90°$ or $\alpha = 45°$. The maximum possible horizontal range for projectiles with initial speed u is

$$R_{\max} = \frac{u^2}{g}.$$

The equation of the path

From equation ③ $t = \dfrac{x}{u \cos \alpha}$

Substitute into equation ④ to give

$$y = u \times \frac{x}{u \cos \alpha} \times \sin \alpha - \tfrac{1}{2}g \times \frac{x^2}{(u \cos \alpha)^2}$$

$$y = x \frac{\sin \alpha}{\cos \alpha} - \frac{gx^2}{2u^2 \cos^2 \alpha}$$

$$y = x \tan \alpha - \frac{gx^2}{2u^2}(1 + \tan^2 \alpha).$$

 It is important that you understand the methods used to derive these formulae and don't rely on learning the results off by heart. They are only true when the given assumptions apply and the variables are as defined in Figure 4.18.

QUESTION 4.7   What are the assumptions on which this work is based?

Chapter 4 Exercises

*In exercises **4.1–4.4** take upwards as positive and use 9.8 ms^{-2} for g unless otherwise stated. All the projectiles start at the origin, unless otherwise stated.*

4.1 In each case find
   a) the time taken for the projectile to reach its highest point
   b) the maximum height.
       i)   Initial velocity 5 ms^{-1} horizontally and 14.7 ms^{-1} vertically.
       ii) Initial velocity 10 ms^{-1} at 30° above the horizontal. Use 10 ms^{-2} for *g*.

4.2 In each case find
   a) the time of flight of the projectile
   b) the horizontal range.
       i)   Initial velocity 20 ms^{-1} horizontally and 19.6 ms^{-1} vertically.
       ii) Initial velocity 5 ms^{-1} at 60° above the horizontal.

4.3 In each case find
   a) the time taken for the projectile to reach its highest point
   b) the maximum height above the origin.
       i)   Initial position (0, 15 m) velocity 5 ms^{-1} horizontally and 14.7 ms^{-1} vertically.
       ii) Initial position (0, 10 m); initial velocity $\begin{pmatrix} 5 \\ 3 \end{pmatrix}$ ms^{-1}.

4.4 Find the horizontal range for these projectiles which start from the origin.

i) Initial velocity $\begin{pmatrix} 2 \\ 7 \end{pmatrix}$ ms^{-1}.   ii) Initial velocity $\begin{pmatrix} 7 \\ 2 \end{pmatrix}$ ms^{-1}.

iii) Sketch the paths of these two projectile using the same axes.

*Use 9.8 ms^{-2} for g in exercises **4.5–4.9** unless otherwise specified.*

4.5 Clare scoops a hockey ball off the ground, giving it an initial velocity of 19 ms^{-1} at 25° to the horizontal.

i) Find the horizontal and vertical components of the ball's initial velocity.

ii) Find the time that elapses before the ball hits the ground.

iii) Find the horizontal distance the ball travels before hitting the ground.

iv) Find how long it takes for the ball to reach maximum height.

v) Find the maximum height reached.

A member of the opposing team is standing 20 m away from Clare in the direction of the ball's flight.

vi) How high is the ball when it passes her? Can she stop the ball?

4.6 A footballer is standing 30 m in front of the goal. He kicks the ball towards the goal with velocity 18 ms^{-1} and angle 55° to the horizontal. The height of the goal's crossbar is 2.5 m. Air resistance and spin may be ignored.

i) Find the horizontal and vertical components of the ball's initial velocity.

ii) Find the time it takes for the ball to cross the goal − line.

iii) Does the ball bounce in front of the goal, go straight into the goal or go over the crossbar?

In fact the goalkeeper is standing 5 m in front of the goal and will stop the ball if its height is less than 2.8 m when it reaches him.

iv) Does the goalkeeper stop the ball?

4.7 A stunt motorcycle rider attempts to jump over a gorge 50 m wide. He uses a ramp at 25° to the horizontal for his take − off and has a speed of 30 ms^{-1} at this time.

i) Assuming that air resistance is negligible, find out whether the rider crosses the gorge successfully.

The stunt man actually believes that in any jump the effect of air resistance is to reduce his distance by 40%.

ii) Calculate his minimum safe take-off speed for this jump.

4.8 A horizontal tunnel has a height of 3 m. A ball is thrown inside the tunnel with an initial speed of 18 ms^{-1}. What is the greatest horizontal distance that the ball can travel before it bounces for the first time?

4.9 Use $g = 10$ ms^{-2} in this question.

A firework is buried so that its top is at ground level and it projects sparks all at a speed of 8 ms^{-1}. Air resistance may be neglected.

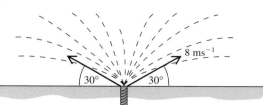

   i) Calculate the height reached by a spark projected vertically and explain why no spark can reach a height greater than this.

   ii) For a spark projected at $30°$ to the horizontal over horizontal ground, show that its height in metres t seconds after projection is $4t - 5t^2$ and hence calculate the distance it lands from the firework.

   iii) For what angle of projection will a spark reach a maximum height of 2 m?

*Use $10\,ms^{-2}$ for g in exercises **4.10**–**4.12** unless otherwise instructed, and use the modelling assumptions that air resistance can be ignored and the ground is horizontal.*

4.10 Jack throws a cricket ball at a wicket 0.7 m high with velocity 10 ms^{-1} at $14°$ above the horizontal. The ball leaves his hand 1.5 m above the origin.

   i) Show that the equation of the path is the parabola $y = 1.5 + 0.25x - 0.053x^2$.

   ii) How far from the wicket is he standing if the ball just hits the top?

4.11 Tennis balls are delivered from a machine at a height of 1 m and with a speed of 25 ms^{-1}. Dan stands 15 m from the machine and can reach to a height of 2.5 m.

   i) Write down the equations for the horizontal and vertical displacements x and y after time t seconds for a ball which is delivered at an angle α to the horizontal.

   ii) Eliminate t from your equations to obtain the equation for y in terms of x and $\tan \alpha$.

   iii) Can Dan reach a ball when $\tan \alpha$ is **a)** 0.5   **b)** 0.2?

   iv) Determine the values of α for which Dan can just hit the ball.

4.12 A golf ball is driven from the tee with speed $30\sqrt{2}\text{ ms}^{-1}$ at an angle α to the horizontal.

   i) Show that during its flight the horizontal and vertical displacements x and y of the ball from the tee satisfy the equation

$$y = x \tan \alpha - \frac{x^2}{360}(1 + \tan^2 \alpha).$$

   ii) The golf ball just clears a tree 5 m high which is 150 m horizontally from the tee. Find the two possible values of $\tan \alpha$.

   iii) Use the discriminant of the quadratic equation in $\tan \alpha$ to find the greatest distance by which the golf ball can clear the tree and find the value of $\tan \alpha$ in this case.

   iv) The ball is aimed at the hole which is on the green immediately behind the tree. The hole is 160 m from the tee. What is the greatest height the tree could be without making it impossible to hit a hole in one?

   (Hint: $2 \sin \alpha \cos \alpha = \sin 2\alpha$)

Personal Tutor Exercise

 4.1 In this question use $10\,\text{ms}^{-2}$ for g.
When attempting to kick a ball from the
ground, a boy finds that the angle of projection
affects the speed at which he can kick the ball.
The horizontal and vertical components of
projection are $10(1 + 2k)\,\text{ms}^{-1}$ and $10(1 - k)\,\text{ms}^{-1}$
respectively, where k is a number such that $0 \leqslant k \leqslant 1$. The ball is kicked over
horizontal ground and air resistance may be neglected.

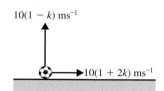

  i) Find the speed and angle of projection in the cases
     a) $k = 0$ **b)** $k = 1$.
  ii) Find expressions in terms of k for the horizontal and vertical positions of the
  ball t seconds after projection.
  iii) Find, in terms of k, how long the ball is in the air before it hits the ground
  again.
  iv) Show that the range is $20(1 + k - 2k^2)$ m.

  Given that $20(1 + k - 2k^2)$ can be written as $\frac{5}{2}(9 - 16(k - \frac{1}{4})^2)$ for all k, write
  down the value of the maximum range.

4.1 Modelling assumptions for projectile motion with acceleration due to gravity
  ● a projectile is a particle
  ● it is not powered
  ● the air has no effect on its motion.

4.2 Projectile motion is usually considered in terms of horizontal and vertical
components.

When the initial position is at O

Angle of projection $= \alpha$

Initial velocity, $\mathbf{u} = \begin{pmatrix} u \cos \alpha \\ u \sin \alpha \end{pmatrix}$

Acceleration, $\mathbf{g} = \begin{pmatrix} 0 \\ -g \end{pmatrix}$

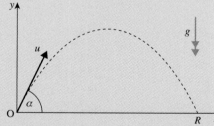

● At time t, velocity, $\mathbf{v} = \mathbf{u} + \mathbf{a}t$

$$\begin{pmatrix} v_x \\ v_y \end{pmatrix} = \begin{pmatrix} u \cos \alpha \\ u \sin \alpha \end{pmatrix} + \begin{pmatrix} 0 \\ -g \end{pmatrix} t$$

$$v_x = u \cos \alpha \qquad \text{①}$$

$$v_y = u \sin \alpha - gt \qquad \text{②}$$

● Displacement, $\mathbf{r} = \mathbf{u}t + \frac{1}{2}\mathbf{a}t^2$

$$\begin{pmatrix} x \\ y \end{pmatrix} = \begin{pmatrix} u\cos\alpha \\ u\sin\alpha \end{pmatrix}t + \frac{1}{2}\begin{pmatrix} 0 \\ -g \end{pmatrix}t^2$$

$$x = ut\cos\alpha \qquad\qquad ③$$

$$y = ut\sin\alpha - \frac{1}{2}gt^2 \qquad\qquad ④$$

4.3 At maximum height $v_y = 0$.

4.4 $y = 0$ when the projectile lands.

4.5 The time to hit the ground is twice the time to maximum height.

4.6 When the point of projection is (x_0, y_0) rather than $(0, 0)$

$$\mathbf{r} = \mathbf{r}_0 + \mathbf{u}t + \frac{1}{2}\mathbf{a}t^2 \qquad\qquad \begin{pmatrix} x \\ y \end{pmatrix} = \begin{pmatrix} x_0 \\ y_0 \end{pmatrix} + \begin{pmatrix} u\cos\alpha \\ u\sin\alpha \end{pmatrix}t + \frac{1}{2}\begin{pmatrix} 0 \\ -g \end{pmatrix}t^2.$$

5 Equilibrium of a particle

Give me matter and motion and I will construct the Universe.

René Descartes

5.1 Finding resultant forces

QUESTION 5.1 This cable car is stationary. Are the tensions in the cable greater than the weight of the car?

A child on a sledge is being pulled up a smooth slope of $20°$ by a rope which makes an angle of $40°$ with the slope. The mass of the child and sledge together is 20 kg and the tension in the rope is 170 N. Draw a diagram to show the forces acting on the child and sledge together. In what direction is the resultant of these forces?

When the child and sledge are modelled as a particle, all the forces can be assumed to be acting at a point. There is no friction force because the slope is smooth. Here is the force diagram.

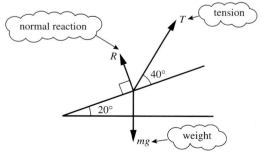

Figure 5.1

QUESTION 5.2 The sledge is sliding along the slope. What direction is the resultant force acting on it?

You can find the normal reaction and the resultant force on the sledge using two methods.

Method 1: Using components

This method involves resolving forces into components in two perpendicular directions as in Chapter 3. It is easiest to use the components of the forces parallel and perpendicular to the slope in the directions of **i** and **j** as shown.

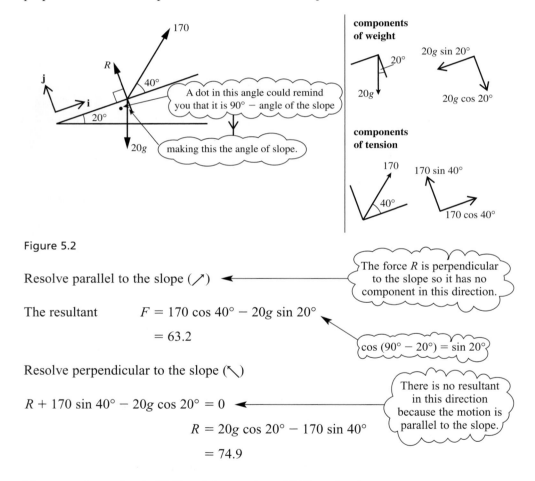

Figure 5.2

Resolve parallel to the slope (↗)

> The force R is perpendicular to the slope so it has no component in this direction.

The resultant $F = 170 \cos 40° - 20g \sin 20°$

$= 63.2$

> $\cos(90° - 20°) = \sin 20°$

Resolve perpendicular to the slope (↖)

> There is no resultant in this direction because the motion is parallel to the slope.

$R + 170 \sin 40° - 20g \cos 20° = 0$

$R = 20g \cos 20° - 170 \sin 40°$

$= 74.9$

The normal reaction is 75 N and the resultant 63 N up the slope.

Alternatively, you could have worked in column vectors as follows.

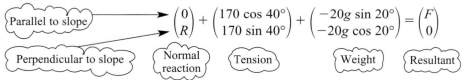

$$\binom{0}{R} + \binom{170 \cos 40°}{170 \sin 40°} + \binom{-20g \sin 20°}{-20g \cos 20°} = \binom{F}{0}$$

Parallel to slope

Perpendicular to slope

Normal reaction   Tension   Weight   Resultant

Note

Try resolving horizontally and vertically. You will obtain two equations in the two unknowns R and F. It is perfectly possible to solve these equations, but quite a lot of work. It is much easier to choose to resolve in directions which ensure that one component of at least one of the unknown forces is zero.

Once you know the resultant force, you can work out the acceleration of the sledge using Newton's second law.

$$F = ma$$
$$63.2 = 20a$$

The acceleration is $3.2\,\text{ms}^{-2}$ (correct to 1 d.p.).

Method 2: Scale drawing

An alternative is to draw a scale diagram with the three forces represented by three of the sides of a quadrilateral taken in order (with the arrows following each other) as shown in Figure 5.3. The resultant is represented by the fourth side AD. This must be parallel to the slope.

QUESTION 5.3   In what order would you draw the lines in the diagram?

From the diagram you can estimate the normal reaction to be about 80 N and the resultant 60 N. This is a reasonable estimate, but components are more precise.

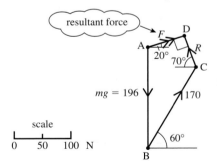

Figure 5.3

QUESTION 5.4   What can you say about the sledge in the cases when

i)   the length AD is not zero?
ii) the length AD is zero so that the starting point on the quadrilateral is the same as the finishing point?
iii) BC is so short that the point D is to the left of A as shown in the diagram?

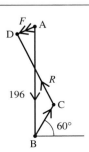

Figure 5.4

5.2 Forces in equilibrium

When forces are in equilibrium their vector sum is zero and the sum of their resolved parts in *any* direction is zero.

EXAMPLE 5.1

A brick of mass 3 kg is at rest on a rough plane inclined at an angle of 30° to the horizontal. Find the friction force F N, and the normal reaction R N of the plane on the brick.

SOLUTION

The diagram shows the forces acting on the brick.

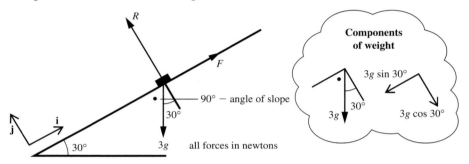

Figure 5.5

Take unit vectors **i** and **j** parallel and perpendicular to the plane as shown.

Since the brick is in equilibrium the resultant of the three forces acting on it is zero.

Resolving in the **i** direction: $F - 29.4 \sin 30° = 0$ ◄──────── $3g = 29.4$ ①

$$F = 14.7$$

Resolving in the **j** direction: $R - 29.4 \cos 30° = 0$ ②

$$R = 25.5$$

Written in vector form the equivalent is

$$\begin{pmatrix} F \\ 0 \end{pmatrix} + \begin{pmatrix} 0 \\ R \end{pmatrix} + \begin{pmatrix} -29.4 \sin 30° \\ -29.4 \cos 30° \end{pmatrix} = \begin{pmatrix} 0 \\ 0 \end{pmatrix}$$

or alternatively

$$F\mathbf{i} + R\mathbf{j} - 29.4 \sin 30° \, \mathbf{i} - 29.4 \cos 30° \, \mathbf{j} = \mathbf{0}.$$

Both these lead to the equations ① and ②.

The triangle of forces

When there are only three (non-parallel) forces acting and they are in equilibrium, the polygon of forces becomes a closed triangle as shown for the brick on the plane.

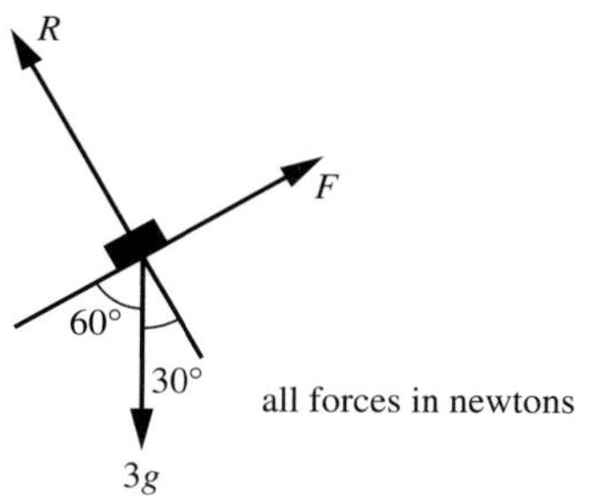

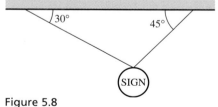

Figure 5.6

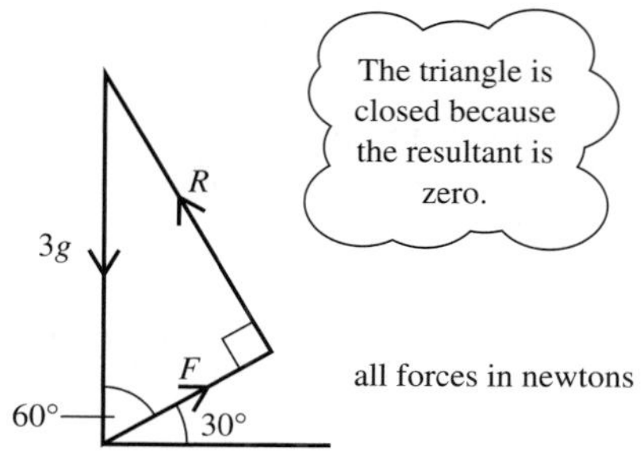

Figure 5.7

Then
$$\frac{F}{3g} = \cos 60°$$

$$F = 29.4 \cos 60° = 14.7\,\text{N}$$

and similarly $\quad R = 29.4 \sin 60° = 25.5\,\text{N}$

This is an example of the theorem known as the *triangle of forces*.

> *When a body is in equilibrium under the action of three non-parallel forces, then*
> **i)** *the forces can be represented in magnitude and direction by the sides of a triangle*
> **ii)** *the lines of action of the forces pass through the same point.*

When more than three forces are in equilibrium the first statement still holds but the triangle is then a polygon. The second is not necessarily true.

EXAMPLE 5.2

This example illustrates two methods for solving problems involving forces in equilibrium. With experience, you will find it easier to judge which method is best for a particular problem.

A sign of mass 10 kg is to be suspended by two strings arranged as shown in the diagram opposite. Find the tension in each string.

Figure 5.8

SOLUTION

The force diagram for this situation is given below.

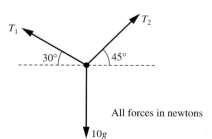

Figure 5.9

Method 1: Resolving forces

Vertically ($\uparrow$): $T_1 \sin 30° + T_2 \sin 45° - 10g = 0$

$$0.5T_1 + 0.707T_2 = 98 \qquad \text{①}$$

Horizontally ($\rightarrow$): $-T_1 \cos 30° + T_2 \cos 45° = 0$

$$-0.866T_1 + 0.707T_2 = 0 \qquad \text{②}$$

Subtracting ② from ① $1.366T_1 = 98$

$$T_1 = 71.74$$

Back substitution gives $T_2 = 87.87$

The tensions are 71.7 N and 87.9 N (to 1 d.p.).

Method 2: Triangle of forces

Since the three forces are in equilibrium they can be represented by the sides of a triangle taken in order.

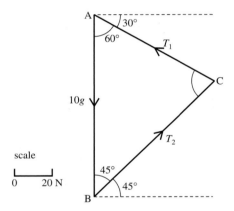

Figure 5.10

QUESTION 5.5   In what order would you draw the three lines in this diagram?

You can estimate the tensions by measurement. This will tell you that $T_1 \approx 72$ and $T_2 \approx 88$ in newtons.

Alternatively, you can use the sine rule to calculate T_1 and T_2 accurately.

In the triangle ABC, $\angle CAB = 60°$, $\angle ABC = 45°$ so $\angle BCA = 75°$ and so

$$\frac{T_1}{\sin 45°} = \frac{T_2}{\sin 60°} = \frac{98}{\sin 75°}$$

giving $$T_1 = \frac{98 \sin 45°}{\sin 75°}$$

and $$T_2 = \frac{98 \sin 60°}{\sin 75°}.$$

As before the tensions are found to be 71.7 N and 87.9 N.

QUESTION 5.6   Lami's theorem states that when three forces acting at a point as shown in the diagram are in equilibrium then

$$\frac{F_1}{\sin \alpha} = \frac{F_2}{\sin \beta} = \frac{F_3}{\sin \gamma}.$$

Sketch a triangle of forces and say how the angles in the triangle are related to α, β and γ.

Hence explain why Lami's theorem is true.

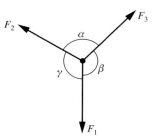

Figure 5.11

EXAMPLE 5.3   The picture shows three men involved in moving a packing case up to the top floor of a warehouse. Brian is pulling on a rope which passes round smooth pulleys at X and Y and is then secured to the point Z at the end of the loading beam.

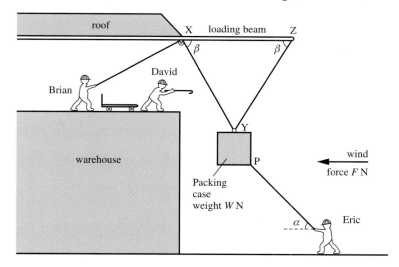

Figure 5.12

The wind is blowing directly towards the building. To counteract this, Eric is pulling on another rope, attached to the packing case at P, with just enough force and in the right direction to keep the packing case central between X and Z.

At the time of the picture the men are holding the packing case motionless.

i)   Draw a diagram showing all the forces acting on the packing case using T_1 and T_2 for the tensions in Brian and Eric's ropes, respectively.

ii)   Write down equations for the horizontal and vertical equilibrium of the packing case.

In one particular situation, $W = 200$, $F = 50$, $\alpha = 45°$ and $\beta = 75°$.

iii) Find the tension T_1.

iv)   Explain why Brian has to pull harder if the wind blows stronger.

SOLUTION

i) The diagram shows all the forces acting on the packing case and the relevant angles.

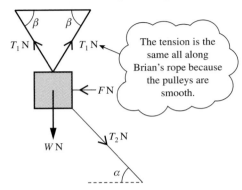

The tension is the same all along Brian's rope because the pulleys are smooth.

Figure 5.13: Force diagram

ii) Equilibrium equations

Resolving horizontally ($\rightarrow$)

$$T_1 \cos \beta + T_2 \cos \alpha - F - T_1 \cos \beta = 0$$

$$T_2 \cos \alpha - F = 0 \qquad ①$$

Resolving vertically ($\uparrow$)

$$T_1 \sin \beta + T_1 \sin \beta - T_2 \sin \alpha - W = 0$$

$$T_1 \sin \beta - T_2 \sin \alpha - W = 0 \qquad ②$$

iii) When $F = 50$ and $\alpha = 45°$ equation ① gives

$$T_2 \cos 45° = 50$$

$$\Rightarrow \quad T_2 \sin 45° = 50$$

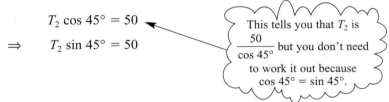

This tells you that T_2 is $\dfrac{50}{\cos 45°}$ but you don't need to work it out because $\cos 45° = \sin 45°$.

Substituting in ② gives $\qquad 2T_1 \sin \beta - 50 - W = 0$

So when $W = 100$ and $\beta = 75°$ $\qquad 2T_1 \sin 75° = 150$

$$T_1 = \frac{150}{2 \sin 75°}$$

The tension in Brian's rope is $77.65 \text{ N} = 78 \text{ N}$ (to the nearest N).

iv) When the wind blows harder, F increases. Given that all the angles remain unchanged, Eric will have to pull harder so the vertical component of T_2 will increase. This means that T_1 must increase and Brian must pull harder.

Or $F = T_2 \cos \alpha$, so as F increases, T_2 increases $\Rightarrow T_2 \sin \alpha + W$ increases $\Rightarrow 2T_1 \sin \beta$ increases. Hence T_1 increases

5.3 Newton's second law in two dimensions

When the forces acting on an object are not in equilibrium it will have an acceleration and you can use Newton's second law to solve problems about its motion.

The equation $\mathbf{F} = m\,\mathbf{a}$ is a vector equation. The resultant force acting on a particle is equal in both magnitude and direction to the mass × acceleration. It can be written in components as

$$\begin{pmatrix} F_1 \\ F_2 \end{pmatrix} = m \begin{pmatrix} a_1 \\ a_2 \end{pmatrix} \quad \text{or} \quad F_1\,\mathbf{i} + F_2\,\mathbf{j} = m\,(a_1\,\mathbf{i} + a_2\,\mathbf{j})$$

so that $F_1 = ma_1$ and $F_2 = ma_2$.

QUESTION 5.7   What direction is the resultant force acting on a child sliding on a sledge down a smooth straight slope inclined at 15° to the horizontal?

EXAMPLE 5.4

Sam and his sister are sledging, but Sam wants to ride by himself. His sister gives him a push at the top of a smooth straight 15° slope and lets go when he is moving at $2\,\text{ms}^{-1}$. He continues to slide for 5 seconds before using his feet to produce a braking force of 95 N parallel to the slope. This brings him to rest. Sam and his sledge have a mass of 30 kg. How far does he travel altogether?

SOLUTION

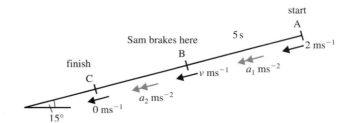

Figure 5.14

To answer this question, you need to know Sam's acceleration for the two parts of his journey. These are constant so you can then use the constant acceleration formulae.

Sliding freely

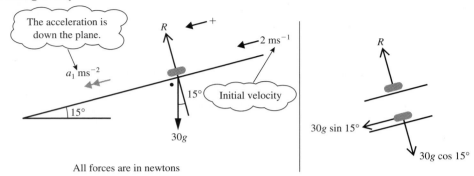

All forces are in newtons

Figure 5.15

Using Newton's second law in the direction of the acceleration gives

$$30g \sin 15° = 30a_1$$
$$a_1 = 2.54$$

Resultant force down the plane = mass × acceleration.

Now you know a_1 you can find how far Sam slides (s_1 m) and his speed (v ms^{-1}) before braking.

Given $u = 2, t = 5, a = 2.54$

$$s = ut + \tfrac{1}{2}at^2$$
$$s_1 = 2 \times 5 + 1.27 \times 25 = 41.75$$
$$v = u + at$$
$$v = 2 + 2.54 \times 5 = 14.7$$

Braking

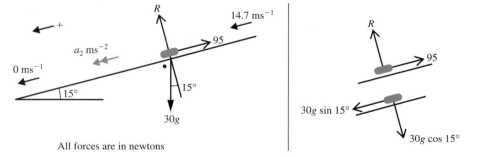

All forces are in newtons

Figure 5.16

By Newton's second law down the plane

$$\text{Resultant force} = \text{mass} \times \text{acceleration}$$
$$30g \sin 15° - 95 = 30a_2$$
$$a_2 = -0.63$$
$$v^2 = u^2 + 2as$$
$$0 = 14.7^2 - 2 \times 0.63 \times s_2$$

Given $u = 14.7, v = 0,$

$$s_2 = \frac{14.7^2}{1.26} = 171.5$$

Sam travels a total distance of $41.75 + 171.5 \text{ m} = 213 \text{ m}$ to the nearest metre.

QUESTION 5.8   Make a list of the modelling assumptions used in this example. What would be the effect of changing these?

EXAMPLE 5.5   A skier is being pulled up a smooth 25° dry ski slope by a rope which makes an angle of 35° with the horizontal. The mass of the skier is 75 kg and the tension in the rope is 350 N. Initially the skier is at rest at the bottom of the slope. The slope is smooth. Find the skier's speed after 5 s and find the distance he has travelled in that time.

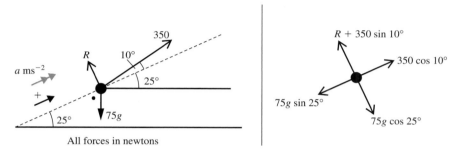

All forces in newtons

Figure 5.17

In the diagram the skier is modelled as a particle. Since the skier moves parallel to the slope consider motion in that direction.

$$\text{Resultant force} = \text{mass} \times \text{acceleration}$$
$$350 \cos 10° - 75g \sin 25° = 75 \times a$$

Taking g as 9.8 ⟶ $a = \dfrac{34.06}{75} = 0.454$ (to 3 d.p.).

This is a constant acceleration so use the constant acceleration formulae.

$$v = u + at$$
$$v = 0 + 0.454 \times 5 \quad \longleftarrow \quad u = 0, a = 0.454, t = 5$$
$$\text{Speed} = 2.27 \text{ ms}^{-1} \text{ (to 2 d.p.).}$$
$$s = ut + \tfrac{1}{2}at^2$$
$$s = 0 + \tfrac{1}{2} \times 0.454 \times 25$$
$$\text{Distance travelled} = 5.68 \text{ m (to 2 d.p.).}$$

EXAMPLE 5.6   A car of mass 1000 kg, including its driver, is being pushed along a horizontal road by three people as indicated in the diagram. The car is moving in the direction PQ.

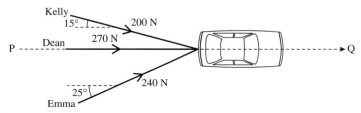

Figure 5.18

i) Calculate the total force exerted by the three people in the direction PQ.
ii) Calculate the force exerted overall by the three people in the direction perpendicular to PQ.
iii) Explain briefly why the car does not move in the direction perpendicular to PQ.

Initially the car is stationary and 5 s later it has a speed of 2 ms^{-1} in the direction PQ.
iv) Calculate the force of resistance to the car's movement in the direction PQ assuming the three people continue to push as described above.

SOLUTION

i) Resolving in the direction PQ, the components in newtons are:

Kelly $\qquad$ 200 cos 15° = 193

Dean $\qquad$ 270

Emma $\qquad$ 240 cos 25° = 218

Total force in the direction PQ = 681 N.

ii) Resolving perpendicular to PQ ($\uparrow$) the components are:

Kelly $\qquad$ -200 sin 15° = -51.8

Dean $\qquad$ 0

Emma $\qquad$ 240 sin 25° = 101.4

Total force in the direction perpendicular to PQ = 49.6 N.

iii) The car does not move perpendicular to PQ because the force in this direction is balanced by a sideways (lateral) friction force between the tyres and the road.

iv) To find the acceleration, a ms^{-2}, of the car:

$$v = u + at$$
$$2 = 0 + 5a \quad \longleftarrow \quad \boxed{u = 0, v = 2, t = 5}$$
$$a = 0.4$$

When the resistance to motion in the direction QP is R N, the diagram shows all the horizontal forces acting on the car and its acceleration.

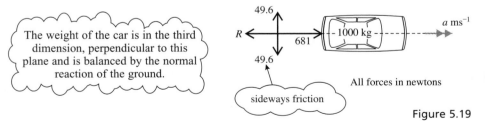

The weight of the car is in the third dimension, perpendicular to this plane and is balanced by the normal reaction of the ground.

49.6

$R \leftarrow$ | 1000 kg | a ms^{-1}

681

49.6

sideways friction

All forces in newtons

Figure 5.19

The resultant force in the direction PQ is $(681 - R)$ N. So by Newton II

$$681 - R = 1000a$$
$$R = 681 - 400$$

The resistance to motion in the direction PQ is 281 N.

Chapter 5 Exercises

For situations **5.1–5.2** below, carry out the following steps. All forces are in newtons.

i)   Draw a scale diagram to show the polygon of the forces and the resultant.

ii)   State whether you think the forces are in equilibrium and, if not, estimate the magnitude and direction of the resultant.

iii)   Write the forces in component form, using the directions indicated and so obtain the components of the resultant.

Hence find the magnitude and direction of the resultant as in Chapter 3, page 75.

iv)   Compare your answers to parts ii) and iii).

5.1

5.2

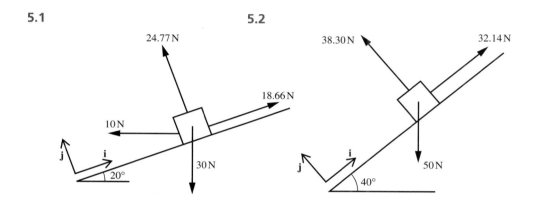

5.3   The diagram shows a girder CD of mass 20 tonnes being held stationary by a crane (which is not shown). The rope from the crane (AB) is attached to a ring at B. Two ropes, BC and BD, of equal length attach the girder to B; the tensions in each of these ropes is T N.

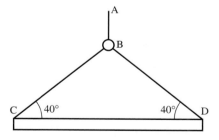

i)   Draw a diagram showing the forces acting on the girder.

ii)   Write down, in terms of T, the horizontal and vertical components of the tensions in the ropes acting at C and D.

iii)   Hence show that the tension in the rope BC is 152.5 kN (to 1 d.p.).

iv)   Draw a diagram to show the three forces acting on the ring at B.

v)   Hence calculate the tension in the rope AB.

vi)   How could you have known the answer to part **v)** without any calculations?

5.4   The diagram shows a simple model
of a crane. The structure is at rest
in a vertical plane. The rod and
cables are of negligible mass and the
load suspended from the joint at
A is 30 N.

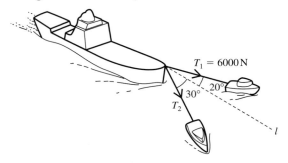

i)   Draw a diagram showing the
forces acting on
   a)   the load
   b)   the joint at A.
ii)   Calculate the forces in the rod and cable 1 and state whether they are in
compression or in tension.

5.5   A ship is being towed by two tugs. Each tug exerts forces on the ship as indicated.
There is also a drag force on the ship.

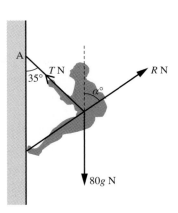

i)   Write down the components of the tensions in the towing cables along and
perpendicular to the line of motion, l, of the ship.
ii)   There is no resultant force perpendicular to the line l. Find T_2.
iii)   The ship is travelling with constant velocity along the line l. Find the magnitude
of the drag force acting on it.

5.6   The diagram shows a man suspended by
means of a rope which is attached at one end
to a peg at a fixed point A on a vertical wall
and at the other to a belt round his waist. The
man has weight $80g$ N, the tension in the rope
is T and the reaction of the wall on the man is
R. The rope is inclined at 35° to the vertical
and R is inclined at $\alpha°$ to the vertical as
shown. The man is in equilibrium.

i)   Explain why $R > 0$.
ii)   By considering his horizontal and vertical
equilibrium separately, obtain two
equations connecting T, R and α.
iii)   Given that $\alpha = 45$, show that T is about 563 N and find R.
iv)   What is the magnitude and the direction of the force on the peg at A?

The peg at A is replaced by a smooth pulley. The rope is passed over the pulley and tied to a hook at B directly below A. Calculate

v) the new value of the tension in the rope section BA

vi) the magnitude of the force on the pulley at A.

5.7 A cyclist of mass 60 kg rides a cycle of mass 7 kg. The greatest forward force that she can produce is 200 N but she is subject to air resistance and friction totalling 50 N.

i) Draw a diagram showing the forces acting on the cyclist when she is going uphill.

ii) What is the angle of the steepest slope that she can ascend?

The cyclist reaches a slope of 8° with a speed of 5 ms^{-1} and rides as hard as she can up it.

iii) Find her acceleration and the distance she travels in 5 s.

iv) What is her speed now?

5.8 A builder is demolishing the chimney of a house and slides the old bricks down to the ground on a straight chute 10 m long inclined at 42° to the horizontal. Each brick has mass 3 kg.

i) Draw a diagram showing the forces acting on a brick as it slides down the chute, assuming the chute to have a flat cross section and a smooth surface.

ii) Find the acceleration of the brick.

iii) Find the time the brick takes to reach the ground.

In fact the chute is not smooth and the brick takes 3 s to reach the ground.

iv) Find the frictional force acting on the brick, assuming it to be constant.

5.9 A tile of mass 2 kg slides 1.2 m from rest down a roof inclined at 30° to the horizontal against a resistance of 5 N. The edge of the roof is 3 m above the ground.

i) Draw a diagram to show the forces acting on the tile as it slides down the roof and use it to find its acceleration.

Calculate also

ii) the velocity of the tile when it reaches the end of the roof.

iii) the time taken for the tile to hit the ground.

iv) the horizontal distance from the end of the roof to the point where the tile lands on the ground.

5.10 Charlotte is sliding down a water chute at the swimming pool and enters the last part at a speed of 4 ms^{-1}. The chute is smooth and straight and the last part has length 2.5 m. It is inclined at 40° to the horizontal and its end is 1.5 m above the water.

i) Find Charlotte's acceleration down the chute. Does it depend on her mass?

ii) What is her speed as she comes to the end of the chute?

iii) What horizontal distance from the end of the chute does Charlotte hit the water?

Personal Tutor Exercise

 5.1 The diagram shows a block of
mass 5 kg on a rough inclined plane. The
block is attached to a 3 kg weight by a
light string which passes over a smooth
pulley and is on the point of sliding up
the slope.

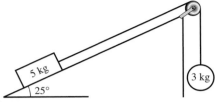

i) Draw a diagram showing the forces
acting on the block.

ii) Resolve these forces into components parallel and perpendicular to the slope.

iii) Find the force of resistance to the block's motion.

The 3 kg weight is replaced by one of mass m kg.

iv) Find the value of m for which the block is on the point of sliding down the
slope, assuming the resistance to motion is the same as before.

KEY POINTS

5.1 The forces acting on a particle can be combined to form a *resultant force* using
scale drawing (trigonometry) or components.

Scale drawing

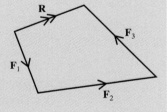

- Draw an accurate diagram, then measure
the resultant. This is less accurate than
calculation.
- To calculate the resultant, find the resultant of
two forces at a time using the trigonometry of
triangles. This is time-consuming for more than
two forces.

Components

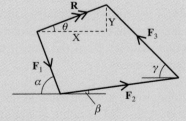

When R is $X\mathbf{i} + Y\mathbf{j}$

$$X = F_1 \cos \alpha + F_2 \cos \beta - F_3 \cos \gamma$$
$$Y = -F_1 \sin \alpha + F_2 \sin \beta + F_3 \sin \gamma$$
$$|\mathbf{R}| = \sqrt{X^2 + Y^2}$$
$$\tan \theta = \frac{Y}{X}$$

5.2   **Equilibrium**

When the resultant **R** is zero, the forces are in equilibrium.

5.3   **Triangle of forces**

If a body is in equilibrium under three non − parallel forces, their lines of action are concurrent and they can be represented by a triangle.

5.4   **Newton's second law**

When the resultant R is not zero there is an acceleration **a** and $\mathbf{R} = m\mathbf{a}$

5.5   When a particle is on a slope, it is usually helpful to resolve in directions parallel and perpendicular to the slope.

6 Friction

Theories do not have to be 'right' to be useful.

Alvin Toffler

This statement about a road accident was offered to a magistrate's court by a solicitor. 'Briefly the circumstances of the accident are that our client was driving his Porsche motor car. He had just left work at the end of the day. He was stationary at the junction with Plymouth Road when a motorcyclist travelling down the Plymouth Road from Tavistock lost control of his motorcycle due to excessive speed and collided with the front offside of our client's motor car.

'The motorcyclist was braking when he lost control and left a 26 metre skid mark on the road. Our advice from an expert witness is that the motorcyclist was exceeding the speed limit of 30 mph.'

QUESTION 6.1   It is the duty of a court to decide whether the motorcyclist was innocent or guilty. Is it possible to deduce his speed from the skid mark? Draw a sketch map and make a list of the important factors that you would need to consider when modelling this situation.

6.1 A model for friction

Clearly the key information is provided by the skid marks. To interpret it, you need a model for how friction works; in this case between the motorcycle's tyres and the road.

As a result of experimental work, Coulomb formulated a model for friction between two surfaces. The following laws are usually attributed to him.

1 Friction always opposes relative motion between two surfaces in contact.
2 Friction is independent of the relative speed of the surfaces.
3 The magnitude of the frictional force has a maximum which depends on the normal reaction between the surfaces and on the roughness of the surfaces in contact.

4 If there is no sliding between the surfaces, then

$$F \leqslant \mu R$$

where F is the force due to friction and R is the normal reaction.
μ is called *the coefficient of friction*.
5 When sliding is just about to occur, friction is said to be *limiting* and $F = \mu R$.
6 When sliding occurs $F = \mu R$.

According to Coulomb's model, μ is a constant for any pair of surfaces. Typical values and ranges of values for the coefficient of friction μ are given in this table.

Surfaces in contact	μ
wood sliding on wood	0.2–0.6
metal sliding on metal	0.15–0.3
normal tyres on dry road	0.8
racing tyres on dry road	1.0
sandpaper on sandpaper	2.0
skis on snow	0.02

How fast was the motorcyclist going?

You can now proceed with the problem. As an initial model, you might make the following assumptions:

1 that the road is level;
2 that the motorcycle was at rest just as it hit the car (Obviously it was not, but this assumption allows you to estimate a minimum initial speed for the motorcycle.);
3 that the motorcycle and rider may be treated as a particle, subject to Coulomb's laws of friction with $\mu = 0.8$ (i.e. dry road conditions).

The calculation then proceeds as follows.

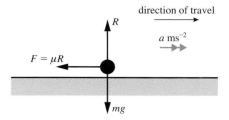

Figure 6.1

Taking the direction of travel as positive, let the motorcycle and rider have acceleration a ms^{-2} and mass m kg. You have probably realised that the acceleration will be negative. The forces (in N) and acceleration are shown in figure 6.1.

Applying Newton's second law:

perpendicular to the road, since there is no vertical acceleration we have

$$R - mg = 0; \qquad \qquad ①$$

parallel to the road, there is a constant force $-\mu R$ from friction, so we have

$$-\mu R = ma. \qquad \qquad ②$$

Solving for a gives

$$a = \frac{-\mu R}{m} = -\frac{\mu m g}{m} = -\mu g.$$

From ①
R = mg

Taking $g = 10 \text{ ms}^{-2}$ and $\mu = 0.8$ gives $a = -8 \text{ ms}^{-2}$.

The constant acceleration equation

$$v^2 = u^2 + 2as$$

can be used to calculate the initial speed of the motorcycle. Substituting $s = 26$, $v = 0$ and $a = -8$ gives

$$u = \sqrt{2 \times 8 \times 26} = 20.4 \text{ ms}^{-1}.$$

This figure can be converted to miles per hour (using the fact that 1 mile $\approx$ 1600m):

$$\text{speed} = \frac{20.4 \times 3600}{1600} \text{ mph}$$
$$= 45.9 \text{ mph}.$$

So this first simple model suggests that the motorcycle was travelling at a speed of at least 45.9 mph before skidding began.

QUESTION 6.2   How good is this model and would you be confident in offering the answer as evidence in court? Look carefully at the three assumptions. What effect do they have on the estimate of the initial speed?

6.2 Modelling with friction

Whilst there is always some frictional force between two sliding surfaces its magnitude is often very small. In such cases we ignore the frictional force and describe the surfaces as *smooth*.

In situations where frictional forces cannot be ignored we describe the surface(s) as *rough*. Coulomb's Law is the standard model for dealing with such cases.

Frictional forces are essential in many ways. For example, a ladder leaning against a wall would always slide if there were no friction between the foot of the ladder and the ground. The absence of friction in icy conditions causes difficulties for road users: pedestrians slip over, cars and motorcycles skid.

Remember that friction always opposes sliding motion.

QUESTION 6.3 In what direction is the frictional force between the back wheel of a cycle and the road?

Historical note

Charles Augustin de Coulomb was born in Angoulême in France in 1736 and is best remembered for his work on electricity rather than for that on friction. The unit for electric charge is named after him.

Coulomb was a military engineer and worked for many years in the West Indies, eventually returning to France in poor health not long before the revolution. He worked in many fields, including the elasticity of metal, silk fibres and the design of windmills. He died in Paris in 1806.

EXAMPLE 6.1 A horizontal rope is attached to a crate of mass 70 kg at rest on a flat surface. The coefficient of friction between the floor and the crate is 0.6. Find the maximum force that the rope can exert on the crate without moving it.

SOLUTION

The forces (in N) acting on the crate are shown in figure 6.2.
Since the crate does not move, it is in equilibrium.

Horizontal forces: $T = F$

Vertical forces: $R = mg$

$$= 70 \times 9.8 = 686$$

The law of friction states that:

$$F \leqslant \mu R \text{ for objects at rest.}$$

So in this case $F \leqslant 0.6 \times 686$

$$F \leqslant 411.6$$

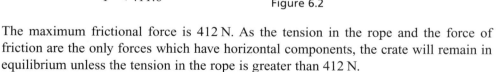

Figure 6.2

The maximum frictional force is 412 N. As the tension in the rope and the force of friction are the only forces which have horizontal components, the crate will remain in equilibrium unless the tension in the rope is greater than 412 N.

EXAMPLE 6.2 Figure 6.3 shows a block of mass 5 kg on a rough table. It is connected by light inextensible strings passing over smooth pulleys to masses of 4 kg and 7 kg which hang vertically. The coefficient of friction between the block and the table is 0.4.

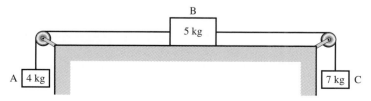

Figure 6.3

i)   Draw a diagram showing the forces acting on the three blocks and the direction of acceleration if the system moves.

ii) Show that acceleration does take place.

iii) Find the acceleration of the system and the tensions in the strings.

SOLUTION

i)

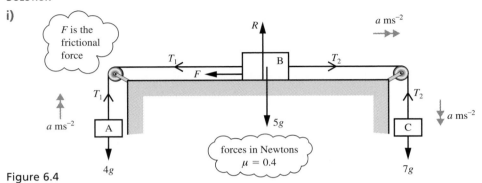

F is the frictional force

forces in Newtons
$\mu = 0.4$

Figure 6.4

If acceleration takes place it is in the direction shown and $a > 0$.

ii) When the acceleration is a ms^{-2} ($\geqslant 0$), Newton's second law gives

for B, horizontally $T_2 - T_1 - F = 5a$ ①

for A, vertically upwards: $T_1 - 4g = 4a$ ②

for C, vertically downwards: $7g - T_2 = 7a$ ③

Adding ①, ② and ③, $3g - F = 16a$ ④

B has no vertical acceleration so $R = 5g$

The maximum possible value of F is $\mu R = 0.4 \times 5g = 2g$

In ④, a can be zero only if $F = 3g$, so $a > 0$ and sliding occurs.

iii) When sliding occurs, you can replace F by $\mu R = 2g$

Then ④ gives $g = 16a$

                                           $a = 0.6125$

Back-substituting gives $T_1 = 41.65$ and $T_2 = 64.3125$

The acceleration is 0.61 ms^{-2} and the tensions are 42 N and 64 N.

EXAMPLE 6.3

Angus is pulling a sledge of mass 12 kg at steady speed across level snow by means of a rope which makes an angle of 20° with the horizontal. The coefficient of friction between the sledge and the ground is 0.15. What is the tension in the rope?

SOLUTION

Since the sledge is travelling at steady speed, the forces acting on it are in equilibrium. They are shown in figure 6.5.

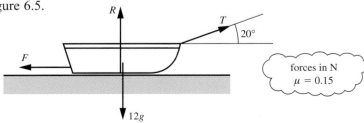

forces in N
$\mu = 0.15$

Figure 6.5

Horizontally: $T \cos 20° = F$
                                       $= 0.15R$ ← 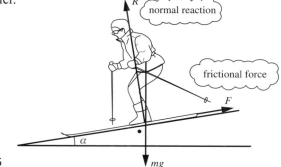 $F = \mu R$ when the sledge slides

Vertically: $T \sin 20° + R = 12g$
                                       $R = 12 \times 9.8 - T \sin 20°$

Combining these gives

$$T \cos 20° = 0.15 (12 \times 9.8 - T \sin 20°)$$
$$T(\cos 20° + 0.15 \sin 20°) = 0.15 \times 12 \times 9.8$$
$$T = 17.8$$

The tension is 17.8 N.

⚠ Notice that the normal reaction is reduced when the rope is pulled in an upwards direction. This has the effect of reducing the friction and making the sledge easier to pull.

EXAMPLE 6.4

A ski slope is designed for beginners. Its angle to the horizontal is such that skiers will either remain at rest on the point of moving or, if they are pushed off, move at constant speed. The coefficient of friction between the skis and the slope is 0.35. Find the angle that the slope makes with the horizontal.

SOLUTION

Figure 6.6 shows the forces on the skier.

Figure 6.6

The weight mg can be resolved into components $mg \cos \alpha$ perpendicular to the slope and $mg \sin \alpha$ parallel to the slope.

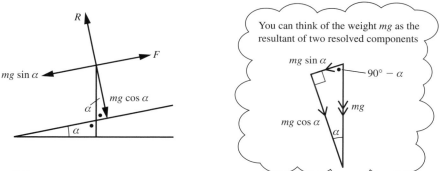

Figure 6.7

Since the skier is in equilibrium (at rest or moving with constant speed) applying Newton's second law:

Parallel to slope: $mg \sin \alpha - F = 0$

$$\Rightarrow \quad F = mg \sin \alpha \qquad \qquad ①$$

Perpendicular to slope: $R - mg \cos \alpha = 0$

$$\Rightarrow \quad R = mg \cos \alpha \qquad \qquad ②$$

In limiting equilibrium or moving at constant speed,

$$F = \mu R$$

$$mg \sin \alpha = \mu \, mg \cos \alpha \quad \longleftarrow \quad \text{Substituting for } F \text{ and } R \text{ from } ① \text{ and } ②$$

$$\Rightarrow \quad \mu = \frac{\sin \alpha}{\cos \alpha} = \tan \alpha.$$

In this case $\mu = 0.35$, so $\tan \alpha = 0.35$ and $\alpha = 19.3$

Note

1 The result is independent of the mass of the skier. This is often found in simple mechanics models. For example, two objects of different mass fall to the ground with the same acceleration. However when such models are refined, for example to take account of air resistance, mass is often found to have some effect on the result.

2 The angle for which the skier is about to slide down the slope is called the angle of friction. The angle of friction is often denoted by λ (lambda) and defined by $\tan \lambda = \mu$.

   When the angle of the slope (α) is equal to the angle of the friction (λ), it is just possible for the skier to stand on the slope without sliding. If the slope is slightly steeper, the skier will slide immediately, and if it is less steep he or she will find it difficult to slide at all without using the ski poles.

Chapter 6 Exercises

You will find it helpful to draw diagrams when doing exercises **6.1**–**6.8**.

6.1 A boy slides a piece of ice of mass 100 g across the surface of a frozen lake. Its initial speed is 10 ms^{-1} and it takes 49 m to come to rest.
   i)   Find the deceleration of the piece of ice.
   ii) Find the frictional force acting on the piece of ice.
   iii) Find the coefficient of friction between the piece of ice and the surface of the lake.
   iv) How far will a 200 g piece of ice travel if it, too, is given an initial speed of 10 ms^{-1}?

6.2 A box of mass 50 kg is being moved across a room. To help it to slide a suitable mat is placed underneath the box.

  i) Explain why the mat makes it easier to slide the box.

  A force of 98 N is needed to slide the mat at a constant velocity.

  ii) What is the value of the coefficient of friction between the mat and the floor?

  A child of mass 20 kg climbs onto the box.

  iii) What force is now needed to slide the mat at constant velocity?

6.3 A car of mass 1200 kg is travelling at 30 ms^{-1} when it is forced to perform an emergency stop. Its wheels lock as soon as the brakes are applied so that they slide along the road without rotating. For the first 40 m the coefficient of friction between the wheels and the road is 0.75 but then the road surface changes and the coefficient of friction becomes 0.8.

  i) Find the deceleration of the car immediately after the brakes are applied.

  ii) Find the speed of the car when it comes to the change of road surface.

  iii) Find the total distance the car travels before it comes to rest.

6.4 In each of the following situations a brick is about to slide down a rough inclined plane. Find the unknown quantity.

  i) The plane is inclined at 30° to the horizontal and the brick has mass 2 kg: find μ.

  ii) The brick has mass 4 kg and the coefficient of friction is 0.7: find the angle of the slope.

  iii) The plane is at 65° to the horizontal and the brick has mass 5 kg: find μ.

  iv) The brick has mass 6 kg and μ is 1.2: find the angle of slope.

6.5 The diagram shows a boy on a simple playground slide. The coefficient of friction between a typically clothed child and the slide is 0.25 and it can be assumed that no speed is lost when changing direction at B. The section AB is 3 m long and makes an angle of 40° with the horizontal. The slide is designed so that a child, starting from rest, stops at just the right moment of arrival at C.

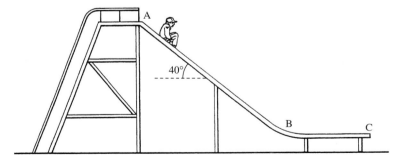

  i) Draw a diagram showing the forces acting on the boy when on the sloping section AB.

  ii) Calculate the acceleration of the boy when on the section AB.

  iii) Calculate the speed on reaching B.

  iv) Find the length of the horizontal section BC.

6.6   The diagram shows a mop being used to clean the floor. The coefficient of friction between the mop and the floor is 0.3.

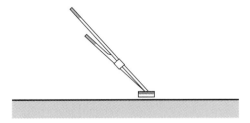

i)   Draw a diagram showing the forces acting on the head of the mop.

In an initial model the weight of the mop is assumed to be negligible.

ii)   Find the angle between the handle and the horizontal when the mop head is moving across the floor at constant velocity. Explain briefly why this angle is independent of how much force is exerted on the mop.

In a more refined model the weight of the head of the mop is taken to be 5 N, but the weight of the handle is still ignored.

iii)   Use this model to calculate the thrust in the handle if it is held at 70° to the horizontal while the head moves at constant velocity across the floor.

iv)   Could the same model be applied to a carpet cleaner on wheels? Explain your reasoning.

6.7

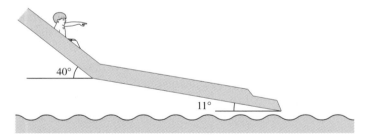

A chute at a water sports centre has been designed so that swimmers first slide down a steep part which is 10 m long and at an angle of 40° to the horizontal. They then come to a 20 m section with a gentler slope, 11° to the horizontal, where they travel at constant speed.

i)   Find the coefficient of friction between a swimmer and the chute.

ii)   Find the acceleration of a swimmer on the steep part.

iii)   Find the speed at the end of the chute of a swimmer who starts at rest. (You may assume that no speed is lost at the point where the slope changes.)

An alternative design of chute has the same starting and finishing points but has a constant gradient.

iv)   With what speed do swimmers arrive at the end of this chute?

6.8   One winter day, Veronica is pulling a sledge up a hill with slope 30° to the horizontal at a steady speed. The weight of the sledge is 40 N. Veronica pulls the sledge with a rope inclined at 15° to the slope of the hill. The tension in the rope is 24 N.

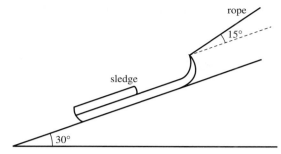

i)   Draw a force diagram showing the forces on the sledge and find the values of the normal reaction of the ground and the frictional force on the sledge.

ii) Show that the coefficient of friction is slightly more than 0.1.

Veronica stops and when she pulls the rope to start again it breaks and the sledge begins to slide down the hill. The coefficient of friction is now 0.1.

iii) Find the new value of the frictional force and the acceleration down the slope.

Coulomb's Law

6.1   The frictional force, F, between two surfaces is given by

$F < \mu R$ when there is no sliding except in limiting equilibrium
$F = \mu R$ in limiting equilibrium
$F = \mu R$ when sliding occurs

where R is the normal reaction of one surface on the other and μ is the coefficient of friction between the surfaces.

6.2   The frictional force always acts in the direction to oppose sliding.

6.3   Remember that the value of the normal reaction is affected by a force which has a component perpendicular to the direction of sliding.

Moments of forces

Give me a firm place to stand and I will move the earth. *Archimedes*

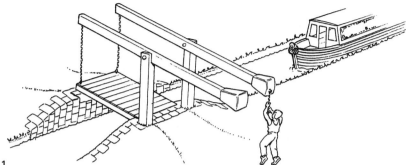

Figure 7.1

The illustration shows a swing bridge over a canal. It can be raised to allow barges and boats to pass. It is operated by hand, even though it is very heavy. How is this possible?

The bridge depends on the turning effects or *moments* of forces. To understand these you might find it helpful to look at a simpler situation.

Two children sit on a simple see-saw, made of a plank balanced on a fulcrum as in figure 7.2. Will the see-saw balance?

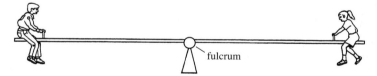

Figure 7.2

If both children have the same mass and sit the same distance from the fulcrum, then you expect the see-saw to balance.

Now consider possible changes to this situation:
i) If one child is heavier than the other then you expect the heavier one to go down;
ii) If one child moves nearer the centre you expect that child to go up.

You can see that both the weights of the children and their distances from the fulcrum are important.

What about this case? One child has mass 35 kg and sits 1.6 m from the fulcrum and the other has mass 40 kg and sits on the opposite side 1.4 m from the fulcrum (see figure 7.3).

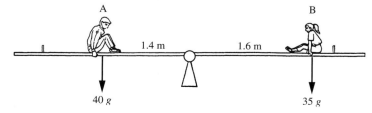

Figure 7.3

Taking the products of their weights and their distances from the fulcrum, gives

A: $\qquad\qquad\qquad\qquad\qquad 40g \times 1.4 = 56g$
B: $\qquad\qquad\qquad\qquad\qquad 35g \times 1.6 = 56g$

So you might expect the see-saw to balance and this indeed is what would happen.

7.1 Rigid bodies

Until now the particle model has provided a reasonable basis for the analysis of the situations you have met. In examples like the see-saw however, where turning is important, this model is inadequate because the forces do not all act through the same point.

In such cases you need the *rigid body model* in which an object, or *body*, is recognised as having size and shape, but is assumed not to be deformed when forces act on it.

Suppose that you push a tray lying on a smooth table with one finger so that the force acts parallel to one edge and through the centre of mass (figure 7.4).

Figure 7.4

The particle model is adequate here: the tray travels in a straight line in the direction of the applied force.

If you push the tray equally hard with two fingers as in figure 7.5, symmetrically either side of the centre of mass, the particle model is still adequate.

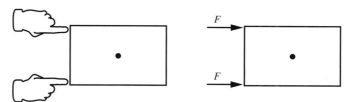

Figure 7.5

However, if the two forces are not equal or are not symmetrically placed, or as in figure 7.6 are in different directions, the particle model cannot be used.

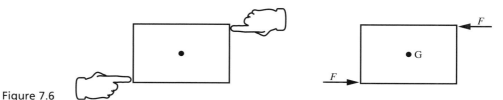

Figure 7.6

The resultant force is now zero, since the individual forces are equal in magnitude but opposite in direction. What happens to the tray? Experience tells us that it starts to rotate about G. How fast it starts to rotate depends, among other things, on the magnitude of the forces and the width of the tray. The rigid body model allows you to analyse the situation.

7.2 Moments

In the example of the see-saw we looked at the product of each force and its distance from a fixed point. This product is called the *moment* of the force about the point.

The see-saw balances because the moments of the forces on either side of the fulcrum are the same magnitude and in opposite directions. One would tend to make the see-saw turn clockwise, the other anti-clockwise. By contrast, the moments about G of the forces on the tray in the last situation do not balance. They both tend to turn it anticlockwise, so rotation occurs.

Conventions and units

The moment of a force F about a point O is defined by

$$\text{moment} = Fd$$

where d is the perpendicular distance from the point O to the line of action of the force (figure 7.7).

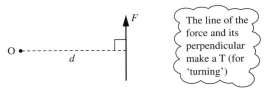

Figure 7.7

In two dimensions, the sense of a moment is described as either positive (anticlockwise) or negative (clockwise) as shown in figure 7.8.

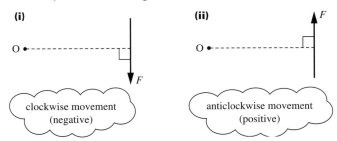

Figure 7.8

If you imagine putting a pin at O and pushing along the line of F, your page would turn clockwise for **i)** and anticlockwise for **ii)**.

In the S.I. system the unit for moment is the newton metre (Nm), because a moment is the product of a force, the unit of which is the newton, and distance, the unit of which is the metre.

Remember that moments are always taken about a point and you must always specify what that point is. A force acting through the point will have no moment about that point because in that case $d = 0$.

QUESTION 7.1   Figure 7.9 shows two tools for undoing wheel nuts on a car. Discuss the advantages and disadvantages of each.

i)   ii)

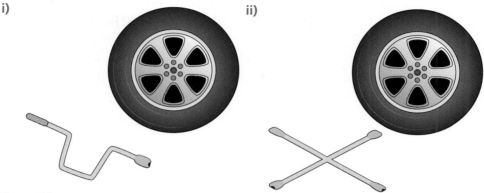

Figure 7.9

When using the spider wrench (the tool with two 'arms'), you apply equal and opposite forces either side of the nut. These produce moments in the same direction. One advantage of this method is that there is no resultant force and hence no tendency for the nut to snap off.

7.3 Couples

Whenever two forces of the same magnitude act in opposite directions along different lines, they have a zero resultant force, but do have a turning effect. In fact the moment will be Fd about any point, where d is the perpendicular distance between the forces. This is demonstrated in figure 7.10.

In each of these situations:

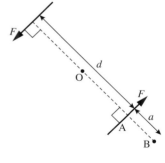

Figure 7.10

Moment about O $F\dfrac{d}{2} + F\dfrac{d}{2} = Fd$ ← anticlockwise is positive

Moment about A $0 + Fd = Fd$

Moment about B $-aF + (a + d)\,F = Fd$

Any set of forces like these with a zero resultant but a non-zero total moment is known as a couple. The effect of a couple on a rigid body is to cause rotation.

7.4 Equilibrium revisited

In earlier chapters we said that an object is in equilibrium if the resultant force on the object is zero. This definition is adequate provided all the forces act through the same point on the object. However, we are now concerned with forces acting at different points, and in this situation even if the forces balance there may be a resultant moment.

Figure 7.11 shows a tray on a smooth surface being pushed equally hard at opposite corners.

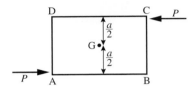

Figure 7.11

The resultant force on the tray is clearly zero, but the resultant moment about its centre point, G, is

$$P \times \frac{a}{2} + P \times \frac{a}{2} = Pa.$$

The tray will start to rotate about its centre and so it is clearly not in equilibrium.

Note

You could have taken moments about any of the corners, A, B, C or D, or any other point in the plane of the paper and the answer would have been the same, Pa anticlockwise.

So we now tighten our mathematical definition of equilibrium to include moments. For an object to remain at rest (or moving at constant velocity) when a system of forces is applied, both the resultant force and the total moment must be zero.

To check that an object is in equilibrium under the action of a system of forces, you need to check two things:
i) that the resultant force is zero;
ii) that the resultant moment about any point is zero. (You only need to check one point.)

EXAMPLE 7.1

Two children are playing with a door. Kerry tries to open it by pulling on the handle with a force 50 N at right angles to the plane of the door, at a distance 0.8 m from the hinges. Peter pushes at a point 0.6 m from the hinges, also at right angles to the door and with sufficient force just to stop Kerry opening it.
i) What is the moment of Kerry's force about the hinges?
ii) With what force does Peter push?
iii) Describe the resultant force on the hinges.

Figure 7.12

SOLUTION

Looking down from above, the line of the hinges becomes a point, H. The door opens clockwise. Anticlockwise is taken to be positive.

i)

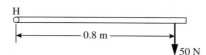

Figure 7.13

Kerry's moment about $H = -50 \times 0.8$

$$= -40\,\text{Nm}$$

The moment of Kerry's force about the hinges is -40 Nm.
(Note that it is a clockwise moment and so negative.)

ii)

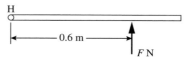

Figure 7.14

Peter's moment about $H = +F \times 0.6$

Since the door is in equilibrium, the total moment on it must be zero, so

$$F \times 0.6 - 40 = 0$$
$$F = \frac{40}{0.6}$$
$$= 66.7$$

Peter pushes with a force of 66.7 N.

iii) Since the door is in equilibrium the overall resultant force on it must be zero.

All the forces are at right angles to the door, as shown in the diagram.

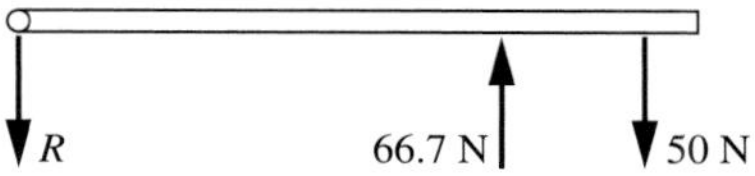

Figure 7.15

Resolve ⊥ to door

$$R + 50 = 66.7$$
$$R = 16.7$$

The reaction at the hinge is a force of 16.7 N in the same direction as Kerry is pulling.

Note

The reaction force at a hinge may act in any direction, according to the forces elsewhere in the system. A hinge can be visualised in cross section as shown in figure 7.16. If the hinge is well oiled, and the friction between the inner and outer parts is negligible, the hinge cannot exert any moment. In this situation the door is said to be 'freely hinged'.

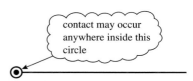

contact may occur anywhere inside this circle

Figure 7.16

EXAMPLE 7.2

The diagram shows a man of weight 600 N standing on a footbridge that consists of a uniform wooden plank just over 2 m long of weight 200 N. Find the reaction forces exerted on each end of the plank.

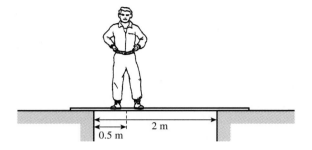

2 m

0.5 m

Figure 7.17

SOLUTION

The diagram shows the forces acting on the plank.

For equilibrium both the resultant force and the total moment must be zero.

As all the forces act vertically we have

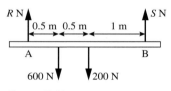

Figure 7.18

$$R + S - 800 = 0 \qquad \text{①}$$

Taking moments about the point A gives

$$(\frown) \qquad R \times 0 - 600 \times 0.5 - 200 \times 1 + S \times 2 = 0 \qquad \text{②}$$

From equation ② $S = 250$ and so equation ① gives $R = 550$.

The reaction forces are 250 N at A and 550 N at B.

Note

1 You cannot solve this problem without taking moments.
2 You can take moments about any point and can, for example, show that by taking moments about B you get the same answer.
3 The whole weight of the plank is being considered to act at its centre.
4 When a force acts through the point about which moments are being taken, its moment about that point is zero.

Levers

A lever can be used to lift or move a heavy object using a relatively small force. Levers depend on moments for their action.

Two common lever configurations are shown below. In both cases a load W is being lifted by an applied force F, using a lever of length l. The calculations assume equilibrium.

Case 1

The fulcrum is at one end of the lever, figure 7.19.

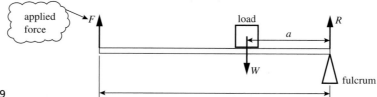

Figure 7.19

Taking moments about the fulcrum

$(\frown)$ $$F \times l - W \times a = 0$$

$$F = W \times \frac{a}{l}$$

Since a is much smaller than l, the applied force F is much smaller than the load W.

Case 2

The fulcrum is within the lever, figure 7.20.

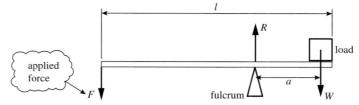

Figure 7.20

Taking moments about the fulcrum

$(\frown)$ $$F \times (l - a) - W \times a = 0$$

$$F = W \times \frac{a}{l - a}$$

Provided that the fulcrum is nearer the end with the load, the applied force is less than the load.

These examples also indicate how to find a single force equivalent to two parallel forces. The force equivalent to F and W should be equal and opposite to R and with the same line of action.

QUESTION 7.2 Describe the single force equivalent to P and Q in each of these cases.

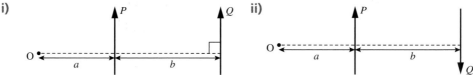

Figure 7.21

In each case state its magnitude and line of action.

QUESTION 7.3 How do you use moments to open a screw-top jar?
Why is it an advantage to press hard when it is stiff?

The moment of a force which acts at an angle

In figure 7.22, the moment of a force about the pivot depends on the perpendicular distance from the pivot to the line of the force.

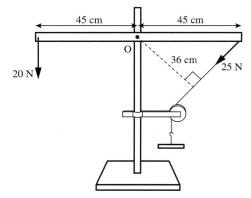

Figure 7.22

In Figure 7.22, where the system remains at rest, the moment about O of the 20 N force is $20 \times 0.45 = 9$ Nm. The moment about O of the 25 N force is $2\,25 \times 0.36 = 2\,9$ Nm. The system is in equilibrium even though unequal forces act at equal distances from the pivot.

The magnitude of the moment of the force F about O in figure 7.23 is given by

$$F \times l = Fd \sin \alpha.$$

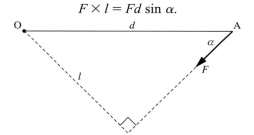

Figure 7.23

Alternatively the moment can be found by noting that the force F can be resolved into components $F \cos \alpha$ parallel to AO and $F \sin \alpha$ perpendicular to AO, both acting through A (Figure 7.24). The moment of each component can be found and then summed to give the total moment.

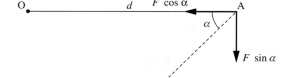

Figure 7.24

The moment of the component along AO is zero because it acts through O. The magnitude of the moment of the perpendicular component is $F \sin \alpha \times d$ so the total moment is $Fd \sin \alpha$, as expected.

EXAMPLE 7.3

A force of 40 N is exerted on a rod as shown. Find the moment of the force about the point marked O.

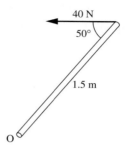

Figure 7.25

SOLUTION

In order to calculate the moment, the perpendicular distance between O and the line of action of the force must be found. This is shown on the diagram.

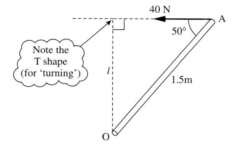

Figure 7.26

Here $l = 1.5 \times \sin 50$.

So the moment about O is

$$F \times l = 40 \times (1.5 \times \sin 50°)$$
$$= 46.0 \text{ Nm.}$$

Alternatively you can resolve the 40 N force into components as in the next diagram.

The component of the force parallel to AO is 40 cos 50 N. The component perpendicular to AO is 40 sin 50 (or 40 cos 40) N.

So the moment about O is
40 sin 50° × 1.5 = 60 sin 50°
$$= 46.0 \text{ Nm as before.}$$

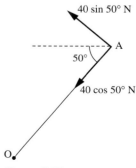

Figure 7.27

EXAMPLE 7.4

A sign outside a pub is attached to a light rod of length 1 m which is freely hinged to the wall and supported in a vertical plane by a light string as in the diagram. The sign is assumed to be a uniform rectangle of mass 10 kg. The angle of the string to the horizontal is 25°.

i) Find the tension in the string.

ii) Find the magnitude and direction of the reaction force of the hinge on the sign.

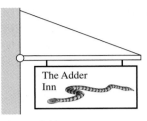

Figure 7.28

SOLUTION

i) The diagram shows the forces acting on the rod, where R_H and R_V are the magnitudes of the horizontal and vertical components of the reaction **R** on the rod at the wall.

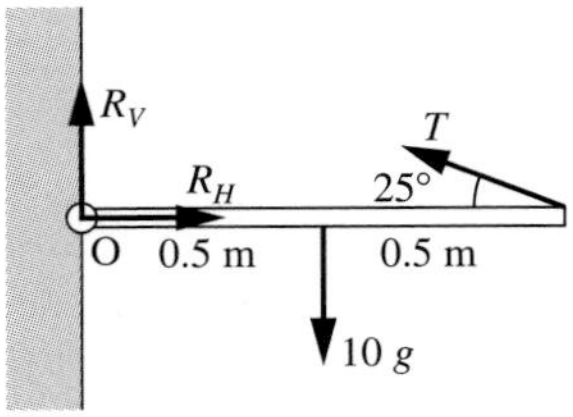

Figure 7.29

Taking moments about O

$$R \times 0 - 10g \times 0.5 + T\sin 25° \times 1 = 0$$

$$\Rightarrow \qquad T\sin 25° = 5g$$

$$T = 116$$

The tension is 116 N.

ii) You can resolve to find the reaction at the wall.

Horizontally: $\qquad\qquad R_H = T\cos 25°$

$$\Rightarrow \qquad R_H = 105$$

Vertically: $\qquad R_V + T\sin 25° = 10g$

$$\Rightarrow \qquad R_V = 10g - 5g = 49$$

$$R = \sqrt{105^2 + 49^2}$$

$$= 116$$

$$\theta = \arctan\left(\tfrac{49}{105}\right) = 25°$$

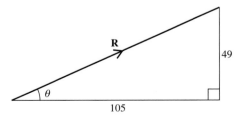

Figure 7.30

The reaction at the hinge has magnitude 116 N and acts at 25 above the horizontal.

QUESTION 7.4

Is it by chance that R and T have the same magnitude and act at the same angle to the horizontal?

EXAMPLE 7.5

A uniform ladder is standing on rough ground and leaning against a smooth wall at an angle of 60° to the ground. The ladder has length 4 m and mass 15 kg. Find the normal reaction forces at the wall and ground and the friction force at the ground.

SOLUTION

The diagram shows the forces acting on the ladder.
The forces are in newtons.

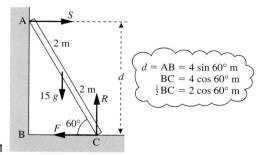

$d = AB = 4 \sin 60°$ m
$BC = 4 \cos 60°$ m
$\frac{1}{2}BC = 2 \cos 60°$ m

Figure 7.31

The diagram shows that there are three unknown forces S, R and F so we need three equations from which to find them. If the ladder remains at rest (in equilibrium) then the resultant force is zero and the resultant moment is zero. These two conditions provide the three necessary equations.

Equilibrium of horizontal components: $S - F = 0$ ①

Equilibrium of vertical components: $R - 15g = 0$ ②

Moments about the foot of the ladder:

$$R \times 0 + F \times 0 + 15g \times 2 \cos 60° - S \times 4 \sin 60° = 0$$

$$\Rightarrow \quad 147 - 4S \sin 60° = 0 \quad ③$$

$$\Rightarrow \quad S = \frac{147}{4 \sin 60°} = 42.4$$

From ① $F = S = 42.4$

From ② $R = 147.$

The force at the wall is 42.4 N those at the ground are 42.4 N horizontally and 147 N vertically.

EXAMPLE 7.6

Figure 7.32 shows a model for a step-ladder standing on a smooth horizontal floor.

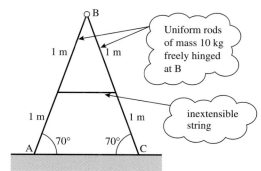

Uniform rods of mass 10 kg freely hinged at B

inextensible string

Figure 7.32

i)   Draw a diagram to show the forces acting on both rods AB and BC.
ii) Explain why the internal forces in the hinge at B are horizontal.
iii) Calculate the tension in the string.

A woman of mass 56 kg stands on a step 0.5 m from A. Calculate

iv) the new reaction at C

v) the tension in the string and the magnitude of the reaction at B.

SOLUTION

i) Figure 7.33 shows the forces acting when the ground is smooth.

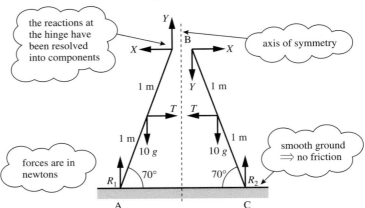

the reactions at the hinge have been resolved into components

axis of symmetry

smooth ground ⇒ no friction

forces are in newtons

Figure 7.33

ii) By Newton's third law, the reactions at B are equal and opposite. Also by the symmetry of the step-ladder and the forces, the vertical components are equal in magnitude and direction. Both conditions can be satisfied only if $Y = 0$ and $R_1 = R_2$.

iii) The two parts of the ladder can be treated separately.

Resolving horizontally for AB or BC $\Rightarrow X = T$

Taking moments about C for BC

$$\Rightarrow 10g \times 1 \cos 70° + T \times 1 \sin 70° = X \times 2 \sin 70°$$

Substituting for X $10g \cos 70° = 2T \sin 70° - T \sin 70°$

$$\Rightarrow T = \frac{10g \cos 70°}{\sin 70°}$$

The tension in the string is 35.7 N.

iv) When a woman stands on a step, the forces are no longer symmetrical so

$Y \neq 0$ and $R_1 \neq R_2$

There are now several possible ways forward. You can treat the ladder as a whole or each part separately but it is best to try to avoid writing down too many equations involving a lot of unknowns.

Take moments about A for the whole system:

$$R_2 \times 4 \cos 70° = 10g \times 3 \cos 70° + 10g \times$$
$$1 \cos 70° + 56g \times 0.5 \cos 70°$$

$$R_2 = 30g + 10g + 28g$$

$$R_2 = 17g$$

Divide by cos 70°

The new reaction at C is 167 N.

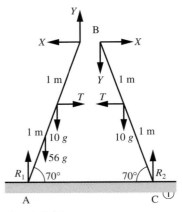

Figure 7.34

v) Take moments about B for BC:

$$R_2 \times 2 \cos 70° = 10g \times 1 \cos 70° + T \times 1 \sin 70°$$

Substituting from ① $\Rightarrow$ $(34g - 10g) \cos 70° = T \sin 70°$

$$T = \frac{24g \cos 70°}{\sin 70°}$$

The tension in the string is 85.6 N.

Resolve vertically for BC: $R_2 = 10g + Y$

$\Rightarrow$ $Y = 7g = 68.6$

horizontally: $X = T = 85.6$

The reaction at B has magnitude $\sqrt{(85.6^2 + 68.6^2)} = 110$ N

 Notice that the extra vertical force due to the woman's weight has an effect on all the forces including those that are horizontal. You cannot assume that any of the reactions remain as they were.

QUESTION 7.5 Suggest an alternative method for part **iii)** of Example 7.6, and check that it gives the same answer.

7.5 Sliding and toppling

The photograph shows a double decker bus on a test ramp. The angle of the ramp to the horizontal is slowly increased.

QUESTION 7.6 What happens to the bus? Would a loaded bus behave differently from the empty bus in the photograph?

QUESTION 7.7   Figure 7.35 shows a cereal packet placed on a slope. Is the box more likely to topple or slide as the angle of the slope to the horizontal increases?

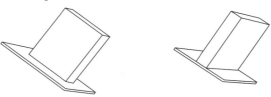

Figure 7.35

To what extent is this situation comparable to that of the bus on the test ramp?

Two critical cases

When an object stands on a surface, the only forces acting are its weight W and the *resultant* of all the contact forces between the surfaces which must act through a point on both surfaces. This resultant contact force is often resolved into two components: the friction, F, parallel to any possible sliding and the normal reaction, R, perpendicular to F as in figures 7.36–8.38.

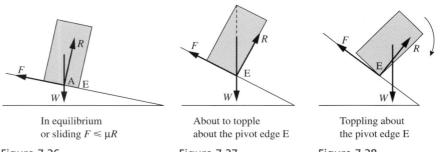

In equilibrium or sliding $F \leqslant \mu R$	About to topple about the pivot edge E	Toppling about the pivot edge E
Figure 7.36	Figure 7.37	Figure 7.38

Equilibrium can be broken in two ways:
i) *The object is on the point of sliding*, then $F = \mu R$ according to our model.
ii) *The object is on the point of toppling*. The pivot is at the lowest point of contact which is the point E in figure 7.37. In this critical case:
  ● the centre of mass is directly above E so the weight acts vertically downwards through E;
  ● the resultant reaction of the plane on the object acts through E, vertically upwards. This is the resultant of F and R.

QUESTION 7.8   Why does the object topple in figure 7.38?

When three non-parallel forces are in equilibrium, their lines of action must be concurrent (they must all pass through one point). Otherwise there is a resultant moment about the point where two of them meet as in figure 7.38.

EXAMPLE 7.7

An increasing force P N is applied to a block, as shown in figure 7.39, until the block moves. The coefficient of friction between the block and the plane is 0.4. Does it slide or topple?

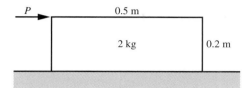

Figure 7.39

SOLUTION

The forces acting are shown in figure 7.40. The normal reaction may be thought of as a single force acting somewhere within the area of contact. When toppling occurs (or is about to occur) the line of action is through the edge about which it topples.

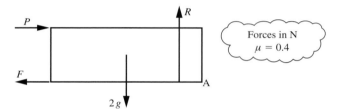

Figure 7.40

Until the block moves, it is in equilibrium.

Horizontally	$P = F$	①
Vertically	$R = 2g$	②

If sliding is about to occur $F = \mu R$

From ① $P = \mu R = 0.4 \times 2g$

$\qquad\qquad = 7.84$

If the block is about to *topple*, then A is the pivot point and the reaction of the plane on the block acts at A. Taking moments about A gives

$(\curvearrowleft)\qquad 2g \times 0.25 - P \times 0.2 = 0$

$\qquad\qquad\qquad P = 24.5$

R acts through A

So to slide P needs to exceed 7.84 N but to topple it needs to exceed 24.5N: the block will slide before it topples.

EXAMPLE 7.8

A rectangular block of mass 3 kg is placed on a slope as shown. The angle α is gradually increased. What happens to the block, given that the coefficient of friction between the block and slope is 0.6?

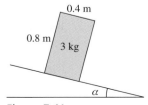

Figure 7.41

SOLUTION

Check for possible sliding

Figure 7.42 shows the forces acting when the block is in equilibrium.

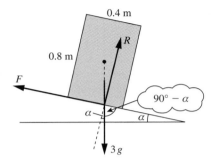

Figure 7.42

Resolve parallel to the slope: $F = 3g \sin \alpha$

Perpendicular to the slope: $R = 3g \cos \alpha$

When the block is on the point of sliding $F = \mu R$ so

$$3g \sin \alpha = \mu \times 3g \cos \alpha$$

$$\Rightarrow \qquad \tan \alpha = \mu = 0.6$$

$$\Rightarrow \qquad \alpha = 31°$$

The block is on the point of sliding when $\alpha = 31°$.

Check for possible toppling

When the block is on the point of toppling about the edge E the centre of mass is vertically above E, as shown in figure 7.43.

Then the angle α is given by:

$$\tan \alpha = \frac{0.4}{0.8}$$

$$\alpha = 26.6$$

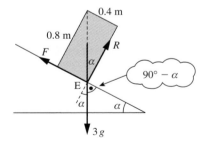

Figure 7.43

The block topples when $\alpha = 26.6°$.

The angle for sliding (31°) is greater than the angle for toppling (26.6°), so the block topples without sliding when $\alpha = 26.6°$.

QUESTION 7.9   Is it possible for sliding and toppling to occur for the same angle?

Chapter 7 Exercises

7.1 The situations below involve several forces acting on each object. For each one, find the total moment.

i)

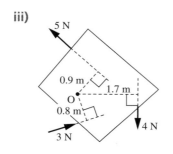

ii)

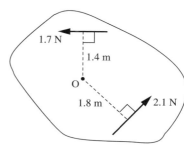

iii)

iv)

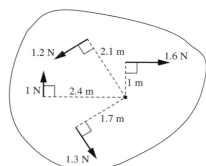

7.2 The diagram shows a motorcycle of mass 250 kg, and its rider whose mass is 80 kg. The centre of mass of the motorcycle lies on a vertical line midway between its wheels. When the rider is on the motorcycle, his centre of mass is 1 m behind the front wheel. Find the vertical reaction forces acting through the front and rear wheels when
i) the rider is not on the motorcycle
ii) the rider is on the motorcycle.

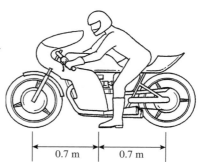

7.3 Find the reaction forces on the hi-fi shelf shown below. The shelf itself has weight 25 N and its centre of mass is midway between A and D.

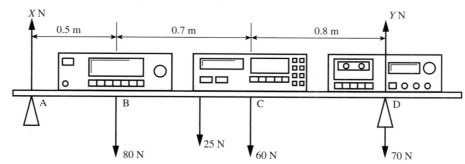

7.4   Karen and Jane are trying to find the positions of their centres of mass. They place a uniform board of mass 8 kg symmetrically on two bathroom scales whose centres are 2 m apart. When Karen lies flat on the board, Jane notes that scale A reads 37 kg and scale B reads 26 kg.

A   B

2 m

i)   Draw a diagram showing the forces acting on Karen and the board and calculate Karen's mass.

ii)   How far from the centre of scale A is her centre of mass?

7.5   The operating instructions for a small crane specify that when the jib is at an angle of 25° above the horizontal, the maximum safe load for the crane is 5000 kg. Assuming that this maximum load is determined by the maximum moment that the pivot can support, what is the maximum safe load when the angle between the jib and the horizontal is:

i)   40   **ii)**   an angle θ?

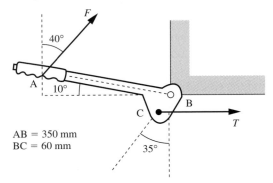

7.6   The diagram shows a car's hand brake. The force F is exerted by the hand in operating the brake, and this creates a tension T in the brake cable. The hand brake is freely pivoted at point B and is assumed to be light.

i)   Draw a diagram showing all the forces acting on the hand brake.

ii)   What is the required magnitude of force F if the tension in the brake cable is to be 1000 N?

iii)   A child applies the hand brake with a force of 10 N. What is the tension in the brake cable?

7.7 The diagram shows four tugs manoeuvring a ship. A and C are pushing it, B and D are pulling it.

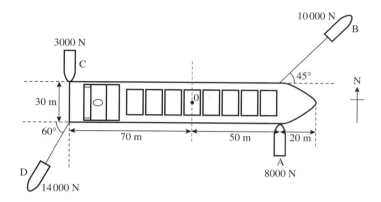

i) Show that the resultant force on the ship is less than 100 N.

ii) Find the overall turning moment on the ship about its centre point, O.

A breeze starts to blow from the South, causing a total force of 2000 N to act uniformly along the length of the ship, at right angles to it.

iii) How (assuming B and D continue to apply the same forces) can tugs A and C counteract the sideways force on the ship by altering the forces with which they are pushing, while maintaining the same overall moment about the centre of the ship?

7.8 A uniform ladder of length 8 m and weight 180 N rests against a smooth, vertical wall and stands on a rough, horizontal surface. A woman of weight 720 N stands on the ladder so that her weight acts at a distance xm from its lower end, as shown in the diagram.

The system is in equilibrium with the ladder at 20° to the vertical.

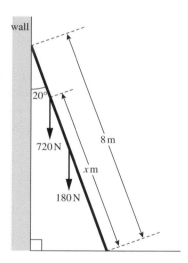

i) Show that the frictional force between the ladder and the horizontal surface is F N, where

$$F = 90(1 + x) \tan 20°$$

ii) Deduce that F increases as x increases and hence find the values of the coefficient of friction between the ladder and the surface for which the woman can stand anywhere on the ladder without it slipping.

7.9 Overhead cables for a tramway are supported by uniform, rigid, horizontal beams of weight 1500 N and length 5 m. Each beam, AB, is freely pivoted at one end A and supports two cables which may be modelled by vertical loads, each of 1000 N, one 1.5 m from A and the other at 1 m from B.

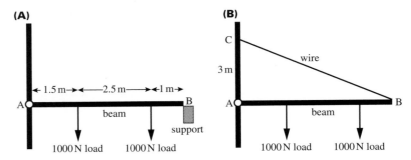

In one situation, the beam is held in equilibrium by resting on a small horizontal support at B, as shown in figure a).

i) Draw a diagram showing all the forces acting on the beam AB. Show that the vertical force acting on the beam at B is 1850 N.

In another situation, the beam is supported by a wire, *instead of the support at* B. The wire is light, attached at one end to the beam at B and at the other to the point C which is 3 m vertically above A, as shown in figure b).

ii) Calculate the tension in the wire.

iii) Find the magnitude and direction of the force on the beam at A.

7.10 A uniform rectangular block of height 30 cm and width 10 cm is placed on a rough plane inclined at an angle α to the horizontal. The block lies on the plane with its length horizontal. The coefficient of friction between the block and the plane is 0.25.

i) Assuming that it does not topple, for what value of α does the block just slide?

ii) Assuming that it does not slide, for what value of α does the block just topple?

iii) The angle α is increased slowly from an initial value of 0. Which happens first, sliding or toppling?

7.11 A man is trying to move a uniform scaffold plank of length 3 m and weight 150 N which is resting on horizontal ground. You may assume that he exerts a slowly increasing force of magnitude P N at a constant angle θ to the vertical and at right angles to the edge CD, as shown in the diagram.

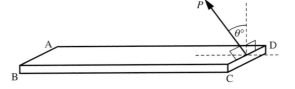

As P increases, the plank will either slide or start to turn about the end AB depending on the values of θ and the coefficient of friction μ between the plank and the ground.

Assume that the plank slides before it turns and is on the point of sliding.

i) Show that the normal reaction of the ground on the plank is $(150 - P \cos \theta)$ N.

ii) Obtain two expressions involving the frictional force acting on the plank and deduce that

$$P = \frac{150\mu}{\sin \theta + \mu \cos \theta}$$

Assume now that the plank starts to turn about the edge AB before it slides and is on the point of turning.

iii) Where is the line of action of the normal reaction of the ground on the plank?

iv) Show that $P = \dfrac{75}{\cos \theta}$

Given that the plank slides before it turns about AB as the force P is gradually increased,

v) find the relationship between μ and θ. Simplify your answer.

7.12 A simple sledge is used to carry goods. The sledge is modelled as a thin, rigid rod, AB, of length 3 m with a weight of 300 N. Its centre of mass is at G, 1 m from the end A. The load is 700 N and is modelled as a point mass 0.75 m from B. Initially the sledge is resting on horizontal ground, as shown in figure (A).

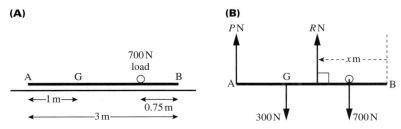

A vertical force of P N is applied at A. This force is insufficient to lift this end of the sledge. The resultant normal reaction, R N, of the ground on the sledge acts through a point of the sledge x m from B, as shown in figure (B).

i) By taking moments about B, show that $3P + Rx = 1125$.

Show also that $P = \dfrac{125(9 - 8x)}{3 - x}$.

ii) The force P is gradually increased. Find the value of P when the end A is just about to come off the ground.

The sledge is now lifted by a force Q N at 60° to the horizontal and it is in equilibrium at 15° to the horizontal, as shown in figure (C). The sledge is on the point of slipping on the ground.

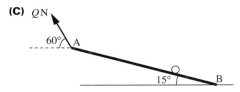

iii) By taking moments about B, show that the value of Q is approximately 512. Calculate the value of the coefficient of friction between the sledge and the ground.

Personal Tutor Exercise

 7.1 Jules is cleaning windows. Her ladder is uniform and stands on rough ground at an angle of 60° to the horizontal and with the top end resting on the edge of a smooth window sill. The ladder has mass 12 kg and length 2.8 m and Jules has mass 56 kg.

i) Draw a diagram to show the forces on the ladder when nobody is standing on it. Show that the reaction at the sill is then $3g$ N.

ii) Find the friction and normal reaction forces at the foot of the ladder.

Jules needs to be sure that the ladder will not slip however high she climbs.

iii) Find the least possible value of μ for the ladder to be safe at 60° to the horizontal.

iv) The value of μ is in fact 0.4. How far up the ladder can Jules stand before it begins to slip?

KEY POINTS

7.1 The moment of a force F about a point O is given by the product Fd where d is the perpendicular distance from O to the line of action of the force.

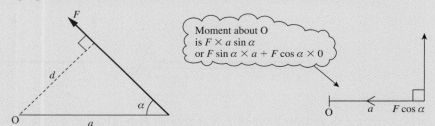

Moment about O is $F \times a \sin \alpha$ or $F \sin \alpha \times a + F \cos \alpha \times 0$

7.2 The S.I. unit for moment is the newton metre (Nm).

7.3 Anticlockwise moments are usually called positive; clockwise negative.

7.4 If a body is in equilibrium the sum of the moments of the forces acting on it, about any point, is zero.

7.5 When three non-parallel forces are in equilibrium, their lines of action are concurrent.

7.6 Two parallel forces P and Q $(P > Q)$ are equivalent to a single force $P + Q$ when P and Q are in the same direction and $P - Q$ when they are in opposite directions. The line of action of the equivalent force is found by taking moments.

8 Centre of mass

Let man then contemplate the whole of nature in her full and grand mystery ... It is an infinite sphere, the centre of which is everywhere, the circumference nowhere.

Blaise Pascal

QUESTION 8.1 Figure 8.1, which is drawn to scale, shows a mobile suspended from the point P. The horizontal rods and the strings are light but the geometrically shaped pieces are made of uniform heavy card. Does the mobile balance? If it does, what can you say about the position of its centre of mass?

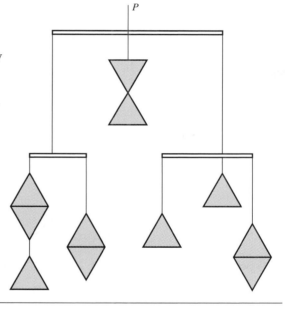

Figure 8.1

QUESTION 8.2 Where is the centre of mass of the gymnast in the picture (right)?

You have met the concept of centre of mass in the context of two general models.

● *The particle model*
The centre of mass is the single point at which the whole mass of the body may be taken to be situated.

● *The rigid body model*
The centre of mass is the balance point of a body with size and shape.

The following examples show how to calculate the position of the centre of mass of a body.

EXAMPLE 8.1

An object consists of three point masses 8 kg, 5 kg and 4 kg attached to a rigid light rod as shown.

Figure 8.2

Calculate the distance of the centre of mass of the object from end O. (Ignore the mass of the rod.)

SOLUTION

Suppose the centre of mass C is $\bar{x}$ m from O. If a pivot were at this position the rod would balance.

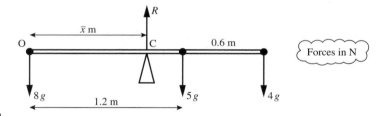

Figure 8.3

For equilibrium $R = 8g + 5g + 4g = 17g$

Taking moments of the forces about O gives:

Total clockwise moment $= (8g \times 0) + (5g \times 1.2) + (4g \times 1.8)$

$$= 13.2g \text{ Nm}$$

Total anticlockwise moment $= R\bar{x}$

$$= 17g\bar{x} \text{ Nm}.$$

The overall moment must be zero for the rod to be in balance, so

$$17g\bar{x} - 13.2g = 0$$

$\Rightarrow$ $\qquad 17\bar{x} = 13.2$

$\Rightarrow$ $\qquad \bar{x} = \Phi \dfrac{13.2}{17} = 0.776.$

The centre of mass is 0.776 m from the end O of the rod.

Note that although *g* was included in the calculation, it cancelled out. The answer depends only on the masses and their distances from the origin and not on the value of *g*. This leads to the following definition for the position of the centre of mass.

Definition

Consider a set of *n* point masses $m_1, m_2, \ldots, m_n$ attached to a rigid light rod (whose mass is neglected) at positions $x_1, x_2, \ldots, x_n$ from one end O. The situation is shown in figure 8.4.

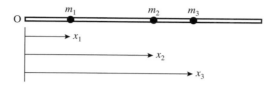

Figure 8.4

The position, $\bar{x}$, of the centre of mass relative to O, is defined by the equation:

moment of whole mass at centre of mass = sum of moments of individual masses

$$(m_1 + m_2 + m_3 + \ldots)\bar{x} = m_1 x_1 + m_2 x_2 + m_3 x_3 + \ldots$$

or

$$M\bar{x} = \sum_{i=1}^{n} m_i x_i$$

where *M* is the total mass (or Σm_i).

EXAMPLE 8.2

A uniform rod of length 2m has mass 5 kg. Masses of 4 kg and 6 kg are fixed at each end of the rod. Find the centre of mass of the rod.

SOLUTION

Since the rod is uniform, it can be treated as a point mass at its centre. Figure 8.5 illustrates this situation.

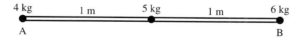

Figure 8.5

Taking the end A as origin,

$$M\bar{x} = \Sigma m_i x_i$$
$$(4 + 5 + 6)\bar{x} = 4 \times 0 + 5 \times 1 + 6 \times 2$$
$$15\bar{x} = 17$$
$$\bar{x} = \tfrac{17}{15}$$
$$= 1\tfrac{2}{15}$$

So the centre of mass is 1.133 m from the 4 kg point mass.

QUESTION 8.3   Check that the rod would balance about a pivot $1\tfrac{2}{15}$ m from A.

EXAMPLE 8.3

A rod AB of mass 1.1 kg and length 1.2 m has its centre of mass 0.48 m from the end A. What mass should be attached to the end B to ensure that the centre of mass is at the mid-point of the rod?

SOLUTION

Let the extra mass be m kg.

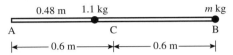

Figure 8.6

Method 1 Refer to the mid-point, C, as origin, so $\bar{x} = 0$. Then

$$(1.1 + m) \times 0 = 1.1 \times (-0.12) + m \times 0.6$$

$\Rightarrow$ $\qquad\qquad 0.6\,m = 1.1 \times 0.12$

$\Rightarrow$ $\qquad\qquad\qquad m = 0.22.$

> The 1.1 mass has negative x referred to C

A mass of 220 grams should be attached to B.

Method 2 Refer to the end A, as origin, so $\bar{x} = 0.6$. Then

$$(1.1 + m) \times 0.6 = 1.1 \times 0.48 + m \times 1.2$$

$\Rightarrow$ $\qquad 0.66 + 0.6m = 0.528 + 1.2m$

$\Rightarrow$ $\qquad\qquad 0.132 = 0.6\,m$

$\qquad\qquad\qquad m = 0.22$ as before.

8.1 Composite bodies

The position of the centre of mass of a composite body such as a cricket bat, tennis racquet or golf club is important to sports people who like to feel its balance. If the body is symmetric then the centre of mass will lie on the axis of symmetry. The next example shows how to model a composite body as a system of point masses so that the methods of the previous section can be used to find the centre of mass.

EXAMPLE 8.4

A squash racquet of mass 200 g and total length 70 cm consists of a handle of mass 150 g whose centre of mass is 20 cm from the end, and a frame of mass 50g, whose centre of mass is 55 cm from the end.

Find the distance of the centre of mass from the end of the handle.

SOLUTION

Figure 8.7 shows the squash racquet and its dimensions.

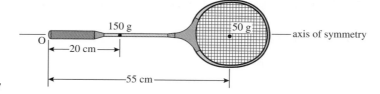

Figure 8.7

The centre of mass lies on the axis of symmetry. Model the handle as a point mass of 0.15 kg a distance 0.2 m from O and the frame as a point mass of 0.05 kg a distance 0.55m from the end O.

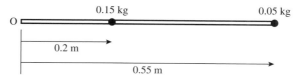

Figure 8.8

The distance, $\bar{x}$, of the centre of mass from O is given by

$$(0.15 + 0.05)\,\bar{x} = (0.15 \times 0.2) + (0.05 \times 0.55)$$
$$\bar{x} = 0.2875.$$

The centre of mass of the squash racquet is 28.75 cm from the end of the handle.

Centres of mass for different shapes

If an object has an axis of symmetry, like the squash racquet in the example above, then the centre of mass lies on it.

The table below gives the position of the centre of mass of some uniform objects that you may encounter, or wish to include within models of composite bodies.

Body	Position of centre of mass	Diagram
Solid cone or pyramid	$\dfrac{1}{4}h$ from base	
Hollow cone or pyramid	$\dfrac{1}{3}h$ from base	
Solid hemisphere	$\dfrac{3}{8}r$ from base	
Hollow hemisphere	$\dfrac{1}{2}r$ from base	
Semi-circular lamina	$\dfrac{4r}{3\pi}$ from base	

8.2 Centre of mass for two- and three-dimensional bodies

The techniques developed for finding the centre of mass using moments can be extended into two and three dimensions.

If a two-dimensional body consists of a set of n point masses $m_1, m_2, \ldots, m_n$ located at positions $(x_1, y_1), (x_2, y_2), \ldots, (x_n, y_n)$ as in figure 8.9 then the position of the centre of mass of the body $(\bar{x}, \bar{y})$ is given by

$$M\bar{x} = \Sigma m_i x_i \quad and \quad M\bar{y} = \Sigma m_i y_i$$

where $M(=\Sigma m_i)$ is the total mass of the body.

In three dimensions, the z co-ordinates are also included; to find $\bar{z}$ use

$$M\bar{z} = \Sigma m_i z_i.$$

The centre of mass of any composite body in two or three dimensions can be found by replacing each component by a point mass at its centre of mass.

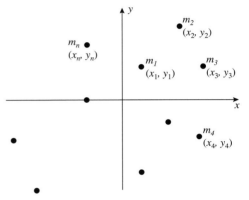

Figure 8.9

Vocabulary

A thin plane surface, like a protractor, is called a *lamina* (plural *laminae*). The laminae considered in this chapter are all uniform; that is they have the same mass per unit area, throughout. The *centroid* of a plane region is the centre of mass of a lamina of the same shape and negligible thickness.

EXAMPLE 8.5

Joanna makes herself a pendant in the shape of a letter J made up of rectangular shapes as shown in figure 8.10.

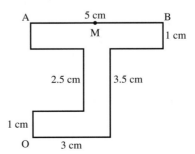

Figure 8.10

i) Find the position of the centre of mass of the pendant.
ii) Find the angle that AB makes with the horizontal if she hangs the pendant from a point, M, in the middle of AB.

She wishes to hang the pendant so that AB is horizontal.

iii) How far along AB should she place the ring that the suspending chain will pass through?

SOLUTION

i) The first step is to split the pendant into three rectangles. The centre of mass of each of these is at its middle, as shown in figure 8.11.

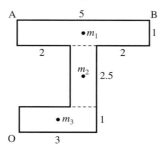

Figure 8.11

You can model the pendant as three point masses m_1, m_2 and m_3, which are proportional to the areas of the rectangular shapes. Since the areas are 5 cm², 2.5 cm² and 3 cm², the masses, in suitable units, are 5, 2.5 and 3, and the total mass is $5 + 2.5 + 3 = 10.5$ (in the same units).

The table below gives the mass and position of m_1, m_2 and m_3.

Mass		m_1	m_2	m_3	M
Mass units		5	2.5	3	10.5
Position of	x	2.5	2.5	1.5	$\bar{x}$
centre of mass	y	4	2.25	0.5	$\bar{y}$

Now it is possible to find $\bar{x}$:

$$M\bar{x} = \Sigma m_i x_i$$
$$10.5\bar{x} = 5 \times 2.5 + 2.5 \times 2.5 + 3 \times 1.5$$
$$\bar{x} = \frac{23.25}{10.5} = 2.2 \text{ cm}$$

Similarly for $\bar{y}$:

$$M\bar{y} = \Sigma m_i y_i$$
$$10.5\bar{y} = 5 \times 4 + 2.5 \times 2.25 + 3 \times 0.5$$
$$\bar{y} = \frac{27.125}{10.5} = 2.6 \text{ cm}$$

The centre of mass is at (2.2, 2.6).

ii) When the pendant is suspended from M, the centre of mass, G, is vertically below M, as shown in figure 8.12.

The pendant hangs like the first diagram but you might find it easier to draw your own diagram like the second.

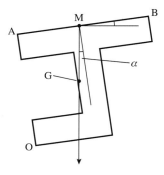

Figure 8.12

$$GP = 2.5 - 2.2 = 0.3$$

$$MP = 4.5 - 2.6 = 1.9$$

$$\therefore \qquad \tan \alpha = \frac{0.3}{1.9} \quad \Rightarrow \quad \alpha = 9°$$

AB makes an angle of 9° with the horizontal (or 8.5° working with unrounded figures).

iii) For AB to be horizontal the point of suspension must be directly above the centre of mass, and so it is 2.2 cm from A.

EXAMPLE 8.6

Find the centre of mass of a body consisting of a square plate of mass 3 kg and side length 2 m, with small objects of mass 1 kg, 2 kg, 4 kg and 5 kg at the corners of the square.

SOLUTION

Figure 8.13 shows the square plate, with the origin taken at the corner at which the 1 kg mass is located. The mass of the plate is represented by a 3 kg point mass at its centre.

In this example the total mass M (in kilograms) is $1 + 2 + 4 + 5 + 3 = 15$.

The two formulae for $\bar{x}$ and $\bar{y}$ can be combined into one using column vector notation:

$$\begin{bmatrix} M\bar{x} \\ M\bar{y} \end{bmatrix} = \begin{bmatrix} \Sigma m_i x_i \\ \Sigma m_i y_i \end{bmatrix}$$

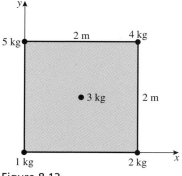

Figure 8.13

which is equivalent to

$$M \begin{pmatrix} \bar{x} \\ \bar{y} \end{pmatrix} = \Sigma m_i \begin{pmatrix} x_i \\ y_i \end{pmatrix}.$$

Substituting our values for M and m_i and x_i and y_i:

$$15\begin{pmatrix} \overline{x} \\ \overline{y} \end{pmatrix} = 1\begin{pmatrix} 0 \\ 0 \end{pmatrix} + 2\begin{pmatrix} 2 \\ 0 \end{pmatrix} + 4\begin{pmatrix} 2 \\ 2 \end{pmatrix} + 5\begin{pmatrix} 0 \\ 2 \end{pmatrix} + 3\begin{pmatrix} 1 \\ 1 \end{pmatrix}$$

$$15\begin{pmatrix} \overline{x} \\ \overline{y} \end{pmatrix} = \begin{pmatrix} 15 \\ 21 \end{pmatrix}$$

$$\begin{pmatrix} \overline{x} \\ \overline{y} \end{pmatrix} = \begin{pmatrix} 1 \\ 1.4 \end{pmatrix}$$

The centre of mass is at the point (1, 1.4).

EXAMPLE 8.7

Figure 8.14 shows a square table of side 2 m which is supported by four legs, each 1 m high, situated at the corners of a 1m square. The mass of each leg is 1 kg and that of the table top 2 kg. The legs and the top are uniform. A small heavy object of mass 6 kg is placed on one corner of the table.

i)   Find the position of the centre of mass relative to the point O shown in the diagram using the axes indicated with unit length 1 m.
ii) What is the significance of your result?

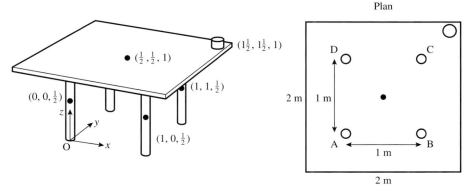

Figure 8.14

SOLUTION

i)   Let the co-ordinates of the centre of mass be $(\overline{x}, \overline{y}, \overline{z})$.

You can put the information in a table and then write the equations in vectors as follows, or deal with each co-ordinate separately.

	leg A	leg B	leg C	leg D	table top	object	whole
mass	1	1	1	1	2	6	12
x	0	1	1	0	$\frac{1}{2}$	$1\frac{1}{2}$	$\overline{x}$
y	0	0	1	1	$\frac{1}{2}$	$1\frac{1}{2}$	$\overline{y}$
z	$\frac{1}{2}$	$\frac{1}{2}$	$\frac{1}{2}$	$\frac{1}{2}$	1	1	$\overline{z}$

Taking moments about O:

$$12 \times \begin{pmatrix} \bar{x} \\ \bar{y} \\ \bar{z} \end{pmatrix} = \overset{A}{\begin{pmatrix} 0 \\ 0 \\ \frac{1}{2} \end{pmatrix}} + 1 \overset{B}{\begin{pmatrix} 1 \\ 0 \\ \frac{1}{2} \end{pmatrix}} + 1 \overset{C}{\begin{pmatrix} 1 \\ 1 \\ \frac{1}{2} \end{pmatrix}} + 1 \overset{D}{\begin{pmatrix} 0 \\ 1 \\ \frac{1}{2} \end{pmatrix}}$$

$$+ 2 \begin{pmatrix} \frac{1}{2} \\ \frac{1}{2} \\ 1 \end{pmatrix} + 6 \begin{pmatrix} 1\frac{1}{2} \\ 1\frac{1}{2} \\ 1 \end{pmatrix}$$

$$\Rightarrow 12 \begin{pmatrix} \bar{x} \\ \bar{y} \\ \bar{z} \end{pmatrix} = \begin{pmatrix} 12 \\ 12 \\ 10 \end{pmatrix}$$

$\Rightarrow$ The centre of mass is at $(1, 1, 0.83)$.

ii) This means that the centre of mass is in the leg nearest the object and so the table is on the point of toppling over. All the weight is being taken by that one leg.

EXAMPLE 8.8

A metal disc of radius 15 cm has a hole of radius 5 cm cut in it as shown in figure 8.15.
Find the centre of mass of the disc.

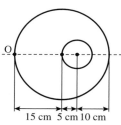

15 cm 5 cm 10 cm

Figure 8.15

SOLUTION

Think of the original uncut disc as a composite body made up of the final body and a disc to fit into the hole. Since the material is uniform the mass of each part is proportional to its area.

The uncut disc = the final body + the cut out disc

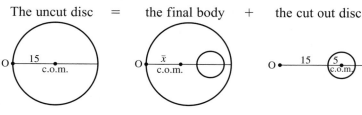

Figure 8.16

Area	$15^2\pi = 225\pi$	$15^2\pi - 5^2\pi = 200\pi$	$5^2\pi = 25\pi$
Distance from O to centre of mass	15 cm	$\bar{x}$ cm	20 cm

Taking moments about O

$$225\pi \times 15 = 200\pi \times \bar{x} + 25\pi \times 20$$

$$\Rightarrow \quad \bar{x} = \frac{225 \times 15 - 25 \times 20}{200}$$

Dived by π

$$= 14.375$$

The centre of mass is 14.4 cm from O, that is 0.6 cm to the left of the centre of the disc.

8.3 Calculating volumes

What do the shapes of the objects have in common? For each one, suggest a way of finding its volume.

How would you find the volume of
i) an egg?
ii) a table tennis ball?
iii) the moon?

One way of finding the volume of an irregular shape is to immerse it in water and measure the apparent increase in the volume of the water. This method is satisfactory for finding the volume of something small and dense like an egg, but you would find it difficult to use for a table tennis ball, and it would be impossible to find the volume of the moon in this way!

You will be familiar with the formula $V = \frac{4}{3}\pi r^3$ for the volume of a sphere. The methods you meet in this chapter will enable you to prove this formula, and also to calculate the volumes of other solids with *rotational symmetry* like the ones illustrated above.

8.4 Volumes of revolution

If a set-square (without a hole) is rotated through 360° about one of its shorter sides its face will sweep out a solid cone (figure 8.17 (a)). A 60° set-square will sweep out different cones depending on the side chosen. In the same way a semi-circular protractor will sweep out a sphere if it is rotated completely about its diameter as shown in figure 8.17 (b).

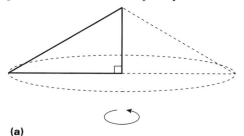
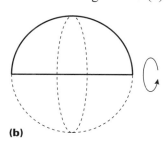

(a)   **(b)**

Figure 8.17

The cone and sphere are examples of *solids of revolution*. If any region of a plane is rotated completely about an axis in the plane it will form a solid of revolution and any solid body with rotational symmetry can be formed in this way.

Imagine that such a solid is cut into thin slices perpendicular to the axis of rotation just as a potato is sliced to make crisps (figure 8.18). Each slice will approximate to a thin disc, but the discs will vary in radius. The volume of the solid can be found by adding the volumes of these discs and then finding the limit of the

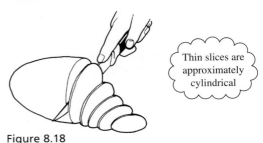

Thin slices are approximately cylindrical

Figure 8.18

volume as their thickness approaches zero. When thin discs are used to find a volume they are called *elementary discs*. (Similarly, when thin strips are used to find an area they are called *elementary strips*.)

This is essentially the same procedure as that developed for finding the area under a curve, and so, integration is required.

Volumes of revolution about the x axis

Look at the solid formed by the rotation about the x axis of the region under the graph of $y = f(x)$ between the values $x = a$ and $x = b$ as shown in figure 8.19.
(You should always start by drawing diagrams when finding areas and volumes.)

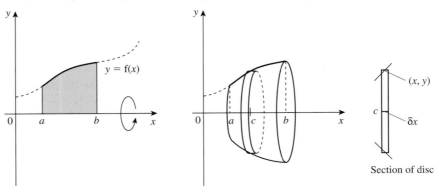

Figure 8.19

The volume of the solid of revolution (which is usually called the volume of revolution) can be found by imagining that it can be sliced into thin discs.

Each disc is a thin cylinder of radius y and thickness δx, so its volume is given by

$$\delta V = \pi y^2 \delta x.$$

The volume of the solid is the limit of the sum of all these elementary volumes as $\delta x \to 0$
i.e. the limit as $\delta x \to 0$ of

$$\sum_{\substack{\text{over all} \\ \text{discs}}} \delta V = \sum_{x=a}^{x=b} \pi y^2 \, \delta x.$$

You need to do this because the discs are assumed to be cylinders.

The limiting values of sums such as these are integrals so

$$V = \int_a^b \pi y^2 \, dx$$

The limits are a and b because x takes values from a to b.

It is essential to realise that this integral is written in terms of dx and so it cannot be evaluated unless the function y is also written in terms of x. For this reason the integral is often written as

$$V = \int_a^b \pi [f(x)]^2 \, dx.$$

EXAMPLE 8.9   Find the volume of revolution formed when the region between the curve $y = x^2 + 2$, the x axis and the lines $x = 1$ and $x = 3$ is rotated completely about the x axis.

SOLUTION

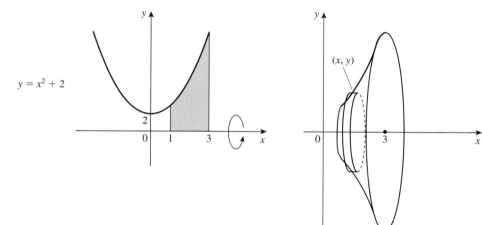

Figure 8.20

The volume is:

$$V = \int_a^b \pi [f(x)]^2 \, dx$$

$$= \int_1^3 \pi (x^2 + 2)^2 \, dx$$

$$= \int_1^3 \pi (x^4 + 4x^2 + 4) \, dx$$

$$= \left[\pi \left(\frac{x^5}{5} + \frac{4x^3}{3} + 4x \right) \right]_1^3$$

$$= \pi \left[\left(\frac{3^5}{5} + 36 + 12 \right) - \left(\frac{1}{5} + \frac{4}{3} + 4 \right) \right].$$

The volume is $\dfrac{1366\pi}{15}$ cubic units or 286 cubic units correct to 3 significant figures.

Note

Unless a decimal answer is required, it is usual to leave π in the answer, which is then exact.

EXAMPLE 8.10   Find the volume of a spherical ball of radius 2 cm.

SOLUTION

The sphere is formed by rotating the circle $x^2 + y^2 = 4$ about the x axis.

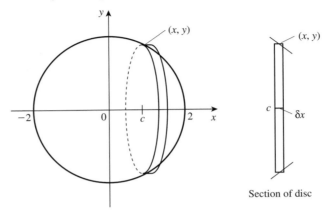

Figure 8.21

The volume of the sphere is found by using

$$V = \int_a^b \pi y^2 \, dx$$

with limits $a = -2$ and $b = 2$.

Since $x^2 + y^2 = 4$, then $y^2 = 4 - x^2$, and the integral may be written as

$$V = \int_{-2}^2 \pi (4 - x^2) \, dx.$$

The constant π can be taken outside the integral so you have

$$V = \pi \left[4x - \frac{x^3}{3} \right]_{-2}^2$$

$$= \pi \left[\left(8 - \frac{8}{3} \right) - \left(-8 + \frac{8}{3} \right) \right]$$

$$= \frac{32\pi}{3} = 33.51\ldots$$

So the volume of a spherical ball with radius 2 cm is 33.5 cm³ correct to 3 significant figures.

The volume of a sphere

The formula for the volume of a sphere of radius r can be found by following the same argument, but using a circle of radius r rather than one of radius 2 cm. Its equation is $x^2 + y^2 = r^2$ and the limits are $-r$ and r.

$$V = \int \pi y^2 \, dx$$

$$= \int_{-r}^r \pi (r^2 - x^2) \, dx$$

Therefore

$$V = \pi\left[r^2 x - \frac{x^3}{3}\right]_{-r}^{r}$$

$$= \pi\left[\left(r^3 - \frac{r^3}{3}\right) - \left(-r^3 - \frac{-r^3}{3}\right)\right]$$

$$\Rightarrow \quad V = \frac{4}{3}\pi r^3$$

Volumes of revolution about the y axis

If the same portion of the curve $y = x^2 + 2$ were to be used to form a region which could be rotated about the y axis, a very different solid would be obtained from that on page 176. The region required would be that which lies between the curve and the y axis and the limits $y = 3$ and $y = 11$ shown in figure 8.22.

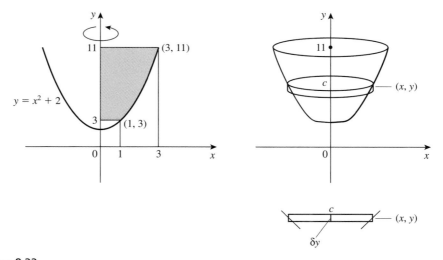

Figure 8.22

An elementary disc has an approximate volume $\delta V = \pi x^2 \delta y$ and the volume of revolution about the y axis is given by

$$V = \int \pi x^2 \, dy.$$

This time it is necessary to write x^2 in terms of y. In this case $x^2 = y - 2$ and so the final integral becomes

$$V = \int_{3}^{11} \pi(y - 2) \, dy$$

$$V = \left[\pi\left(\frac{y^2}{2} - 2y\right)\right]_{3}^{11}$$

$$\Rightarrow \quad V = \pi\left[\left(\frac{11^2}{2} - 22\right) - \left(\frac{3^2}{2} - 6\right)\right]$$

The volume is 40π cubic units.

8.5 Finding a mathematical model for a solid of revolution

A hen's egg has rotational symmetry about its longest axis so it can be represented by a solid of revolution. But what curve should be revolved around the x axis to obtain a reasonable egg shape?

The cross-section looks circular at the 'blunt' end, but the 'sharp' end is longer and looks more like half an ellipse. It may be possible to model the egg as part of a circle plus part of an ellipse, rotated about the x axis. The equation of an ellipse of length $2a$ and width $2b$ as in figure 8.24 can be written as

$$\frac{x^2}{a^2} + \frac{y^2}{b^2} = 1.$$

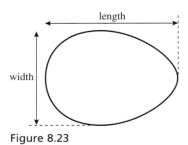

Figure 8.23

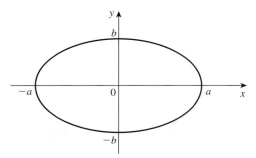

Figure 8.24

A particular egg is 6.3 cm long and 4.6 cm wide. Figure 8.25 shows the regions which are rotated about the x axis.

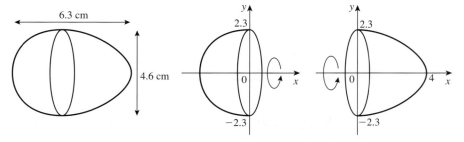

Figure 8.25

The left part is a hemisphere of radius 2.3 cm so its volume is

$$V_1 = \tfrac{2}{3}\pi r^3 = \tfrac{2}{3}\pi(2.3)^3$$
$$= 25.48 \text{ cm}^3 \text{ (to 2 d.p.).}$$

The equation of the elliptical part to the right is

$$\frac{x^2}{4^2} + \frac{y^2}{2.3^2} = 1.$$

On rearrangement this becomes

$$y^2 = 2.3^2\left(1 - \frac{x^2}{4^2}\right).$$

The volume of the elliptical part is therefore given by

$$V_2 = \int_0^4 \pi y^2 \, dx$$

$$= \pi \times 2.3^2 \int_0^4 \left(1 - \frac{x^2}{4^2}\right) dx$$

$$= \pi \times 2.3^2 \left[x - \frac{x^3}{48}\right]_0^4$$

$$= \pi \times 2.3^2 \left[\left(4 - \frac{4^3}{48}\right) - 0\right]$$

$$= 44.32 \text{ cm}^3 \text{ (to 2 d.p.).}$$

The total volume of the egg can now be found as

$$V_1 + V_2 = 25.48 + 44.32$$

$$= 70 \text{ cm}^3 \text{ (correct to 2 s.f.).}$$

8.6 Centres of mass

The position of the centre of mass of a solid of revolution will affect its stability and the way it will roll. Some toys are designed so that they can never be knocked over. Many birds that live on cliffs lay pointed eggs which roll round in circles so that they do not fall over the cliff edge. The position of the centre of mass is also important in the design of a lot of sporting equipment.

8.7 Using integration to find centres of mass

The calculus methods you have been using to determine the volumes of solids of revolution can be extended to find their centres of mass (assuming they are of uniform density).

Notice that by symmetry, the centre of mass of a solid of revolution must lie on its axis, so provided you choose either the x axis or the y axis to be the axis of symmetry, there is only one co-ordinate to determine. For a solid of revolution about the x axis, divide it into thin discs as before. An elementary disc situated at the point (x, y) on the curve, has a centre of mass at the point $(x, 0)$ on the x axis as shown in figure 8.26.

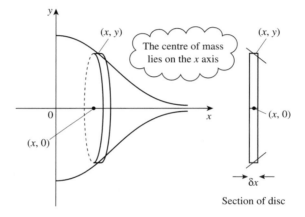

Figure 8.26

The volume of this disc is
$$\delta V = \pi y^2 \, \delta x$$

and so its mass is
$$\delta M = \rho \pi y^2 \, \delta x$$

where ρ is the density.

The sum of such discs forms a solid which approaches the original as $\delta x \to 0$ and the discs become thinner. Both the mass and the position of the centre of mass are known for each disc, so you can use the result from detailed earlier in this chapter. For a composite body, the position, $\bar{x}$, of the centre of mass is given by:

Moment of whole mass at centre of mass = Sum of moments of individual masses

$$M\bar{x} = \sum_i m_i x_i.$$

In this case:
$$\left(\sum_{\substack{\text{All} \\ \text{discs}}} \delta M\right) \bar{x} = \sum_{\substack{\text{All} \\ \text{discs}}} (\delta Mx).$$

Substituting the expression for δM obtained above gives

$$\sum \rho \pi y^2 \, \delta x) \bar{x} = \sum \rho \pi y^2 x \, \delta x.$$

In the limit as $\delta x \to 0$ these sums may be represented by integrals and so

$$(\int \rho \pi y^2 \, dx)\bar{x} = \int \rho \pi y^2 x \, dx.$$

Assuming that ρ is uniform, you can divide through by $\rho \pi$ to give

$$(\int y^2 \, dx)\bar{x} = \int y^2 x \, dx \quad \text{or} \quad \bar{x} = \frac{\int y^2 x \, dx}{\int y^2 \, dx}.$$

This result is valid for any solid of revolution of *uniform density* which is formed by rotation about the x axis.

It may also be written as

$$V\bar{x} = \int \pi y^2 x \, dx$$

where V is the volume of the solid.

In the next example, the centre of mass of a solid hemisphere is found using the above result.

EXAMPLE 8.11 Find the centre of mass of a solid hemisphere of radius r.

SOLUTION

With the hemisphere oriented as in the diagram, the point (x, y) lies on the curve $x^2 + y^2 = r^2$ and the limits for the integration are $x = 0$ and $x = r$. You also know that the volume, V, is $\frac{2}{3}\pi r^3$ but the equations will look simpler if this is substituted later.

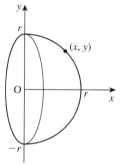

$$V\bar{x} = \int_0^r \pi y^2 x \, dx$$

Substituting $y^2 = r^2 - x^2$ gives $V\bar{x} = \int_0^r \pi(r^2 - x^2)x \, dx$

$$= \int_0^r \pi(r^2 x - x^3) \, dx$$

$$= \pi \left[\frac{r^2 x^2}{2} - \frac{x^4}{4} \right]_0^r$$

Figure 8.27

So $V\bar{x} = \frac{1}{4}\pi r^4$

Substituting for V gives: $\frac{2}{3}\pi r^3 \bar{x} = \frac{1}{4}\pi r^4$

$$\Rightarrow \quad \bar{x} = \frac{3}{8}r$$

The centre of mass of a solid hemisphere is $\frac{3}{8}r$ from the centre of its base.

When finding centres of mass in this way it is often necessary to use *integration by parts*. The following example illustrates this.

EXAMPLE 8.12 Find the position of the centre of mass of the solid of revolution formed when the region between the curve $y = 2e^x$, the lines $x = 0$ and $x = 2$, and the x axis is rotated through $360°$ about the x axis.

SOLUTION

The diagrams show the region and the solid obtained

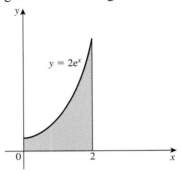

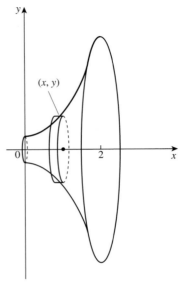

Figure 8.28

The volume, V, of the solid is given by

$$V = \int \pi y^2 \, dx$$

$$= \pi \int_0^2 (2e^x)^2 \, dx$$

$$V = 4\pi \int_0^2 e^{2x} \, dx$$

$$= 4\pi \left[\frac{e^{2x}}{2} \right]_0^2$$

$$= 2\pi(e^4 - 1).$$

By symmetry, the y co-ordinate of the centre of mass is 0. The x co-ordinate is given by $\bar{x}$ in

$$V\bar{x} = \int \pi y^2 x \, dx$$

$$= \pi \int_0^2 (2e^x)^2 x \, dx$$

Therefore

$$V\bar{x} = 4\pi \int_0^2 x e^{2x} \, dx. \qquad ①$$

This integral requires the technique of integration by parts, expressed by the general formula

$$\int_a^b u \frac{dv}{dx} \, dx = \left[uv \right]_a^b - \int_a^b v \frac{du}{dx} \, dx.$$

To find $\qquad \int x e^{2x} \, dx$

let $\qquad u = x \qquad \Rightarrow \qquad \dfrac{du}{dx} = 1,$

and let $\qquad \dfrac{dv}{dx} = e^{2x} \qquad \Rightarrow \qquad v = \dfrac{e^{2x}}{2}.$

The limits are $a = 0$ and $b = 2$.

Substituting these in the general formula, you obtain

$$\int_0^2 xe^{2x}\,dx = \left[x\frac{e^{2x}}{2}\right]_0^2 - \int_0^2 \frac{e^{2x}}{2}\times 1\times dx$$

$$= \left[\tfrac{1}{2}xe^{2x}\right]_0^2 - \left[\tfrac{1}{4}e^{2x}\right]_0^2$$

$$= (e^4 - 0) - \left(\tfrac{1}{4}e^4 - \tfrac{1}{4}\right)$$

$$= \tfrac{1}{4}(3e^4 + 1).$$

Substituting this and $V = 2\pi(e^4 - 1)$ in ①

$$V\bar{x} = 4\pi\int_0^2 xe^{2x}\,dx$$

gives

$$2\pi(e^4 - 1)\bar{x} = 4\pi\times\tfrac{1}{4}(3e^4 + 1)$$

$$\Rightarrow \quad \bar{x} = \frac{(3e^4 + 1)}{2(e^4 - 1)}$$

$$= 1.54 \text{ (correct to 2 d.p.)}.$$

The centre of mass is at $(1.54, 0)$.

Note

If the answer is required in decimal form, it is best to substitute the numerical values of constants such as e and π later rather than sooner. Notice that π cancels in this example.

8.8 Centres of mass of composite bodies

Once the positions of the centres of mass of standard solids have been found, it is possible to use these results to find the positions of the centres of mass of composite bodies. Some useful results are summarised in the following table.

Diagram	Body	Volume	Height of c.o.m. above base
	Solid sphere, radius r	$\frac{4}{3}\pi r^3$	r
	Solid hemisphere, radius r	$\frac{2}{3}\pi r^3$	$\frac{3}{8}r$
	Solid cone, height h, radius r	$\frac{1}{3}\pi r^2 h$	$\frac{1}{4}h$

You may also find it helpful to know the position of the centre of mass for some shells (hollow bodies with negligible wall thickness), and these are given below.

Diagram	Body	Curved surface area	Height of c.o.m. above base
	Hollow hemisphere, radius r	$2\pi r^2$	$\frac{1}{2}r$
	Hollow cone, height h, radius r	$\pi r l$ $(l = \sqrt{r^2 + h^2})$	$\frac{1}{3}h$

It is possible to find the position of the centres of mass of thin shells by using formulae already obtained for solid bodies. A hemispherical shell, for example, can be formed by removing a small hemisphere (say H_1) from a larger one (H_2). Suppose that the radii of the two hemispheres are a and b respectively.

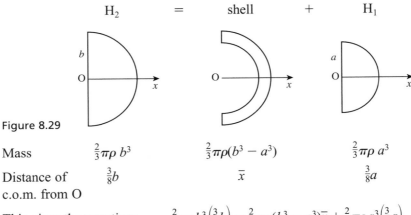

Figure 8.29

	H_2	shell	H_1
Mass	$\frac{2}{3}\pi\rho\, b^3$	$\frac{2}{3}\pi\rho(b^3 - a^3)$	$\frac{2}{3}\pi\rho\, a^3$
Distance of c.o.m. from O	$\frac{3}{8}b$	$\bar{x}$	$\frac{3}{8}a$

This gives the equation:

$$\frac{2}{3}\pi\rho b^3 \left(\frac{3}{8}b\right) = \frac{2}{3}\pi\rho(b^3 - a^3)\bar{x} + \frac{2}{3}\pi\rho a^3 \left(\frac{3}{8}a\right)$$

$$(b^3 - a^3)\bar{x} = b^3\left(\frac{3}{8}b\right) - a^3\left(\frac{3}{8}a\right)$$

$$\Rightarrow \quad \bar{x} = \frac{3(b^4 - a^4)}{8(b^3 - a^3)} \qquad \qquad \text{①}$$

This is the position of the centre of mass of a shell of any thickness. The position of the centre of mass of a thin shell of radius a, can be found by letting $b = a + h$, and then letting $h \to 0$.

It is necessary to expand $(a + h)^4$ and $(a + h)^3$ using the binomial theorem.

$$\text{Then} \quad b^4 - a^4 = (a + h)^4 - a^4$$
$$= (a^4 + 4a^3 h + 6a^2 h^2 + 4ah^3 + h^4) - a^4$$
$$= h(4a^3 + 6a^2 h + 4ah^2 + h^3)$$

and similarly:
$$b^3 - a^3 = (a + h)^3 - a^3$$
$$= h(3a^2 + 3ah + h^2).$$

Then
$$\bar{x} = \frac{3(b^4 - a^4)}{8(b^3 - a^3)}$$
$$= \frac{3h(4a^3 + 6a^2h + 4ah^2 + h^3)}{8h(3a^2 + 3ah + h^2)}.$$

Cancelling h and then letting $h \to 0$ gives $\bar{x}$ for the shell:
$$\bar{x} = \frac{3}{8} \times \frac{4a^3}{3a^2} = \frac{a}{2}.$$

EXAMPLE 8.13

i) Find the position of the centre of mass of a uniform hemispherical bowl of thickness 1 cm and inside radius 9 cm.

ii) A light handle AB of length 18 cm is attached to a point A on the rim of the bowl in such a way that the line AB passes through the centre of the hemisphere. What angle does AB make with the vertical when the bowl is hung from the end B?

SOLUTION

i) Think of the bowl as a solid hemisphere of radius 10 cm from which another solid hemisphere of radius 9 cm has been removed. Then the original 10 cm hemisphere can be treated as a composite body, consisting of the bowl and the 9 cm hemisphere.

You can put all the information in a table.

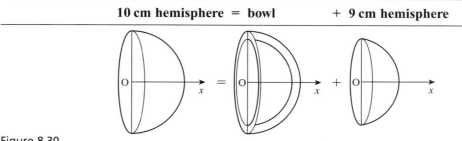

	10 cm hemisphere	**= bowl**	**+ 9 cm hemisphere**
Mass	$\frac{2}{3}\pi\rho \times 10^3$	$\frac{2}{3}\pi\rho(10^3 - 9^3)$	$\frac{2}{3}\pi\rho \times 9^3$
Distance from O to c.o.m.	$\frac{3}{8} \times 10$	$\bar{x}$	$\frac{3}{8} \times 9$

Figure 8.30

Taking moments gives
$$\left(\tfrac{2}{3}\pi\rho 10^3\right) \times \left(\tfrac{3}{8} \times 10\right) = \left(\tfrac{2}{3}\pi\rho(10^3 - 9^3) \times \bar{x}\right) + \left(\tfrac{2}{3}\pi\rho 9^3\right) \times \left(\tfrac{3}{8} \times 9\right).$$

This can be simplified by dividing through by $\frac{2}{3}\rho\pi$ to give
$$\tfrac{3}{8} \times 10^4 = (10^3 - 9^3)\bar{x} + \tfrac{3}{8} \times 9^4$$
$$\bar{x} = \frac{3(10^4 - 9^4)}{8(10^3 - 9^3)}$$
$$= 4.76 \text{ (to 3 s.f.).}$$

ii) The handle is light so the position of the centre of mass, G, is unchanged when it is attached. For equilibrium G must be vertically below B and the angle required is the angle ABG = α.

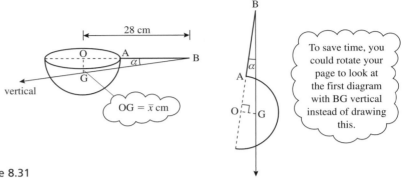

Figure 8.31

From the diagram
$$\tan\alpha = \frac{OG}{OB} = \frac{\bar{x}}{28}$$
$$\Rightarrow \quad \alpha = 9.65°.$$

The angle between AB and the vertical is about 10°.

Note

An alternative, but equivalent, method for finding $\bar{x}$ is to regard the bowl as the sum of a solid hemisphere of radius 10 cm and a solid hemisphere of radius 9 cm which has negative mass.

In this case the moments equation would be written
$$\tfrac{2}{3}\pi\rho(10^3 - 9^3)\bar{x} = \left(\tfrac{2}{3}\pi\rho 10^3\right)\times\left(\tfrac{3}{8}\times 10\right) + \left(-\tfrac{2}{3}\pi\rho 9^3\right)\times\left(\tfrac{3}{8}\times 9\right).$$

EXAMPLE 8.14

A traffic cone is modelled as a hollow cone of height 45 cm standing on a heavy cylindrical base of radius 20 cm and height 5 cm. The mass of the base is twice that of the cone.
i) Find the position of the centre of mass of the traffic cone.
ii) What is the angle of the steepest slope it can be placed on without toppling?

SOLUTION

i) You can put the information in a table.

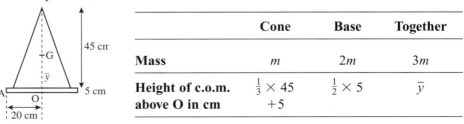

	Cone	Base	Together
Mass	m	$2m$	$3m$
Height of c.o.m. above O in cm	$\tfrac{1}{3}\times 45$ $+5$	$\tfrac{1}{2}\times 5$	$\bar{y}$

Figure 8.32

Then
$$20m \quad + \quad 5m \quad = \quad 3m\bar{y}$$
$$\Rightarrow \qquad \bar{y} = \tfrac{25}{3}$$

ii) The cone begins to topple when the line GA is vertical.

Then $\tan\alpha = \dfrac{20}{y} = 2.4$

The steepest slope is $\alpha = 67.4°$ to the horizontal.

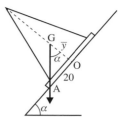

Figure 8.33

8.9 Centres of mass of plane regions

Calculus methods for determining the centre of mass of a lamina

When a lamina or plane region is bounded by curves for which the equations are known, calculus methods can be used to find the position of its centre of mass in a similar manner to that used for volumes of revolution.

When a lamina occupies a region between a curve and the x axis, as shown in figure 8.34, it can be divided into elementary strips of width δx and length y. One such strip, situated at the point (x, y) on the curve, is shown. The dot indicates the centre of mass of this elementary strip. Let σ (the Greek letter sigma), be the mass per unit area of the lamina.

The length of the strip is y and its width is δx, so its mass δM is $\sigma y \delta x$.

The co-ordinates of the centre of mass of the strip are $\left(x, \dfrac{y}{2}\right)$ so $\bar{x}$ can be found using

$$(\Sigma \, \delta M)\bar{x} = \Sigma(\delta Mx).$$

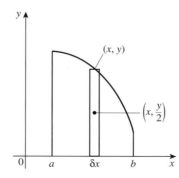

Figure 8.34

Substituting $\delta M = \sigma y \delta x$ in

$$(\Sigma \delta M)\bar{x} = \Sigma(\delta Mx)$$

gives

$$(\Sigma \sigma y \delta x)\bar{x} = \Sigma(\sigma y \delta x \times x).$$

If the material is uniform, the mass per unit area is constant and so the equation can be divided through by σ. In that case,

$$(\Sigma y \delta x)\bar{x} = \Sigma(yx\delta x)$$

and in the limit as $\delta x \to 0$:

$$\left(\int_a^b y \, dx\right)\bar{x} = \int_a^b xy \, dx. \qquad \text{①}$$

Notice that the first integral gives the area of the lamina.

You can find $\bar{y}$ in a similar way. Using the y co-ordinate, $\dfrac{y}{2}$, of the centre of the strip and dividing by σ as before:

$$(\Sigma y \delta x)\bar{y} = \Sigma\left(y\delta x \times \frac{y}{2}\right).$$

In the limit as $\delta x \to 0$ this becomes:

$$\left(\int_a^b y \, dx\right)\bar{y} = \int_a^b \frac{y^2}{2} \, dx. \qquad \text{②}$$

The equations ① and ② can be combined to give the co-ordinates of the centre of mass of the lamina in the form of a position vector:

$$A\binom{\bar{x}}{\bar{y}} = \begin{pmatrix} \displaystyle\int_a^b xy \, dx \\ \displaystyle\int_a^b \frac{y^2}{2} \, dx \end{pmatrix}$$

where A is the total area of the lamina or $\displaystyle\int_a^b y \, dx$.

⚠ Before deciding which formula to use, draw a diagram such as figure 8.34 so that you can see whether the formula you are using is sensible.

EXAMPLE 8.15   Find the position of the centre of mass of a uniform semi-circular lamina of radius r.

SOLUTION

The diagram below shows the lamina.

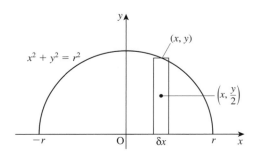

Figure 8.35

By symmetry, $\bar{x} = 0$

To find $\bar{y}$, use $A\bar{y} = \displaystyle\int_a^b \frac{y^2}{2}\,dx.$

In this case, $A = \frac{1}{2}\pi r^2$ (because the lamina is a semi-circle), and the limits are $-r$ and $+r$.

Since the equation of the curve is $x^2 + y^2 = r^2$, you can write y^2 as $r^2 - x^2$.

This gives
$$\tfrac{1}{2}\pi r^2 \bar{y} = \int_{-r}^{r} \tfrac{1}{2}(r^2 - x^2)\,dx$$

$$= \tfrac{1}{2}\left[r^2 x - \frac{x^3}{3} \right]_{-r}^{r}$$

$$\Rightarrow \quad \tfrac{1}{2}\pi r^2 \bar{y} = \tfrac{1}{2}\left\{ \left(r^3 - \frac{r^3}{3} \right) - \left(-r^3 - \frac{(-r)^3}{3} \right) \right\}$$

$$= \tfrac{2}{3} r^3$$

$$\Rightarrow \quad \bar{y} = \frac{4r}{3\pi}.$$

This is the distance of the centre of mass of the semi-circular lamina from its centre. It is roughly 0.4 times the radius.

EXAMPLE 8.16   Find the co-ordinates of the centroid of the region between the curve $y = 4 - x^2$ and the positive x and y axes.

SOLUTION

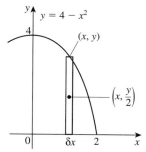

Figure 8.36

It is best to find the area, A, first. In this case:

$$A = \int y\,dx$$

$$= \int_0^2 (4 - x^2)\,dx$$

$$= \left[4x - \frac{x^3}{3} \right]_0^2$$

$$= \left(8 - \frac{8}{3} \right) - 0$$

$$= \frac{16}{3}.$$

Then use
$$A\left(\frac{\bar{x}}{\bar{y}}\right) = \left(\begin{array}{c} \int_0^2 xy \, dx \\ \int_0^2 \frac{y^2}{2} \, dx \end{array}\right) \quad \text{to give}$$

$$A\bar{x} = \int_0^2 (4 - x^2)x \, dx \quad \text{and} \quad A\bar{y} = \int_0^2 \frac{(4 - x^2)^2}{2} \, dx$$

$$= \int_0^2 (4x - x^3) \, dx \qquad\qquad = \int_0^2 \left(\frac{16 - 8x^2 + x^4}{2}\right) dx$$

$$= \left[\frac{4x^2}{2} - \frac{x^4}{4}\right]_0^2 \qquad\qquad = \frac{1}{2}\left[16x - \frac{8x^3}{3} + \frac{x^5}{5}\right]_0^2$$

$$= 4 \qquad\qquad\qquad = \frac{128}{15}.$$

Substituting $A = \frac{16}{3}$ gives

$$\bar{x} = 4 \times \frac{3}{16} \quad \text{and} \quad \bar{y} = \frac{128}{15} \times \frac{3}{16}$$

$$\bar{x} = \frac{3}{4} = 0.75 \qquad \bar{y} = \frac{8}{5} = 1.6.$$

The co-ordinates of the centroid are (0.75, 1.6).

Using strips parallel to the *x* axis

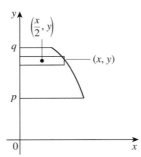

Figure 8.37

Sometimes it is necessary (or easier) to divide the region into strips parallel to the *x* axis as shown in figure 8.37. The area A is then $\int_p^q x \, dy$ and

$$A\left(\frac{\bar{x}}{\bar{y}}\right) = \left(\begin{array}{c} \int_p^q \frac{x^2}{2} \, dy \\ \int_p^q yx \, dy \end{array}\right).$$

Check that you can see the reason for these.

In this case the *x* in each integral should be written in terms of *y*. Where there is a choice between methods you should choose the one which gives the easiest integral to evaluate.

Chapter 8 Exercises

8.1 A rod has length 2 m and mass 3 kg. The centre of mass should be in the middle but due to a fault in the manufacturing process it is not. This error is corrected by placing a 200 g mass 5 cm from the centre of the rod. Where is the centre of mass of the rod itself?

8.2 The masses of the earth and the moon are 5.98×10^{24} kg and 7.38×10^{22} kg, and the distance between their centres is 3.84×10^5 km. How far from the centre of the earth is the centre of mass of the earth–moon system?

8.3 A uniform scaffold pole of length 5 m has brackets bolted to it as shown in the diagram below. The mass of each bracket is 1 kg.

0.5 m   0.5 m   0.5 m   0.5 m   1 m   1 m   1 m

The centre of mass is 2.44 m from the left-hand end. What is the mass of the pole?

8.4 An object of mass m_1 is placed at one end of a light rod of length l. An object of mass m_2 is placed at the other end. Find the position of the centre of mass.

8.5 The diagram illustrates a mobile tower crane. It consists of the main vertical section (mass M tonnes), housing the engine, winding gear and controls, and the boom. The centre of mass of the main section is on its centre line. The boom, which has negligible mass, supports the load (L tonnes) and the counterweight (C tonnes). The main section stands on supports at P and Q, distance $2d$ m apart. The counterweight is held at a fixed distance a m from the centre line of the main section and the load at a variable distance l m.

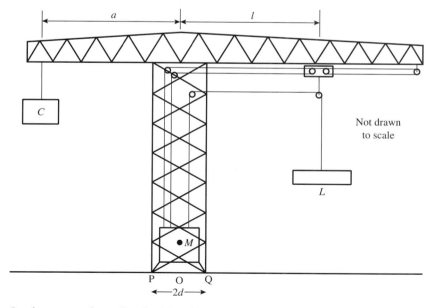

i) In the case when $C = 3$, $M = 10$, $L = 7$, $a = 8$, $d = 2$ and $l = 13$, find the horizontal position of the centre of mass and say what happens to the crane.

   ii) Show that for these values of C, M, a, d and l the crane will not fall over when it has no load, and find the maximum safe load that it can carry.

   iii) Formulate two inequalities in terms of C, M, L, a, d and l that must hold if the crane is to be safe loaded or unloaded.

   iv) Find, in terms of M, a, d and l, the maximum load that the crane can carry.

8.6 A filing cabinet has the dimensions shown in the diagram. The body of the cabinet has mass 20 kg and its construction is such that its centre of mass is at a height of 60 cm, and is 25 cm from the back of the cabinet. The mass of a drawer and its contents may be taken to be 10 kg and its centre of mass to be 10 cm above its base and 10 cm from its front face.

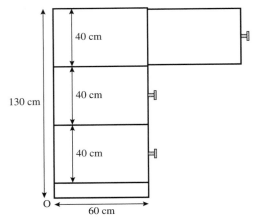

   i) Find the position of the centre of mass when all the drawers are closed.

   ii) Find the position of the centre of mass when the top two drawers are fully open.

   iii) Show that when all three drawers are fully opened the filing cabinet will tip over.

   iv) Two drawers are fully open. How far can the third one be opened without the cabinet tipping over?

8.7 Co-ordinates refer to the axes shown in the figures and the units are centimetres.

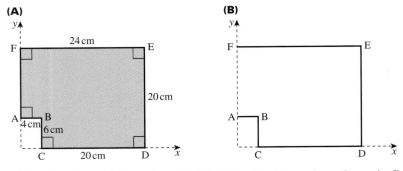

A uniform lamina with boundary ABCDEF has the dimensions shown in figure (A).

   i) Calculate the co-ordinates of the centre of mass of this lamina, showing that the x co-ordinate is 12.5 correct to one decimal place.

A length of uniform thin wire is bent into the same shape ABCDEF shown in figure (B); note that there is no join FA.

ii) Calculate the co-ordinates of the centre of mass of this wire, showing that the x co-ordinate is 14.6 correct to one decimal place.

A small animal feeder with cross-section in the shape of ABCDEF and length 60 cm is shown in figures (C) and (D). All the faces are made from uniform, thin sheet metal except for AFLG which is left open to allow access to the food inside.

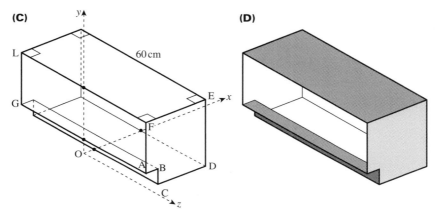

iii) Using your answers to parts i) and ii), or otherwise, calculate the x and y co-ordinates of the centre of mass of the feeder.

8.8 A simple lift bridge (commonly used on canals) which stays open or closed without the use of any restraining forces is shown in figure (A). A mathematical model of the structure is shown in figure (B), where the bridge is represented by two thin rods AO and OB, rigidly joined and freely pivoted at a fixed point O. The rod AO has length 2.5 m, mass 400 kg and centre of mass 1.8 m from O. The rod OB has length 4 m, mass 350 kg and centre of mass 2 m from O. In the closed position AO is horizontal and OB is inclined at an angle of 40° with the horizontal. In the open position OB is horizontal.

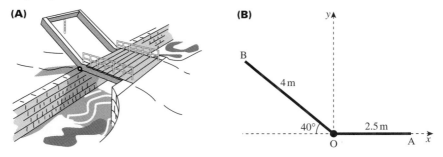

i) Calculate the position of the centre of mass of the bridge in the closed position, referred to the co-ordinate axes indicated in figure (B). Give your answers correct to three significant figures. Draw a sketch showing the position of the centre of mass.

ii) Describe how the position of the centre of mass of the bridge changes as the bridge is opened.

iii) Establish that this bridge will stay open or closed once moved to that position. Show any further calculations you find necessary.

iv) When closed, the bridge is opened by pulling on a chain attached to the point B. The chain is pulled perpendicular to OB. What is the tension in the chain when the bridge just begins to turn?

8.9 A mathematical model for a large garden pot is obtained by rotating through 360° about the y axis the part of the curve $y = 0.0001x^4$ which is between $x = 10$ and $x = 25$, and then adding a flat base. Units are in centimetres.

i) Draw a sketch of the curve and shade in the cross-section of the pot, indicating which line will form its base.

ii) Garden compost is sold in litres. How many litres will be required to fill the pot to a depth of 35 cm? (Ignore the thickness of the pot.)

8.10 The circle $x^2 + (y + 1)^2 = 5$ meets the x axis at the points A and B. The minor arc AB is rotated about the x axis to form a solid of revolution which can be used to model a rugby ball.

i) Find the co-ordinates of A and B and show that

$$y^2 = 6 - x^2 - 2\sqrt{5 - x^2}$$

ii) Hence calculate the volume of the solid of revolution.

$$\left(\text{Hint: } \int_0^2 \sqrt{5 - x^2} \, dx = \frac{5}{2} \arcsin \frac{2}{\sqrt{5}} + 1 \right)$$

iii) A similar rugby ball is 30 cm long. Find its volume.

8.11 Water fills a light hemispherical bowl of radius 12 cm to a depth of 6 cm. Find the height of the centre of mass of the water above the base of the bowl.

8.12 The shape of the bowl of a wine glass is produced by rotating through 360° about the y axis the part of the curve $y = kx^2$ which lies between the origin and the line $y = h$.

i) Draw a diagram of the bowl and, by integration, find the volume of the solid of revolution contained within it.

ii) Show that the centre of mass of this solid is at the point $(0, \frac{2}{3}h)$ and hence write down its distance below the top of the glass. (Notice that this is independent of the value of k.)

iii) The bowl of the glass can be modelled as a solid block of depth 9 cm with another similar block of depth 8.8 cm scooped out. Make out a table similar to that on page 186 using the results obtained in part i). Show that the centre of mass of the bowl is 4.45 cm below the rim.

iv) The complete wine glass can be modelled by a circular base of negligible thickness and radius 3 cm, a cylindrical stem of height 8 cm and the bowl (as above) of depth 9 cm. The masses of these three parts are in the ratio 1:1:2. Find the distance of the centre of mass of the whole glass from the circular base.

v) If the glass is carried on a tray, what is the maximum angle to the horizontal that the tray can be held without the glass toppling over. You may assume that the glass does not slide.

8.13 In the diagram, OAB is a uniform right-angled triangular lamina. OA = 6 cm and OB = 12 cm.

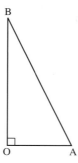

i)   Write down the area of the triangle.

ii) Using OA as the x axis and OB as the y axis, find the equation of the line AB.

iii) Find, by integration, the co-ordinates $(\bar{x}, \bar{y})$ of the centre of mass, G, of the triangle and verify that $\bar{x} = \frac{1}{3}$OA and $\bar{y} = \frac{1}{3}$OB.

8.14 A letter P is made up from a rectangle and a semi-circle with a smaller semi- circle cut out as shown. In this exercise, you can use the result for the position of the centre of mass of a semi-circle, found on page 186. It is at a distance $\frac{4r}{3\pi}$ from the centre.

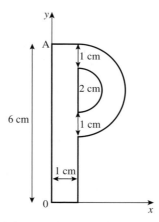

i)   Taking O as the origin, write down the areas of the component parts of the letter, and the co-ordinates of their centres of mass.

ii) Find the co-ordinates of the centre of mass of the letter.

iii) What angle will AO make with the vertical if the letter is hung from the corner A?

Personal Tutor Exercise

 8.1 The diagram shows the hyperbola $\dfrac{x^2}{a^2} - \dfrac{y^2}{b^2} = 1$.

When it is rotated through 360° about the y axis a hyperboloid is formed.

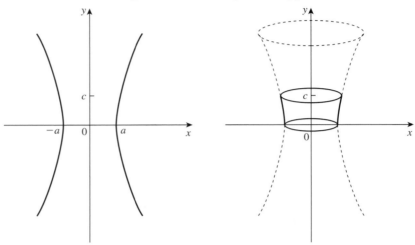

i) Show that the volume enclosed by this hyperboloid between $y = 0$ and $y = c$ is given by $V = \dfrac{\pi a^2 c}{3b^2}(3b^2 + c^2)$.

The cooling towers of power stations can be modelled as hyperboloids and the dimensions of one such tower, at Didcot power station in Oxfordshire, are shown on the picture.

A = 51.3 m
B = 47.5 m
C = 24.5 m
D = 89.5 m
E = 85.5 m

The equation $\dfrac{x^2}{560} - \dfrac{y^2}{3600} = 1$ is a suitable mathematical model for the curve which is rotated to form the hyperboloid.

ii) Use the formula above to find the volume occupied by the cooling tower, and show that the six towers of the power station occupy over 1.9 million cubic metres of space.

8.1   The centre of mass of a body has the property that:

the moment, about any point, of the whole mass of the body taken at the centre of mass is equal to the sum of the moments of the various particles comprising the body.

$$M\overline{\mathbf{r}} = \Sigma m_i\, \mathbf{r}_i \quad \text{where} \quad M = \Sigma m_i$$

8.2   In one dimension

$$M\overline{x} = \Sigma m_i\, x_i$$

8.3   In two dimensions

$$M\begin{pmatrix} \overline{x} \\ \overline{y} \end{pmatrix} = \Sigma m_i \begin{pmatrix} x_i \\ y_i \end{pmatrix}$$

8.4   In three dimensions

$$M\begin{pmatrix} \overline{x} \\ \overline{y} \\ \overline{z} \end{pmatrix} = \Sigma m_i \begin{pmatrix} x_i \\ y_i \\ z_i \end{pmatrix}$$

8.5   Volumes of revolution

- About the *x* axis:

$$V = \int_a^b \pi y^2\, dx$$

- About the *y* axis:

$$V = \int_p^q \pi x^2\, dx$$

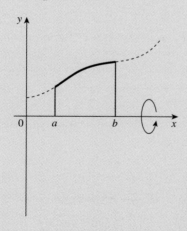

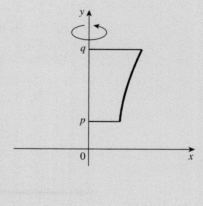

8.6   Centres of mass of uniform bodies

For a volume of revolution

- About the *x* axis:

$$\left(\int_a^b \pi y^2\, dx \right) \overline{x} = \int_a^b \pi x y^2\, dx$$

and $\overline{y} = 0$

- About the *y* axis:

$$\overline{x} = 0 \quad \text{and}$$

$$\left(\int_p^q \pi x^2\, dy \right) \overline{y} = \int_p^q \pi y x^2\, dy$$

For a uniform plane lamina, area A, as shown in the diagrams

- $A = \int_a^b y \, dx$

- $A\begin{pmatrix} \bar{x} \\ \bar{y} \end{pmatrix} = \begin{pmatrix} \int_a^b xy \, dx \\ \int_a^b \dfrac{y^2}{2} \, dx \end{pmatrix}$

- $A = \int_p^q x \, dy$

- $A\begin{pmatrix} \bar{x} \\ \bar{y} \end{pmatrix} = \begin{pmatrix} \int_p^q \dfrac{x^2}{2} \, dy \\ \int_p^q yx \, dy \end{pmatrix}$

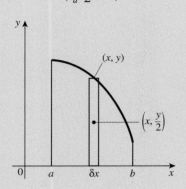

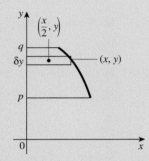

9 Energy, work and power

I like work: it fascinates me. I can sit and look at it for hours.

Jerome K. Jerome

M C Escher's 'Waterfall'
© 2000 Cordon Art B.V. –
Baarn – Holland.
All rights reserved.

QUESTION 9.1   This is a picture of a perpetual motion machine. What does this term mean and will this one work?

9.1 Energy and momentum

When describing the motion of objects in everyday language the words *energy* and *momentum* are often used quite loosely and sometimes no distinction is made between them. In mechanics they must be defined precisely.

For an object of mass m moving with velocity $\mathbf{v}$:

- *Kinetic energy* $= \frac{1}{2}mv^2$ (this is the energy it has due to its motion)
- *Momentum* $= m\mathbf{v}$.

Notice that kinetic energy is a scalar quantity with magnitude only, but momentum is a vector in the same direction as the velocity.

Both the kinetic energy and the momentum are liable to change when a force acts on a body and you will learn more about how the energy is changed in this chapter. You will meet momentum again in Chapter 10.

9.2 Work and energy

In everyday life you encounter many forms of energy such as heat, light, electricity and sound. You are familiar with the conversion of one form of energy to another: from chemical energy stored in wood to heat energy when you burn it; from electrical energy to the energy of a train's motion, and so on. The S.I. unit for energy is the joule, J.

Mechanical energy and work

In mechanics two forms of energy are particularly important.

Kinetic energy is the energy which a body possesses because of its motion.

● *The kinetic energy of a moving object* $= \frac{1}{2} \times mass \times (speed)^2$.

Potential energy is the energy which a body possesses because of its position. It may be thought of as stored energy which can be converted into kinetic or other forms of energy. You will meet this again on page 208.

The energy of an object is usually changed when it is acted on by a force. When a force is applied to an object which moves in the direction of its line of action, the force is said to do work. For a constant force this is defined as follows.

● *The work done by a constant force = force × distance moved in the direction of the force.*

The following examples illustrate how to use these ideas.

EXAMPLE 9.1

A brick, initially at rest, is raised by a force averaging 40 N to a height 5 m above the ground where it is left stationary. How much work is done by the force?

SOLUTION

The work done by the force raising the brick is

$$40 \times 5 = 200 \text{ J}.$$

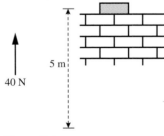

Figure 9.1

Examples 9.2 and 9.3 show how the work done by a force can be related to the change in kinetic energy of an object.

EXAMPLE 9.2

A train travelling on level ground is subject to a resisting force (from the brakes and air resistance) of 250 kN for a distance of 5 km. How much kinetic energy does the train lose?

SOLUTION

The forward force is $-250\,000$ N.

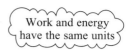

Work and energy have the same units

The work done by it is $-250\,000 \times 5000 = -1\,250\,000\,000$ J.

Hence $-1\,250\,000\,000$ J of kinetic energy are gained by the train, in other words $+1\,250\,000\,000$ J of kinetic energy are lost and the train slows down. This energy is converted to other forms such as heat and perhaps a little sound.

EXAMPLE 9.3

A car of mass m kg is travelling at u ms^{-1} when the driver applies a constant driving force of F N. The ground is level and the road is straight and air resistance can be ignored. The speed of the car increases to v ms^{-1} in a period of t s over a distance of s m. Find the relationship between F, S, m, u and v.

SOLUTION

Treating the car as a particle and applying Newton's second law:

$$F = ma$$

$$a = \frac{F}{m}.$$

Since F is assumed constant, the acceleration is constant also so using $v^2 = u^2 + 2as$

$$v^2 = u^2 + \frac{2Fs}{m}$$

$$\Rightarrow \tfrac{1}{2}mv^2 = \tfrac{1}{2}mu^2 + Fs$$

$$Fs = \tfrac{1}{2}mv^2 - \tfrac{1}{2}mu^2.$$

Thus

- *work done by force = final kinetic energy – initial kinetic energy of car.*

The work-energy principle

Examples 9.4 and 9.5 illustrate the *work-energy* principle which states that:

- *The total work done by the forces acting on a body is equal to the increase in the kinetic energy of the body.*

EXAMPLE 9.4

A sledge of total mass 30 kg, initially moving at 2 ms^{-1}, is pulled 14 m across smooth horizontal ice by a horizontal rope in which there is a constant tension of 45 N. Find its final velocity.

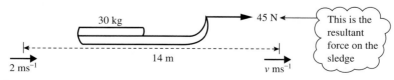

Figure 9.2

SOLUTION

Since the ice is smooth, the work done by the force is all converted into kinetic energy and the final velocity can be found using

work done by the force = final kinetic energy − initial kinetic energy

$$45 \times 14 = \tfrac{1}{2} \times 30 \times v^2 - \tfrac{1}{2} \times 30 \times 2^2.$$

Giving $v^2 = 46$ and the final velocity of the sledge as 6.8 ms^{-1}.

EXAMPLE 9.5

The combined mass of a cyclist and her bicycle is 65 kg. She accelerated from rest to 8 ms^{-1} in 80 m along a horizontal road.
i)   Calculate the work done by the net force in accelerating the cyclist and her bicycle.
ii) Hence calculate the net forward force (assuming the force to be constant).

SOLUTION

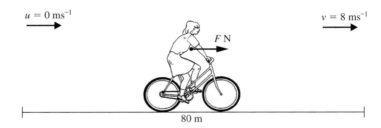

$u = 0 \text{ ms}^{-1}$ $F \text{ N}$ $v = 8 \text{ ms}^{-1}$

80 m

Figure 9.3

i)   The work done by the net force F is given by

$$\text{work} = \text{final K.E.} - \text{initial K.E.}$$
$$= \tfrac{1}{2}mv^2 - \tfrac{1}{2}mu^2$$
$$= \tfrac{1}{2} \times 65 \times 8^2 - 0$$
$$= 2080 \text{ J}.$$

The work done is 2080 J.

ii)   Work done $= Fs$
$$= F \times 80$$
So $80F = 2080$
$$F = 26$$

The net forward force is 26 N.

Work

It is important to realise that:

● work is done by a force
● work is only done when there is movement
● a force only does work on an object when it has a component in the direction of motion of the object.

It is quite common to speak of the work done by a person, say in pushing a lawn mower. In fact this is the work done by the force of the person on the lawn mower.

Notice that if you stand holding a brick stationary above your head, painful though it may be, the force you are exerting on it is doing no work. Nor is this vertical force doing any work if you walk round the room keeping the brick at the same height. However, once you start climbing the stairs, a component of the brick's movement is in the direction of the upward force that you are exerting on it, so the force is now doing some work.

When applying the work–energy principle, you have to be careful to include all the forces acting on the body. In the example of a brick of weight 40 N being raised 5 m vertically, starting and ending at rest, the change in kinetic energy is clearly 0.

This seems paradoxical when it is clear that the force which raised the brick has done $40 \times 5 = 200$ J of work. However, the brick was subject to another force, namely its weight, which did $-40 \times 5 = -200$ J of work on it, giving a total of $200 + (-200) = 0$ J.

Conservation of mechanical energy

The net forward force on the cyclist in Example 9.5 is the girl's driving force minus resistive forces such as air resistance and friction in the bearings. In the absence of such resistive forces, she would gain more kinetic energy; also the work she does against them is lost, it is dissipated as heat and sound. Contrast this with the work a cyclist does against gravity when going uphill. This work can be recovered as kinetic energy on a downhill run. The work done against the force of gravity is conserved and gives the cyclist potential energy (see page 208).

Forces such as friction which result in the dissipation of mechanical energy are called *dissipative forces*. Forces which conserve mechanical energy are called *conservative forces*. The force of gravity is a conservative force and so is the tension in an elastic string; you can test this using an elastic band.

EXAMPLE 9.6

A bullet of mass 25 g is fired at a wooden barrier 3 cm thick. When it hits the barrier it is travelling at 200 ms⁻¹. The barrier exerts a constant resistive force of 5000 N on the bullet.

i) Does the bullet pass through the barrier and if so with what speed does it emerge?
ii) Is energy conserved in this situation?

SOLUTION

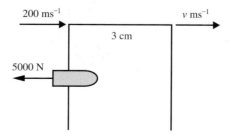

Figure 9.4

i) The work done *by* the force is defined as the product of the force and the distance moved *in the direction of the force*. Since the bullet is moving in the direction opposite to the net resistive force, the work done by this force is negative.

$$\text{Work done} = -5000 \times 0.03 \text{ J}$$
$$= -150 \text{ J}$$

The initial kinetic energy of the bullet is

$$\text{Initial K.E.} = \tfrac{1}{2}mu^2$$
$$= \tfrac{1}{2} \times 0.025 \times 200^2$$
$$= 500 \text{ J}$$

A loss in energy of 150 J will not reduce kinetic energy to zero, so the bullet will still be moving on exit.

Since the work done is equal to the change in kinetic energy,

$$-150 = \tfrac{1}{2}mv^2 - 500$$

Solving for v

$$\tfrac{1}{2}mv^2 = 500 - 150$$
$$v^2 = \frac{2 \times (500 - 150)}{0.025}$$
$$v = 167 \text{ (to nearest whole number).}$$

So the bullet emerges from the barrier with a speed of 167 ms^{-1}.

ii) Total energy is conserved but there is a loss of mechanical energy of $\tfrac{1}{2}mu^2 - \tfrac{1}{2}mv^2 = 150$ J. This energy is converted into non-mechanical forms such as heat and sound.

EXAMPLE 9.7   An aircraft of mass m kg is flying at a constant velocity v ms^{-1} horizontally. Its engines are providing a horizontal driving force F N.

i) Draw a diagram showing the driving force, the lift force L N, the air resistance (drag force) R N and the weight of the aircraft.

ii) State which of these forces are equal in magnitude.

iii) State which of the forces are doing no work.

iv) In the case when $m = 100\,000$, $v = 270$ and $F = 350\,000$, find the work done in a 10-second period by those forces which are doing work, and show that the work–energy principle holds in this case.

At a later time the pilot increases the thrust of the aircraft's engines to 400 000 N. When the aircraft has travelled a distance of 30 km, its speed has increased to 300 ms^{-1}.

v) Find the work done against air resistance during this period, and the average resistance force.

SOLUTION

i)

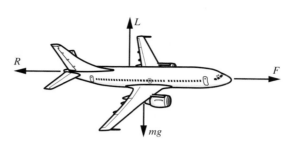

Figure 9.5

ii) Since the aircraft is travelling at constant velocity it is in equilibrium.

Horizontal forces: $F = R$

Vertical forces: $L = mg$

iii) Since the aircraft's velocity has no vertical component, the vertical forces, L and mg, are doing no work.

iv) In 10 s at 270 ms^{-1} the aircraft travels 2700 m.

Work done by force $F = 350\,000 \times 2700 \quad = 94\,500\,000$ J

Work done by force $R = 350\,000 \times -2700 = -94\,500\,000$ J

The work–energy principle states that in this situation

work done by F + work done by R = change in kinetic energy.

Now work done by F + work done by $R = (94\,500\,000 - 94\,500\,000) = 0$ J, and change in kinetic energy $= 0$ (since velocity is constant), so the work–energy principle does indeed hold in this case.

v)

$$\text{Final K.E.} - \text{initial K.E.} = \tfrac{1}{2}mv^2 - \tfrac{1}{2}mu^2$$

$$= \tfrac{1}{2} \times 100\,000 \times 300^2 - \tfrac{1}{2} \times 100\,000 \times 270^2$$

$$= 855 \times 10^6 \text{ J}$$

$$\text{Work done by driving force} = 400\,000 \times 30\,000$$

$$\text{Total work done} = \text{K.E. gained}$$

$$= 12\,000 \times 10^6 \text{ J}$$

$$\text{Work done by resistance force} + 12\,000 \times 10^6 = 855 \times 10^6$$

$$\text{Work done by resistance force} = -11\,145 \times 10^6 \text{ J}$$

$$\text{Average force} \times \text{distance} = \text{work done by force}$$

$$\text{Average force} \times 30\,000 = -11\,145 \times 10^6$$

$\Rightarrow$ The average resistance force is 371 500 N (in the negative direction).

Note

When an aircraft is in flight, most of the work done by the resistance force results in air currents and the generation of heat. A typical large jet cruising at 35 000 feet has a body temperature about 30°C above the surrounding air temperature. For supersonic flight the temperature difference is much greater. Concorde flew with a skin temperature more than 200°C above that of the surrounding air.

9.3 Gravitational potential energy

As you have seen, kinetic energy (K.E.) is the energy that an object has because of its motion. Potential energy (P.E.) is the energy an object has because of its position. The units of potential energy are the same as those of kinetic energy or any other form of energy, namely joules.

One form of potential energy is *gravitational potential energy*. The gravitational potential energy of the object in Figure 9.6 of mass m kg at height h m above a fixed reference level, O, is mgh J. If it falls to the reference level, the force of gravity does mgh J of work and the body loses mgh J of potential energy.

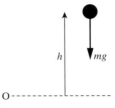

Figure 9.6

A loss in gravitational potential energy is an alternative way of accounting for the work done by the force of gravity.

If a mass m kg is *raised* through a distance h m, the gravitational potential energy *increases* by mgh J. If a mass m kg is *lowered* through a distance h m the gravitational potential energy *decreases* by mgh J.

EXAMPLE 9.8

Calculate the gravitational potential energy, relative to the ground, of a ball of mass 0.15 kg at a height of 2 m above the ground.

SOLUTION

Mass $m = 0.15$, height $h = 2$.

Gravitational potential energy $= mgh$
$$= 0.15 \times 9.8 \times 2$$
$$= 2.94 \text{ J}.$$

Note

If the ball falls:

$$\text{loss in P.E.} = \text{work done by gravity}$$
$$= \text{gain in K.E.}$$

There is no change in the total energy (P.E. + K.E.) of the ball.

Using conservation of mechanical energy

When gravity is the only force which does work on a body, mechanical energy is conserved. When this is the case, many problems are easily solved using energy. This is possible even when the acceleration is not constant.

EXAMPLE 9.9

A skier slides down a smooth ski slope 400 m long which is at an angle of 30° to the horizontal. Find the speed of the skier when he reaches the bottom of the slope.

At the foot of the slope the ground becomes horizontal and is made rough in order to help him to stop. The coefficient of friction between his skis and the ground is $\frac{1}{4}$.

i)   Find how far the skier travels before coming to rest.

ii) In what way is your model unrealistic?

SOLUTION

The skier is modelled as a particle.

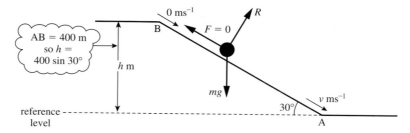

Figure 9.7

i)   Since in this case the slope is smooth, the frictional force is zero. The skier is subject to two external forces, his weight mg and the normal reaction from the slope.

The normal reaction between the skier and the slope does no work because the skier does not move in the direction of this force. The only force which does work is gravity, so mechanical energy is conserved.

$$\text{Total mechanical energy at B} = mgh + \tfrac{1}{2}mu^2$$
$$= m \times 9.8 \times 400 \sin 30° + 0$$
$$= 1960m \text{ J}$$

Total mechanical energy at A $= (0 + \tfrac{1}{2}mv^2)$ J

Since mechanical energy is conserved,

$$\tfrac{1}{2}mv^2 = 1960m \qquad\qquad ①$$
$$v^2 = 3920$$
$$v = 62.6$$

The skier's speed at the bottom of the slope is 62.6 ms^{-1}.

Notice that the mass of the skier cancels out. Using this model, all skiers should arrive at the bottom of the slope with the same speed. Also the slope could be curved so long as the total height lost is the same.

For the horizontal part there is some friction. Suppose that the skier travels a further distance s m before stopping.

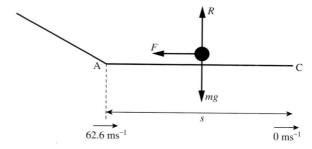

Figure 9.8

Coulomb's law of friction gives $\quad F = \mu R = \frac{1}{4}R$.

Since there is no vertical acceleration we can also say $R = mg$

So $$F = \frac{1}{4}mg.$$

Work done by the friction force $F \times (-s) = -\frac{1}{4}mgs$.

> Negative because the motion is in the opposite direction to the force

The increase in kinetic energy between A and C $= (0 - \frac{1}{2}mv^2)$ J.

Using the work–energy principle

$$-\frac{1}{4}mgs = -\frac{1}{2}mv^2 = -1960m \text{ from } ①$$

Solving for s gives $s = 800$.

So the distance the skier travels before stopping is 800 m.

ii)   The assumptions made in solving this problem are that friction on the slope and air resistance are negligible, and that the slope ends in a smooth curve at A. Clearly the speed of 62.6 ms^{-1} is very high, so the assumption that friction and air resistance are negligible must be suspect.

EXAMPLE 9.10   Ama, whose mass is 40 kg, is taking part in an assault course. The obstacle shown in figure 9.9 is a river at the bottom of a ravine 8 m wide which she has to cross by swinging on a rope 5 m long secured to a point on the branch of a tree, immediately above the centre of the ravine.

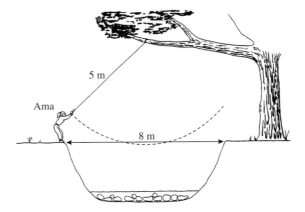

Figure 9.9

Find how fast Ama is travelling at the lowest point of her crossing:

i)   if she starts from rest
ii)   if she launches herself off at a speed of 1 ms^{-1}.

Will her speed be 1 ms^{-1} faster throughout her crossing?

SOLUTION

i) The vertical height Ama loses is HB in the diagram.

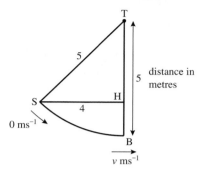

Figure 9.10

Using Pythagoras

$$TH = \sqrt{5^2 - 4^2} = 3$$
$$HB = 5 - 3 = 2$$
$$\text{P.E. lost} = mgh$$
$$= 40g \times 2$$
$$\text{K.E. gained} = \tfrac{1}{2}mv^2 - 0$$
$$= \tfrac{1}{2} \times 40 \times v^2$$

By conservation of energy, K.E. gained = P.E. lost

$$\tfrac{1}{2} \times 40 \times v^2 = 40 \times 9.8 \times 2$$
$$v = 6.26$$

Ama is travelling at $6.26\,\text{ms}^{-1}$.

ii) If she has initial speed $1\,\text{ms}^{-1}$ at S and speed $v\,\text{ms}^{-1}$ at B, her initial K.E. is $\tfrac{1}{2} \times 40 \times 1^2$ J and her K.E. at B is $\tfrac{1}{2} \times 40 \times v^2$.

Using conservation of energy,

$$\tfrac{1}{2} \times 40 \times v^2 - \tfrac{1}{2} \times 40 \times 1^2 = 40 \times 9.8 \times 2$$

This gives $v = 6.34$, so Ama's speed at the lowest point is now $6.34\,\text{ms}^{-1}$, only $0.08\,\text{ms}^{-1}$ faster than in part **i)**, so she clearly will not travel $1\,\text{ms}^{-1}$ faster throughout.

Historical note

James Joule was born in Salford in Lancashire on Christmas Eve 1818. He studied at Manchester University at the same time as the famous chemist, Dalton.

Joule spent much of his life conducting experiments to measure the equivalence of heat and mechanical forms of energy to ever increasing degrees of accuracy. Working with William Thomson, he also discovered that a gas cools when it expands without doing work against external forces. It was this discovery that paved the way for the development of refrigerators.

Joule died in 1889 but his contribution to science is remembered with the S.I. unit for energy named after him.

9.4 Work and kinetic energy for two-dimensional motion

QUESTION 9.2 Imagine that you are cycling along a level winding road in a strong wind. Suppose that the strength and direction of the wind are constant, but because the road is winding sometimes the wind is directly against you but at other times it is from your side.

How does the work you do in travelling a certain distance – say 1 m – change with your direction?

Work done by a force at an angle to the direction of motion

You have probably deduced that as a cyclist you would do work against the component of the wind force that is directly against you. The sideways component does not resist your forward progress.

Suppose that you are sailing and the angle between the force, F, of the wind on your sail and the direction of your motion is θ. In a certain time you travel a distance d in the direction of F, see figure 9.11, but during that time you actually travel a distance s along the line OP.

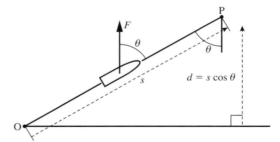

Figure 9.11

Work done by $F = Fd$

Since $d = s \cos \theta$, the work done by the force F is $Fs \cos \theta$. This can also be written as the product of the component of F along OP, $F \cos \theta$, and the distance moved along OP, s.

$$F \times s \cos \theta = F \cos \theta \times s$$

[Notice that the direction of F is not necessarily the same as the direction of the wind, it depends on how you have set your sails.]

EXAMPLE 9.11 As a car of mass m kg drives up a slope at an angle α to the horizontal it experiences a constant resistive force F N and a driving force D N. What can be deduced about the work done by D as the car moves a distance d m uphill if:
i)   the car moves at constant speed?
ii) the car slows down?
iii) the car gains speed?

The initial and final speeds of the car are denoted by $u\,\mathrm{ms}^{-1}$ and $v\,\mathrm{ms}^{-1}$ respectively.

iv) Write v^2 in terms of the other variables.

SOLUTION

The diagram shows the forces acting on the car. The table shows the work done by each force. The normal reaction, R, does no work as the car moves no distance in the direction of R.

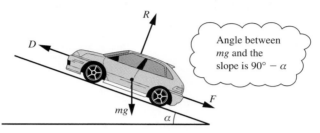

Angle between mg and the slope is $90° - \alpha$

Figure 9.12

Force	Work done
Resistance F	$-Fd$
Normal reaction R	0
Force of gravity mg	$-mgd \cos (90° - \alpha) = -mgd \sin a$
Driving force D	Dd
Total work done	$Dd - Fd - mgd \sin \alpha$

i) If the car moves at a constant speed there is no change in kinetic energy so the total work done is zero, giving

Work done by D is
$$Dd = Fd + mgd \sin \alpha.$$

ii) If the car slows down the total work done by the forces is negative, hence

Work done by D is
$$Dd < Fd + mgd \sin \alpha.$$

iii) If the car gains speed the total work done by the forces is positive

Work done by D is
$$Dd > Fd + mgd \sin \alpha.$$

iv)
$$\text{Total work done} = \text{final K.E.} - \text{initial K.E.}$$

$$\Rightarrow \qquad Dd - Fd - mgd \sin \alpha = \tfrac{1}{2}mv^2 - \tfrac{1}{2}mu^2$$

Multiplying by $\dfrac{2}{m}$

$$\Rightarrow \qquad v^2 = u^2 + \frac{2d}{m}(D - F) - 2gd \sin \alpha$$

9.5 Power

It is claimed that a motorcycle engine can develop a maximum *power* of 26.5 kW at a top *speed* of 103 mph. This suggests that power is related to speed and this is indeed the case.

Power is the rate at which work is being done. A powerful car does work at a greater rate than a less powerful one.

You might find it helpful to think in terms of a force, F, acting for a very short time t over a small distance s. Assume F to be constant over this short time.

Power is the rate of working so

$$\text{power} = \frac{\text{work}}{\text{time}}$$

$$= \frac{Fs}{t}$$

$$= Fv.$$

> This gives you the power at an *instant* of time. The result is true whether or not F is constant.

The power of a vehicle moving at speed v under a driving force F is given by Fv.

For a motor vehicle the power is produced by the engine, whereas for a bicycle it is produced by the cyclist. They both make the wheels turn, and the friction between the rotating wheels and the ground produces a forward force on the machine.

The unit of power is the watt (W), named after James Watt. The power produced by a force of 1 N acting on an object that is moving at $1 \, \text{ms}^{-1}$ is 1 W. Because the watt is such a small unit you will probably use kilowatts more often (1 kW = 1000 W).

EXAMPLE 9.12 A car of mass 1000 kg can produce a maximum power of 45 kW. Its driver wishes to overtake another vehicle. Ignoring air resistance, find the maximum acceleration of the car when it is travelling at
i) $12\,\text{ms}^{-1}$
ii) $28\,\text{ms}^{-1}$ (these are about 27 mph and 63 mph).

SOLUTION

i) Power = force × velocity
The driving force at $12\,\text{ms}^{-1}$ is F_1 N where

$$45\,000 = F_1 \times 12$$

$$\Rightarrow \quad F_1 = 3750.$$

By Newton's second law $F = ma$

$$\Rightarrow \quad \text{acceleration} = \frac{3750}{1000} = 3.75\,\text{ms}^{-2}.$$

ii) Now the driving force F_2 is given by

$$45\,000 = F_2 \times 28$$

$$\Rightarrow \quad F_2 = 1607$$

$$\Rightarrow \quad \text{acceleration} = \frac{1607}{1000} = 1.61\,\text{ms}^{-2}.$$

This example shows why it is easier to overtake a slow moving vehicle.

EXAMPLE 9.13 A car of mass 900 kg produces power 45 kW when moving at a constant speed. It experiences a resistance of 1700 N.
i) What is its speed?
ii) The car comes to a downhill stretch inclined at 2° to the horizontal. What is its maximum speed downhill if the power and resistance remain unchanged?

SOLUTION

i) As the car is travelling at a constant speed, there is no resultant force on the car. In this case the forward force of the engine must have the same magnitude as the resistance forces, i.e. 1700 N.
Denoting the speed of the car by $v\,\text{ms}^{-1}$, $P = Fv$ gives

$$v = \frac{P}{F}$$

$$= \frac{45\,000}{1700}$$

$$= 26.5.$$

The speed of the car is $26.5\,\text{ms}^{-1}$ (approximately 60 mph).

ii) The diagram shows the forces acting.

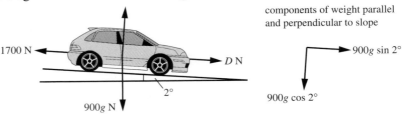

Figure 9.13

At maximum speed there is no acceleration so the resultant force down the slope is zero.

When the driving force is D N

$$D + 900g \sin 2° - 1700 = 0$$
$$\Rightarrow \qquad D = 1392$$
But power is Dv so $45\,000 = 1392v$
$$\Rightarrow \qquad v = \frac{45\,000}{1392}$$

The maximum speed is 32.3 ms^{-1} (about 73 mph).

Historical note

James Watt was born in 1736 in Greenock in Scotland, the son of a house- and ship-builder. As a boy James was frail and he was taught by his mother rather than going to school. This allowed him to spend time in his father's workshop where he developed practical and inventive skills.

As a young man he manufactured mathematical instruments: quadrants, scales, compasses and so on. One day he was repairing a model steam engine for a friend and noticed that its design was very wasteful of steam. He proposed an alternative arrangement, which was to become standard on later steam engines. This was the first of many engineering inventions which made possible the subsequent industrial revolution. James Watt died in 1819, a well known and highly respected man. His name lives on as the S.I. unit for power.

Chapter 9 Exercises

9.1 A sprinter of mass 60 kg is at rest at the beginning of a race and accelerates to 12 ms^{-1} in a distance of 30 m. Assume air resistance to be negligible.

   i) Calculate the kinetic energy of the sprinter at the end of the 30 m.

   ii) Write down the work done by the sprinter over this distance.

   iii) Calculate the forward force exerted by the sprinter, assuming it to be constant, using work = force × distance.

   iv) Using force = mass × acceleration and the constant acceleration formulae, show that this force is consistent with the sprinter having speed 12 ms^{-1} after 30 m.

9.2 The forces acting on a hot air balloon of mass 500 kg are its weight and the total uplift force.
   i) Find the total work done when the speed of the balloon changes from
   a) $2\,\text{ms}^{-1}$ to $5\,\text{ms}^{-1}$ **b)** $8\,\text{ms}^{-1}$ to $3\,\text{ms}^{-1}$
   ii) If the balloon rises 100 m vertically while its speed changes calculate in each case the work done by the uplift force.

9.3 A bullet of mass 20 g, found at the scene of a police investigation, had penetrated 16 cm into a wooden post. The speed for that type of bullet is known to be $80\,\text{ms}^{-1}$.
   i) Find the kinetic energy of the bullet before it entered the post.
   ii) What happened to this energy when the bullet entered the wooden post?
   iii) Write down the work done in stopping the bullet.
   iv) Calculate the resistive force on the bullet, assuming it to be constant.

   Another bullet of the same mass and shape had clearly been fired from a different and unknown type of gun. This bullet had penetrated 20 cm into the post.
   v) Estimate the speed of this bullet before it hit the post.

9.4 A bricklayer carries a hod of bricks of mass 25 kg up a ladder of length 10 m inclined at an angle of $60°$ to the horizontal.
   i) Calculate the increase in the gravitational potential energy of the bricks.
   ii) If instead he had raised the bricks vertically to the same height, using a rope and pulleys, would the increase in potential energy be **a)** less, **b)** the same, or **c)** more than in part **i)**?

9.5 A girl of mass 45 kg slides down a smooth water chute of length 6 m inclined at an angle of $40°$ to the horizontal.
   i) Find
   a) the decrease in her potential energy
   b) her speed at the bottom.
   ii) How are answers to part **i)** affected if the slide is not smooth?

9.6 A hockey ball of mass 0.15 kg is hit from the centre of a pitch. Its position vector (in m), t s later is modelled by

$$\mathbf{r} = 10t\,\mathbf{i} + (10t - 4.9t^2)\,\mathbf{j}$$

where the unit vectors $\mathbf{i}$ and $\mathbf{j}$ are in directions along the line of the pitch and vertically upwards.
   i) What value of g is used in this model?
   ii) Find an expression for the gravitational potential energy of the ball at time t. For what values of t is your answer valid?
   iii) What is the maximum height of the ball? What is its velocity at that instant?
   iv) Find the initial velocity, speed and kinetic energy of the ball.
   v) Show that according to this model mechanical energy is conserved and state what modelling assumption is implied by this. Is it reasonable in this context?

9.7 Akosua draws water from a well 12 m below the ground. Her bucket holds 5 kg of water and by the time she has pulled it to the top of the well it is travelling at $1.2\,\text{ms}^{-1}$.
   i) How much work does Akosua do in drawing the bucket of water?

On an average day 150 people in the village each draw six such buckets of water. One day a new electric pump is installed that takes water from the well and fills an overhead tank 5 m above ground level every morning. The flow rate through the pump is such that the water has speed 2 ms^{-1} on arriving in the tank.

ii) Assuming that the villagers' demand for water remains unaltered, how much work does the pump do in one day?

It takes the pump one hour to fill the tank each morning.

iii) At what rate does the pump do work, in joules per second (watts)?

9.8 Sacks of fertiliser, each of mass 30 kg, are being unloaded into a store. They slide from A to B in contact with a slope at 40° to the horizontal. From B they fall to the floor of the store.

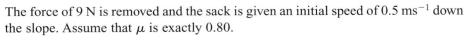

It is found that a sack at rest on the slope is just prevented from sliding down by a force of 9 N acting up the slope and parallel to it.

i) Show that the coefficient of friction, μ, between the sack and the slope, is about 0.80.

The force of 9 N is removed and the sack is given an initial speed of 0.5 ms^{-1} down the slope. Assume that μ is exactly 0.80.

ii) Calculate the work done against friction when the sack slides the 2 m from A to B.

iii) There are negligible resistances to the motion of the sack after it reaches B. Calculate its speed as it hits the floor, having fallen a vertical distance of 3 m from B.

9.9 A weightlifter takes 2 seconds to lift 120 kg from the floor to a position 2 m above it, where the weight has to be held stationary.

i) Calculate the work done by the weightlifter.

ii) Calculate the average power developed by the weightlifter.

The weightlifter is using the 'clean and jerk' technique. This means that in the first stage of the lift he raises the weight 0.8 m from the floor in 0.5 s. He then holds it stationary for 1 s before lifting it up to the final position in another 0.5 s.

iii) Find the average power developed by the weightlifter during each of the stages of the lift.

9.10 A mouse of mass 15 g is stationary 2 m below its hole when it sees a cat. It runs to its hole, arriving 1.5 seconds later with a speed of 3 ms^{-1}.

i) Show that the acceleration of the mouse is not constant.

ii) Calculate the average power of the mouse.

9.11 A tractor and its plough have a combined mass of 6000 kg. When developing a power of 5 kW, the tractor is travelling at a steady speed of 2.5 ms^{-1} along a horizontal field.

i) Calculate the resistance to the motion.

The tractor comes to a patch of wet ground where the resistance to motion is different. The power developed by the tractor during the next 10 seconds has an average value of 8 kW over this time. During this time, the tractor accelerates uniformly from 2.5 ms^{-1} to 3 ms^{-1}.

ii) Show that the work done against the resistance to motion during the 10 seconds is 71 750 J.

Assuming that the resistance to the motion is constant, calculate its value.

The tractor now comes to a slope at $\sin^{-1}\left(\frac{1}{20}\right)$ to the horizontal. The non-gravitational resistance to motion on this slope is 2000 N. The tractor accelerates uniformly from 3 ms^{-1} to 3.25 ms^{-1} over a distance of 100 m while climbing the slope.

iii) Calculate the time taken to travel this distance of 100 m and the average power required over this time period.

9.12 A winch pulls a crate of mass 1500 kg up a slope at 20° to the horizontal. The light wire attached to the winch and the crate is parallel to the slope, as shown in figure (A).

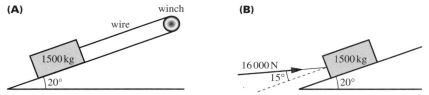

The crate takes 50 seconds to move 25 m up the slope at a constant speed when the power supplied by the winch is 6 kW.

i) How much work is done by the tension in the wire in the 50 seconds?

ii) Calculate the resistance to the motion of the crate up the slope.

iii) Show that the coefficient of friction between the crate and the slope is 0.50 (correct to two significant figures).

The winch breaks down and the crate is then *pushed* up the slope by a mechanical shovel by means of a constant force of 16000 N inclined at 15° to the slope, as shown in figure (B). You may assume that the crate does not tip up.

iv) Calculate the distance travelled by the crate up the slope as it speeds up from rest to 2.5 ms^{-1}. [You may assume the coefficient of friction between the crate and the slope is exactly 0.5.]

Personal Tutor Exercise

 9.1 A cyclist with her bicycle has a total mass of 84 kg.

Initially she travels for 100 m at a steady speed of 5 ms^{-1} along a horizontal road against a constant resistance of 49 N.

i) Calculate the work done against the resistance.

She now travels at a steady speed of 5 ms^{-1} up a constant slope at an angle of $\arcsin\left(\frac{1}{21}\right)$ to the horizontal; there is still a constant resistance of 49 N.

ii) Show that the power being developed by the cyclist is 441 W.

Travelling at 5 ms^{-1}, she now comes to another horizontal section and maintains her power of 441 W. After three seconds her speed has increased to 6 ms^{-1}.

iii) Explain briefly why the driving force on the bicycle is not constant over the three seconds.

iv) Calculate the total work done against the resistance to motion during the three seconds.

The cyclist returns along the same road. As she begins to descend the hill she stops pedalling.

v) Calculate the distance she travels while her speed reduces from 6 ms^{-1} to 3 ms^{-1} against an average resistance of 49 N.

KEY POINTS

9.1 The work done by a constant force F is given by Fs where s is the distance moved in the direction of the force.

9.2 The kinetic energy (K.E.) of a body of mass m moving with speed v is given by $\frac{1}{2}mv^2$. Kinetic energy is the energy a body possesses on account of its motion.

9.3 The work–energy principle states that the total work done by all the forces acting on a body is equal to the increase in the kinetic energy of the body.

9.4 The gravitational potential energy of a body of mass m at height h above a given reference level is given by mgh. It is the work done against the force of gravity in raising the body.

9.5 Mechanical energy (K.E. and P.E.) is conserved when no forces other than gravity do work.

9.6 Power is the rate of doing work, and is given by Fv.

9.7 The S.I. unit for energy is the joule and that for power is the watt.

10 Impulse and momentum

I collided with a stationary truck coming the other way.

Statement on an insurance form reported in the Toronto News.

The karate expert in the picture has just broken a pile of roof − tiles with a single blow from his head. Forces in excess of 3000 N have been measured during karate chops. How is this possible?

10.1 Impulse

Although the karate expert produces a very large force, it acts for only a short time. This is often the case in situations where impacts occur, as in the following example involving a tennis player.

EXAMPLE 10.1

A tennis player hits the ball as it is travelling towards her at $10 \, \mathrm{ms}^{-1}$ horizontally. Immediately after she hits it, the ball is travelling away from her at $20 \, \mathrm{ms}^{-1}$ horizontally. The mass of the ball is 0.06 kg. What force does the tennis player apply to the ball?

SOLUTION

You cannot tell unless you know how long the impact lasts, and that will vary from one shot to another.

QUESTION 10.1

Show that the average force she applies to the ball in the cases where the impact lasts 0.1 s and 0.015 s are 18 N and 120 N respectively. What does 'average' mean in this context?

While you cannot calculate the force unless you know the time for which it acts, you can work out the product force × time. This is called the *impulse*.

When a constant force acts for a time *t* the impulse of the force is defined as

$$\text{impulse} = \text{force} \times \text{time}.$$

The impulse is a vector in the direction of the force. When the force and time cannot be known separately, as in the case of the tennis ball, an impulse is often denoted by **J** and its magnitude by *J*. The S.I. unit for impulse is the newton second (Ns).

Impulse and momentum in one dimension

When the motion is in one dimension and the velocity of an object of mass m is changed from u to v by a constant force F you can use Newton's second law and the equations for motion with constant acceleration.

$$F = ma$$

and
$$v = u + at$$

$$\Rightarrow \quad mv = mu + mat$$

Substituting F for ma gives
$$mv = mu + Ft$$

$$\Rightarrow \quad Ft = mv - mu \qquad \textcircled{1}$$

The quantity 'mass $\times$ velocity' is defined as the momentum of the moving object.

The equation $\textcircled{1}$ can then be written as

So impulse = change in momentum

$$\textit{impulse of force} = \textit{final momentum} - \textit{initial momentum.} \qquad \textcircled{2}$$

This equation also holds for any large force acting for a short time even when it cannot be assumed to be constant. The force on the tennis ball will increase as it embeds itself into the strings and then decrease as it is catapulted away, but you can calculate the impulse of the tennis racket on the ball as

$$0.06 \times 20 - 0.06 \times (-10) = 1.8 \text{ Ns} \qquad \text{(the } -10 \text{ takes account of the change in direction)}$$

Equation $\textcircled{2}$ is also true for a variable force but then calculus is used to work out the impulse. It is also true, but less often used, when a longer time is involved.

QUESTION 10.2 The magnitude of the momentum of an object is often thought of as its resistance to being stopped. Compare the momentum and kinetic energy of a cricket ball of mass 0.15 kg bowled very fast at 40 ms^{-1} and a 20 tonne railway truck moving at a very slow speed of 1 cm per second.

Which would you rather be hit by, an object with high momentum and low energy, or one with high energy and low momentum?

EXAMPLE 10.2 A ball of mass 50 g hits the ground with a speed of 4 ms^{-1} and rebounds with an initial speed of 3 ms^{-1}. If the ball is in contact with the ground for 0.01 s,
i) find the average force exerted on the ball
ii) find the loss of kinetic energy during the impact.

SOLUTION

i) The impulse is given by:
$$J = mv - mu$$
$$= 0.05 \times 3 - 0.05 \times (-4)$$
$$= 0.35$$

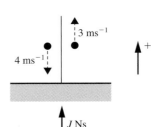

Figure 10.1

The impulse J is also given by

$$J = Ft$$

where F is the average force, i.e. the constant force which, acting for the same time interval, would have the same effect as the variable force which actually acted.

$\therefore$
$$0.35 = F \times 0.01$$
$$F = 35$$

So the ground exerts an average upward force of 35 N.

ii)   Initial K.E. $= \frac{1}{2} \times 0.05 \times 4^2$

$$= 0.400 \text{ J}$$

Final K.E. $= \frac{1}{2} \times 0.05 \times 3^2$

$$= 0.225 \text{ J}$$

Loss in K.E. $= 0.175 \text{ J}$

(This is converted into heat and sound.)

Note

Example 10.2 demonstrates the important point that mechanical energy is not conserved during an impact.

Although the force of gravity acts during the impact, its impulse is negligible over such a short time.

EXAMPLE 10.3   A car of mass 800 kg is pushed with a constant force of magnitude 200 N for 10 s. If the car starts from rest, find its speed at the end of the ten-second interval. Resistance to motion may be ignored.

SOLUTION

The force of 200 N acts for 10 s, so the impulse on the car is

$$J = 200 \times 10 = 2000 \text{ (in Ns)}.$$

(The impulse is in the direction of the force.)

Hence the change in momentum (in Ns) is

$$mv = 2000$$

$\therefore$
$$v = \frac{2000}{800} = 2.5$$

The speed at the end of the time interval is 2.5 ms^{-1}.

Impulse and momentum in more than one dimension

Both impulse and momentum are vectors. The impulse of a force is in the direction of the force and the momentum of a moving object is in the direction of its velocity. When an impulse $\mathbf{J}$ changes the velocity of a mass m from $\mathbf{u}$ to $\mathbf{v}$, the impulse–momentum equation is

$$\mathbf{J} = m\mathbf{v} - m\mathbf{u}.$$

The diagram shows how this applies to a ball which changes direction when it is hit by a bat.

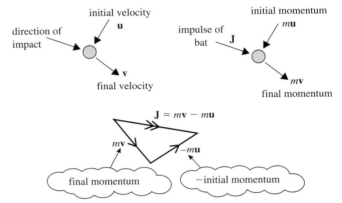

Figure 10.2

EXAMPLE 10.4

In a game of snooker the cue ball (W) of mass 0.2 kg is hit towards a stationary red ball (R) at $0.8\ \text{ms}^{-1}$. After the collision the cue ball is moving at $0.6\ \text{ms}^{-1}$ having been deflected through $30°$.

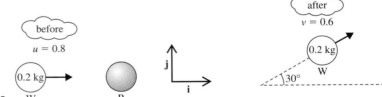

Figure 10.3

Find the impulse on the cue ball and show this in a vector diagram.

SOLUTION

In terms of unit vectors $\mathbf{i}$ and $\mathbf{j}$ the velocities of the cue ball before and after the collision are given by

$$\mathbf{u} = 0.8\mathbf{i}$$
$$\mathbf{v} = 0.6 \cos 30°\,\mathbf{i} + 0.6 \sin 30°\,\mathbf{j}.$$

Then impulse = final momentum − initial momentum

$$\mathbf{J} = m\mathbf{v} - m\mathbf{u}$$
$$= 0.2(0.6 \cos 30°\,\mathbf{i} + 0.6 \sin 30°\,\mathbf{j}) - 0.2(0.8\mathbf{i})$$
$$= -0.056\,\mathbf{i} + 0.06\,\mathbf{j}$$

$$\text{Magnitude of impulse} = \sqrt{0.056^2 + 0.06^2}$$
$$= 0.082$$
$$\text{Direction: } \tan \alpha = \frac{0.06}{0.056}$$
$$\alpha = 47°$$
$$\theta = 133°$$

The impulse has magnitude 0.082 N at an angle of 133° to the initial motion of the ball.

This is shown on the vector diagram below. Note that the impulse–momentum equation shows the direction of the impulsive force acting on the cue ball.

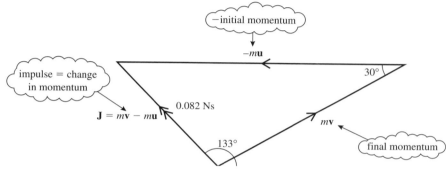

Figure 10.4

QUESTION 10.3   What happens to the red ball in Example 10.4?

EXAMPLE 10.5   A hockey ball of mass 0.15 kg is moving at 4 ms^{-1} parallel to the side of a pitch when it is struck by a blow from a hockey stick that exerts an impulse of 4 Ns at an angle of 120° to its direction of motion. Find the final velocity of the ball.

SOLUTION

The vector diagram shows the motion of the ball.

Figure 10.5

In terms of unit vectors $\mathbf{i}$ and $\mathbf{j}$ we have
$$\mathbf{u} = 4\mathbf{i}$$
and
$$\mathbf{J} = -4 \cos 60° \, \mathbf{i} + 4 \sin 60° \, \mathbf{j}$$
$$= -2\mathbf{i} + 3.46\mathbf{j}$$
Using
$$\mathbf{J} = m\mathbf{v} - m\mathbf{u}$$
$$-2\mathbf{i} + 3.46\mathbf{j} = 0.15\mathbf{v} - 0.15 \times 4\mathbf{i}$$
$$\Rightarrow \qquad 0.15\mathbf{v} = -2\mathbf{i} + 0.6\mathbf{i} + 3.46\mathbf{j}$$
$$\Rightarrow \qquad 0.15\mathbf{v} = -1.4\mathbf{i} + 3.46\mathbf{j}$$
$$\Rightarrow \qquad \mathbf{v} = -9.33\mathbf{i} + 23.1\mathbf{j}$$

The magnitude of the velocity is given by $v = \sqrt{9.33^2 + 23.1^2}$
$$= 24.9 \text{ ms}^{-1}.$$

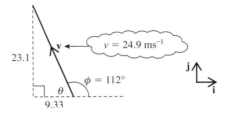

Figure 10.6

The angle, θ, is given by $\theta = \tan^{-1}\left(\frac{23.1}{9.33}\right)$
$$= 68°$$
$$\phi = 180° - 68°$$
$$= 112°.$$

After the blow, the ball has a velocity of magnitude 24.9 ms^{-1} at an angle of 112° to the original direction of motion.

The motion of a particle under a variable force

Very often the forces that change the motion of objects are not constant. When a trampolinist hits the bed of a trampoline, the upward force acting on her is linked to the deformation of the bed and the extension of the springs.

The greater the deformation and extension, the greater the force. While it is possible to deal with situations like this by considering the average force, it is also possible to deal precisely with variable forces.

Motion under a variable force is also covered by the impulse–momentum equation. Think of a particle of mass m moving along a straight line and subject to a variable force F for a time T.

Suppose that during this time the velocity of the particle changes from U to V.

Applying Newton's second law,

$$F = ma = m\frac{dv}{dt}.$$

Integrating between $t = 0$ and $t = T$ we have

$$\int_0^T F \, dt = \int_0^T m\frac{dv}{dt} \, dt$$

$$= \int_{v=U}^{v=V} m \, dv$$

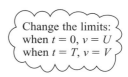

Change the limits:
when $t = 0$, $v = U$
when $t = T$, $v = V$

$$= mV - mU.$$

The right-hand side of the equation is the change in momentum and the left-hand side, $\int_0^T F \, dt$, is the impulse. The impulse is therefore the area under the graph of F against t.

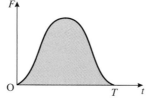

The shaded area represents the impulse, $\int F \, dt$

Figure 10.7

Note that if the force, F, is constant then

$$\text{impulse} = \int_0^T F \, dt = F\int_0^T dt = FT \text{ as before.}$$

10.2 Conservation of momentum

Collisions

QUESTION 10.4   In an experiment to investigate car design two vehicles were made to collide head-on. How would you investigate this situation? Can you find a relationship between the change in momentum of the van and that of the car?

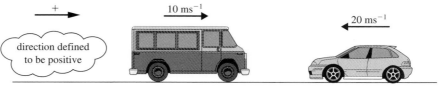

direction defined to be positive

10 ms^{-1}

20 ms^{-1}

Figure 10.8

2500 kg

1000 kg

The first thing to remember is Newton's third law. The force that body A exerts on body B is equal to the force that B exerts on A, but in the opposite direction.

Suppose that once the van is in contact with the car, it exerts a force F on the car for a time t. Newton's third law tells us that the car also exerts a force F on the van for a time t. (This applies whether F is constant or variable.) So both vehicles receive equal impulses, but in opposite directions. Consequently the increase in momentum of the car in the positive direction is exactly equal to the increase in momentum of the van in the negative direction. For the two vehicles together, the total change in momentum is zero.

This example illustrates the *law of conservation of momentum*.

> *The law of conservation of momentum states that when there are no external influences on a system, the total momentum of the system is constant.*

Since momentum is a vector quantity, this applies to the magnitude of the momentum in any direction.

For a collision, you can say

total momentum before collision = total momentum after collision.

 It is important to remember that although momentum is conserved in a collision, mechanical energy is not conserved. Some of the work done by the forces is converted into heat and sound.

EXAMPLE 10.6

The two vehicles in the previous discussion collide head − on, and as a result the van comes to rest.

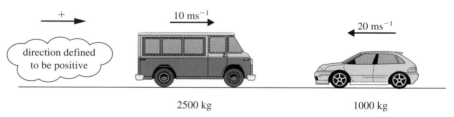

Figure 10.9

Find
i)   the final velocity of the car, $v\,\text{ms}^{-1}$
ii) the impulse on each vehicle
iii) the kinetic energy lost.
iv) If it is assumed that the impact lasts for one − twentieth of a second, find the force on each vehicle and its acceleration.

SOLUTION

i)

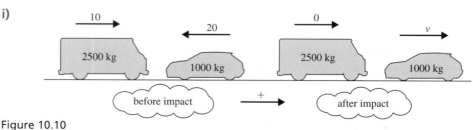

Figure 10.10

Using conservation of momentum, and taking the positive direction as being to the right:

$$2500 \times 10 + 1000 \times (-20) = 2500 \times 0 + 1000 \times v$$
$$5000 = 1000v$$
$$v = 5$$

The final velocity of the car is $5\ \text{ms}^{-1}$ in the positive direction (i.e. the car travels backwards).

ii)

Impulse = final momentum − initial momentum

For the van,   impulse = $2500 \times 0 - 2500 \times 10$
$$= -25\,000\ \text{Ns}.$$

For the car,   impulse = $1000 \times 5 - 1000 \times (-20)$
$$= +25\,000\ \text{Ns}.$$

The van experiences an impulse of 25000Ns in the negative direction, the car an equal and opposite impulse.

iii)

Total initial K.E. $= \frac{1}{2} \times 2500 \times 10^2 + \frac{1}{2} \times 1000 \times 20^2$
$$= 325\,000\ \text{J}$$

Total final K.E. $= \frac{1}{2} \times 2500 \times 0^2 + \frac{1}{2} \times 1000 \times 5^2$
$$= 12\,500\ \text{J}$$

Loss in K.E. $= 312\,500\ \text{J}$

iv)

Impulse = average force × time
$$25\,000 = F \times \tfrac{1}{20}$$
$$F = 500\,000\ \text{N}\quad \text{(acting to the right on the car and to the left on the van).}$$

Using $F = ma$ on each vehicle gives an average acceleration of $500\ \text{ms}^{-2}$ for the car and $-200\ \text{ms}^{-2}$ for the van.

QUESTION 10.5   These accelerations ($500\ \text{ms}^{-2}$ and $-200\ \text{ms}^{-2}$) seem very high. Are they realistic for a head − on collision?

Work out the distance each car travels during the time interval of one − twentieth of a second between impact and separation. This will give you an idea of the amount of damage there would be.

Is it better for cars to be made strong so that there is little damage, or to be designed to crumple under impact?

EXAMPLE 10.7   In an experiment on lorry bumper design, the Transport Research Laboratory arranged for a car and a lorry, of masses 1 and 3.5 tonnes, to travel towards each other, both with speed $9\ \text{ms}^{-1}$. After colliding both vehicles moved together and the total momentum had been conserved.

What was their combined velocity after the collision?

SOLUTION

The situation before the collision is illustrated below.

9 ms^{-1} 9 ms^{-1}

+

1 tonne   3.5 tonnes

Figure 10.11

Taking the positive direction to be to the right, before the collision

momentum of the car in Ns: $1000 \times 9 = 9000$

momentum of the lorry in Ns: $3500 \times (-9) = -31\,500$

total momentum in Ns: $9000 - 31\,500 = -22\,500$.

After the collision assume they move as a single object of mass 4.5 tonnes with velocity v ms^{-1} in the positive direction so the total momentum is now $4500v$ Ns.

Momentum is conserved so
$$4500v = -22\,500$$
$$v = -5.$$

The car and lorry move at 5 ms^{-1} in the direction the lorry was moving.

EXAMPLE 10.8 A child of mass 30 kg running through a supermarket at 4 ms^{-1} leaps on to a stationary shopping trolley of mass 15 kg. Find the speed of the child and trolley together, assuming that the trolley is free to move easily.

SOLUTION

The diagram shows the situation before the child hits the trolley.

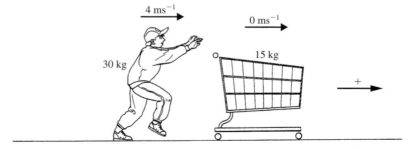

4 ms^{-1}

0 ms^{-1}

15 kg

30 kg

+

Figure 10.12

Taking the direction of the child's velocity as positive, total momentum (in Ns) before impact

$$= 4 \times 30 + 0 \times 15$$
$$= 120.$$

The situation after impact is shown below.

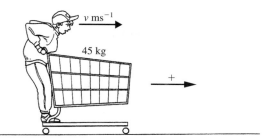

Figure 10.13

The total mass of child and trolley is 45 kg, so the total momentum after is $45v$ Ns.

Conservation of momentum gives:

$$45v = 120$$
$$v = 2\tfrac{2}{3}.$$

The child and the trolley together move at $2\tfrac{2}{3}$ ms^{-1}.

Explosions

Conservation of momentum also applies when explosions take place provided there are no external forces, for example when a bullet is fired from a rifle, or a rocket is launched.

EXAMPLE 10.9 A rifle of mass 8 kg is used to fire a bullet of mass 80 g at a speed of 200 ms^{-1}. Calculate the initial recoil speed of the rifle.

SOLUTION

Before the bullet is fired the total momentum of the system is zero.

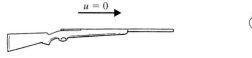

Figure 10.14

> Before firing: rifle and bullet have zero momentum

After the firing the situation is as illustrated below.

Figure 10.15

The total momentum in the positive direction after the firing is $8v + 0.08 \times 200$.

For momentum to be conserved,

$$8v + 0.08 \times 200 = 0$$

so that

$$8v = -200 \times 0.08$$
$$v = -2$$

> You have probably realised that v would turn out to be negative

The recoil speed of the rifle is 2 ms^{-1}.

10.3 Newton's law of impact

Question 10.6 If you drop two different balls, say a tennis ball and a cricket ball, from the same height, will they both rebound to the same height? How will the heights of the second bounces compare with the heights of the first ones?

Your own experience probably tells you that different balls will rebound to different heights. For example, a tennis ball will rebound to a greater height than a cricket ball. Furthermore, the surface on which the ball is dropped will affect the bounce. A tennis ball dropped onto a concrete floor will rebound higher than if dropped onto a carpeted floor.

Coefficient of restitution

Newton's experiments on collisions led him to formulate a simple law relating to the speeds before and after a direct collision between two bodies, called *Newton's law of impact*.

- $$\frac{\text{speed of separation}}{\text{speed of approach}} = \text{constant}$$

This can also be written as

$$\text{speed of separation} = \text{constant} \times \text{speed of approach}$$

This constant is called the *coefficient of restitution* and is conventionally denoted by the letter e. For two particular surfaces, e is a constant between 0 and 1. It does not have units, being the ratio of two speeds.

For very bouncy balls, e is close to 1, and for balls that do not bounce, e is close to 0. A collision for which $e = 1$ is called perfectly elastic, and a collision for which $e = 0$ is called perfectly inelastic.

Direct impact with a fixed surface

The value of e for the ball that you used in the experiment is given by $\frac{v_s}{v_a}$, and you should have found that this had approximately the same value each time for any particular ball. When a moving object hits a fixed surface which is perpendicular to its motion it rebounds in the opposite direction. If the speed of approach is v_a and the speed of separation is v_s Newton's law of impact gives

$$\frac{v_s}{v_a} = e$$

$$\Rightarrow \qquad v_s = ev_a.$$

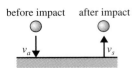

Figure 10.16

Collisions between bodies moving in the same straight line

Figure 10.17 shows two objects that collide while moving along a straight line. Object A is catching up with B, and after the collision either B moves away from A or they continue together.

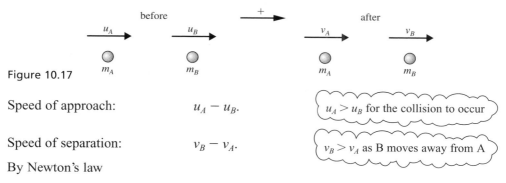

Figure 10.17

Speed of approach: $u_A - u_B$.

$u_A > u_B$ for the collision to occur

Speed of separation: $v_B - v_A$.

$v_B > v_A$ as B moves away from A

By Newton's law

$$\text{speed of separation} = e \times \text{speed of approach}$$

$$\Rightarrow \qquad v_B - v_A = e\,(u_A - u_B). \qquad \qquad \text{①}$$

A second equation relating the velocities follows from the law of conservation of momentum in the positive direction ($\rightarrow$):

$$\text{momentum after collision} = \text{momentum before collision}$$

$$m_A v_A + m_B v_B = m_A u_A + m_B u_B. \qquad \qquad \text{②}$$

These two equations, ① and ②, allow you to calculate the final velocities, v_A and v_B, after any collision as shown in the next two examples.

EXAMPLE 10.10 A direct collision takes place between two snooker balls. The cue ball travelling at $2\ \text{ms}^{-1}$ hits a stationary red ball. After the collision the red ball moves in the direction in which the cue ball was moving before the collision. Assume that the balls have equal mass, and that the coefficient of restitution between the two balls is 0.6. Predict the velocities of the two balls after the collision.

SOLUTION

Let the mass of each ball be m, and call the (white) cue ball 'W' and the red ball 'R'. The situation is summarised in figure 10.18.

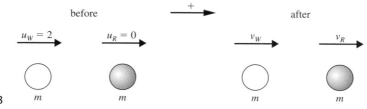

Figure 10.18

$$\text{Speed of approach} = 2 - 0 = 2$$
$$\text{Speed of separation} = v_R - v_W$$

By Newton's law of impact,

$$\text{speed of separation} = e \times \text{speed of approach}$$

$$\Rightarrow \quad v_R - v_W = 0.6 \times 2$$

$$\Rightarrow \quad v_R - v_W = 1.2. \qquad \text{①}$$

Conservation of momentum gives

$$mv_W + mv_R = mu_W + mu_R.$$

Dividing through by m, and substituting $u_W = 2$, $u_R = 0$, this becomes

$$v_W + v_R = 2. \qquad \text{②}$$

Adding equations ① and ② gives $2v_R = 3.2$,

so $v_R = 1.6$, and from equation ②, $v_W = 0.4$.

After the collision both balls move forward, the red ball at a speed of $1.6\ \text{ms}^{-1}$ and the cue ball at a speed of $0.4\ \text{ms}^{-1}$.

EXAMPLE 10.11

An object A of mass m moving with speed $2u$ hits an object B of mass $2m$ moving with speed u in the opposite direction from A.

i) Show that the ratio of speeds remains unchanged whatever the value of e.

ii) Find the loss of kinetic energy in terms of m, u and e.

SOLUTION

i) Let the velocities of A and B after the collision be v_A and v_B respectively.

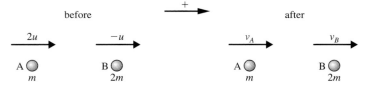

Figure 10.19

$$\text{Speed of approach} = 2u - (-u) = 3u$$

$$\text{Speed of separation} = v_B - v_A$$

Using Newton's law of impact

$$\text{speed of separation} = e \times \text{speed of approach}$$

$$\Rightarrow \quad v_B - v_A = e \times 3u \qquad \text{①}$$

Conservation of momentum gives

$$mv_A + 2mv_B = m(2u) + 2m(-u)$$

Dividing by m gives

$$v_A + 2v_B = 0 \qquad \text{②}$$

Equation ① is

$$v_B - v_A = 3eu$$

Adding ① and ②

$$3v_B = 3eu$$
$$v_B = eu$$

From ②,

$$v_A = -2eu$$

The ratio of speeds was initially $2u : u$ and finally $2eu : eu$ so the ratio of speeds is unchanged at $2 : 1$ (providing $e \neq 0$).

ii) Initial K.E. of A $\frac{1}{2}m \times (2u)^2 = 2mu^2$

Initial K.E. of B $\frac{1}{2}(2m) \times u^2 = mu^2$

Total K.E. before impact $= 3mu^2$.

Final K.E. of A $\frac{1}{2}m \times 4e^2u^2 = 2me^2u^2$

Final K.E. of B $\frac{1}{2}(2m) \times e^2u^2 = me^2u^2$

Total K.E. after impact $= 3me^2u^2$

Loss of K.E. $= 3mu^2(1 - e^2)$.

Note

In this case, A and B lose *all* their energy when $e = 0$, but this is not true in general. Only when $e = 1$ is there no loss in K.E. Kinetic energy is lost in any collision in which the coefficient of restitution is not equal to 1.

10.4 Oblique impacts

When an object hits a smooth plane there can be no impulse parallel to the plane so the component of momentum, and hence velocity, is unchanged in this direction. Perpendicular to the plane, the momentum is changed but Newton's law of impact still applies.

The diagrams show the components of the velocity of a ball immediately before and after it hits a smooth plane with coefficient of restitution e.

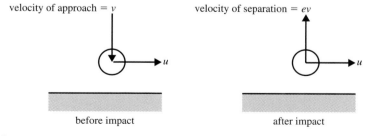

Figure 10.20

When the ball is travelling with speed U at an angle α to the plane, the components of the final velocity are $U \cos \alpha$ parallel to the plane and $eU \sin \alpha$, perpendicular to the plane.

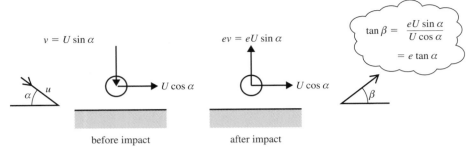

Figure 10.21

The *impulse* on the ball is equal to final momentum − initial momentum. This is perpendicular to the plane because there is no change in momentum parallel to the plane.

In figure 10.20 the impulse is

$$mev - m(-v) = (1 + e)\, mv \text{ upwards.}$$

In figure 10.21 the impulse is

$$meU \sin \alpha - m(-U \sin \alpha) = (1 + e)\, mU \sin \alpha \text{ upwards.}$$

Whenever an impact takes place, energy is likely to be lost. In the cases illustrated in the diagrams the *loss in kinetic energy* is

$$\tfrac{1}{2}m\,(u^2 + v^2) - \tfrac{1}{2}m\,(u^2 + e^2 v^2) = \tfrac{1}{2}m\,(1 - e^2)\, v^2$$

or $\tfrac{1}{2}m\,(1 - e^2)\, U^2 \sin^2 \alpha.$

QUESTION 10.7   What happens to the ball when $e = 0$ and $e = 1$?

EXAMPLE 10.12   A ball of mass 0.2 kg moving at $12\ \text{ms}^{-1}$ hits a smooth horizontal plane at an angle of $75°$ to the horizontal. The coefficient of restitution is 0.5. Find
i)   the impulse on the ball
ii)   the impulse on the plane
iii)   the kinetic energy lost by the ball.

SOLUTION

i)   The diagram shows the velocities before and after impact.

Figure 10.22

Parallel to the plane $u = 12 \cos 75$
Perpendicular to the plane: $v = 0.5 \times 12 \sin 75°$ by Newton's law of impact
                                 $= 6 \sin 75°$

The impulse on the ball = final momentum − initial momentum

$$= \begin{pmatrix} 12 \cos 75° \\ 6 \sin 75° \end{pmatrix} - \begin{pmatrix} 12 \cos 75° \\ -12 \sin 75° \end{pmatrix} \text{ (using directions } \mathbf{i}, \mathbf{j} \text{ as shown)}$$

$$= \begin{pmatrix} 0 \\ 18 \sin 75° \end{pmatrix}$$

The impulse on the ball is $18 \sin 75°\,\mathbf{j}$ that is 17.4 Ns perpendicular to the plane and upwards in the $\mathbf{j}$ direction.

ii) By Newton's third law, the impulse on the plane is equal and opposite to the impulse on the ball. It is 17.4 Ns perpendicular to the plane in the direction of $-\mathbf{j}$.

iii) The initial kinetic energy $= \frac{1}{2} \times 0.2 \times 12^2 = 14.4$ J
   Final kinetic energy $= \frac{1}{2} \times 0.2 \times [(12 \cos 75°)^2 + (6 \sin 75°)^2]$
   $= 4.32$ J
   Kinetic energy lost $= 14.4 - 4.32 = 10.1$ J (3 sf)

EXAMPLE 10.13

During a game of snooker two balls, A and B, both of mass 0.17 kg, collide. Before the collision ball A was travelling with velocity $4\mathbf{i} - 2.9\mathbf{j}$ ms^{-1} and ball B was travelling with velocity $4\mathbf{i} + 3.9\mathbf{j}$ ms^{-1}. If after the collision ball A had a velocity of $3\mathbf{i} + 2.1\mathbf{j}$ ms^{-1}, what is the velocity of B after the collision?

SOLUTION

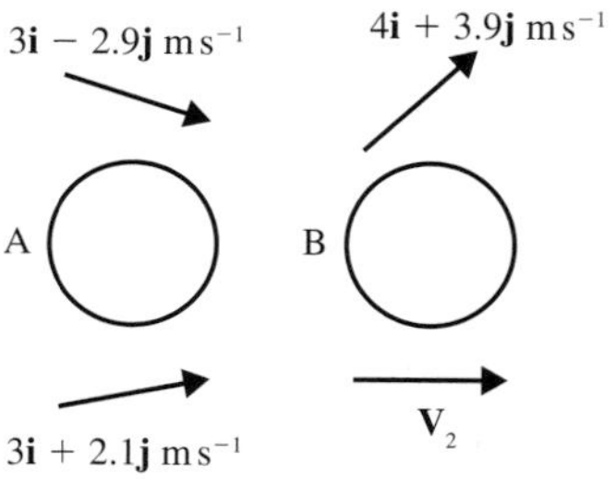

Figure 10.23

Using conservation of momentum, in two dimensions:

total momentum before collision = total momentum after collision
$$0.17\,(4\mathbf{i} - 2.9\mathbf{j}) + 0.17\,(4\mathbf{i} - 3.9\mathbf{j}) = 0.17\,(3\mathbf{i} + 2.1\mathbf{j}) + 0.17\mathbf{v}_2$$
$$0.17\,(4\mathbf{i} + 4\mathbf{i} - 3\mathbf{i}) + 0.17\,(-2.9\mathbf{j} + 3.9\mathbf{j} - 2.1\mathbf{j}) = 0.17\mathbf{v}_2$$
$$0.17\,(5\mathbf{i}) + 0.17\,(-1.1\mathbf{j}) = 0.17\mathbf{v}_2$$
$$\mathbf{v}_2 = 5\mathbf{i} - 1.1\mathbf{j}$$

The velocity of ball B after the collision is $5\mathbf{i} - 1.1\mathbf{j}$.

Chapter 10 Exercises

10.1 A car of mass 0.9 tonnes is travelling at 13.2 ms^{-1} when it crashes head-on into a wall. The car is brought to rest in a time of 0.12 s. Find
  i)   the impulse acting on the car
  ii) the average force acting on the car
  iii) the average deceleration of the car in terms of g (taken to be 10 ms^{-2}).
  iv) Explain why many cars are designed with crumple zones rather than with completely rigid construction.

10.2 A hockey ball of mass 0.15 kg is travelling with velocity $12\mathbf{i} - 8\mathbf{j}$ (in ms^{-1}), where the unit vectors $\mathbf{i}$ and $\mathbf{j}$ are in horizontal directions parallel and perpendicular to the length of the pitch, and the vector $\mathbf{k}$ is vertically upwards. The ball is hit by Jane with an impulse $-4.8\mathbf{i} + 1.2\mathbf{j}$.
  i)   What is the velocity of the ball immediately after Jane has hit it?

  The ball goes straight, without losing any speed, to Fatima in the opposite team who hits it without stopping it. Its velocity is now $14\mathbf{i} + 4\mathbf{j} + 3\mathbf{k}$.
  ii) What impulse does Fatima give the ball?
  iii) Which player hits the ball harder?

10.3 A hailstone of mass 4 g is travelling with speed 20 ms^{-1} when it hits a window as shown in the diagram. It bounces off the window; the vertical component of the velocity is unaltered, but the horizontal component is now 2 ms^{-1} away from the window.
  i)   State the magnitude and direction of the impulse
      a) of the window on the hailstone
      b) of the hailstone on the window.

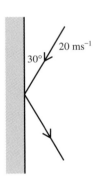

  At the peak of the storm, hailstones like this are hitting the window at the rate of 540 per minute.
  ii) Find the average force of the hail on the window.

10.4 The driver of a car of mass 1000 kg falls asleep while it is travelling at 30 ms^{-1}. The car runs into the back of the car in front which has mass 800 kg and is travelling in the same direction at 20 ms^{-1}. The bumpers of the two cars become locked together and they continue as one vehicle.
  i)   What is the final speed of the cars?
  ii) What impulse does the larger car give to the smaller one?
  iii) What impulse does the smaller car give to the larger one?

10.5 Katherine (mass 40 kg) and Elisabeth (mass 30 kg) are on a sledge (mass 10 kg) which is travelling across smooth horizontal ice at 5 ms^{-1}. Katherine jumps off the back of the sledge with speed 4 ms^{-1} backwards relative to the sledge.
  i)   What is Katherine's absolute speed when she jumps off?
  ii) With what speed does Elisabeth, still on the sledge, then go?

Elisabeth then jumps off in the same manner, also with speed 4 ms^{-1} relative to the sledge.

iii) What is the speed of the sledge now?

iv) What would the final speed of the sledge have been if Katherine and Elisabeth had both jumped off at the same time, with speed 4 ms^{-1} backwards relative to the sledge?

You will find it helpful to draw diagrams before answering the next four questions.

10.6 A trapeze artist of mass 50 kg falls from a height of 20 m into a safety net.

i) Find the speed with which she hits the net. (You may ignore air resistance and should take the value of g to be 10 ms^{-2}.)

Her speed on leaving the net is 15 ms^{-1}.

ii) What is the coefficient of restitution between her and the net?

iii) What impulse does the trapeze artist receive?

iv) How much mechanical energy is absorbed in the impact?

v) If you were a trapeze artist would you prefer a safety net with a high coefficient of restitution or a low one?

10.7 The diagram shows two snooker balls and one edge cushion. The coefficient of restitution between the balls and the cushion is 0.5 and that between the balls is 0.75. Ball A (the cue ball) is hit directly towards the stationary ball B with speed 8 ms^{-1}. Find the speeds and directions of the two balls after their second impact with each other.

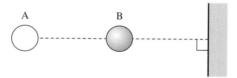

10.8 A spacecraft has two parts which may separate in flight, the body with mass 15000 kg and the nose-cone with mass 5000 kg. They rejoin by means of a 'docking' procedure which requires the nose-cone and body to approach along the axis shown in the diagram with a speed of approach of no more than 3 ms^{-1}. If the speed of approach is too great, the linking mechanism will not engage and the two parts will bounce off each other with a speed of separation which is $\frac{1}{4}$ of the speed of approach (i.e. a coefficient of restitution of 0.25). At all times both parts may be taken as travelling in a straight line.

On one occasion the body is travelling forwards at 105 ms^{-1} and the nose-cone forwards at 103 ms^{-1}.

i) Draw diagrams showing the velocities of the parts before and after docking.

ii) Calculate the final velocity of the spacecraft.

On another occasion the nose-cone is moving backwards at 5.5 ms^{-1} and the body forwards at 2.5 ms^{-1}.

iii) Show that the body comes to rest and find the direction and speed of the nose-cone.

After this has happened, before a further attempt to dock is made, the speed of the body is changed in its line of motion by expelling a mass m kg of fuel. The fuel is expelled at a speed of 2000 ms^{-1} and you may assume that all the fuel is expelled instantaneously (i.e. before the body changes speed).

iv) What value of m would cause the body to have a forward speed of 2.5 ms^{-1} after the fuel is burnt?

10.9 A simple pile-driver for use with fence posts consists of a metal cylinder which fits over the post to be driven into the ground. The pile-driver is lifted by two people using the handles and dropped onto the post, as shown in the diagram.

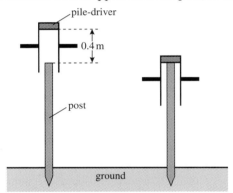

This process is modelled by assuming that the post is vertical, air resistance is negligible, the resistance of the ground is constant and that linear momentum is conserved in the impact between the pile-driver and the post.

The post has a mass of 6 kg and the pile-driver has a mass of 50 kg. The pile-driver is dropped 0.4 m onto the post at every attempt to drive in the post.

i) Show that the speed of the pile-driver when it reaches the post is 2.8 ms^{-1}.

On the first attempt to drive in the post, the post becomes jammed in the pile-driver at the moment of contact so that they then move as a single body.

ii) Draw a diagram indicating the velocities of the pile-driver and post before and after the impact. Calculate the common velocity of the pile-driver and the post immediately after the impact.

On the next attempt to drive in the post, the post does not become jammed and the pile-driver and post separate. The coefficient of restitution between the pile-driver and the post is 0.2.

iii) Calculate the velocities of the post and pile-driver immediately after the impact.

After this impact, the pile-driver is free to move under gravity with negligible resistance to its motion and the post is subject to a constant resistance from the ground of 600 N.

iv) Show that the pile-driver hits the post a second time before the post comes to rest.

10.10 A ball of mass m kg moving at u ms^{-1} hits a smooth horizontal plane at an angle of $\alpha°$ to the horizontal. The coefficient of restitution is 0.
    i)   Find the impulse on the ball.
    ii) Show that the kinetic energy lost is $\frac{1}{2}mu^2 \sin^2 \alpha$.

10.11 The wind blowing against a sail can be modelled as a series of particles hitting the smooth sail at an angle of $30°$ and with zero coefficient of restitution. The wind blows at 20 ms^{-1} and the density of the air is 1.4 k gm^{-3}. Calculate
    i)   the mass of air hitting one square metre of sail each second
    ii) the impulse of this mass on the sail and hence the force acting on a 2 m^2 sail.

10.12 A smooth snooker ball moving at 2 ms^{-1} hits a cushion at an angle of $30°$ to the cushion. The ball then rebounds and hits a second cushion which is perpendicular to the first. The coefficient of restitution for both impacts is 0.8.

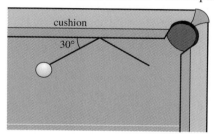

    i)   Find the direction of motion after each impact.
    ii) Find the magnitude of the velocity after the second impact.
    iii) Repeat parts **i)** and **ii)** for a ball moving at u ms^{-1} which hits the first cushion at an angle α. Assume the coefficient of restitution is e. Hence show that the direction of a ball is always reversed after hitting two perpendicular cushions and state the factor by which its speed is reduced.

Personal Tutor

10.1 Small seed pods, each of mass 0.007 kg, are falling vertically at 2.2 ms^{-1}.
    i)   One pod collides with a stationary, hovering fly of mass 0.004 kg and sticks to it. Calculate the velocity of the fly immediately after the collision.
    ii) Another pod collides with a stationary, hovering fly, also of mass 0.004 kg. In this case the pod moves downwards with a speed of 1 ms^{-1} immediately after the collision. Find the velocity of the fly immediately after the collision and the coefficient of restitution.

Another small seed pod is falling vertically at a speed of 1.5 ms^{-1} when it bursts into two parts. One part, P, has a mass of 0.005 kg and immediately after the pod bursts it has velocity $(2.4\mathbf{i} + 0.7\mathbf{j})$ ms^{-1}, where $\mathbf{i}$ and $\mathbf{j}$ are unit vectors with $\mathbf{i}$ horizontal and $\mathbf{j}$ vertically downwards. The other part of the pod, Q, has a mass of 0.002 kg.
    iii) Explain briefly why the two parts of the pod must move in the same vertical plane immediately after the pod bursts.
    iv) Calculate the velocity of part Q immediately after the pod bursts, giving your answer as a vector in terms of $\mathbf{i}$ and $\mathbf{j}$.

10.1   The impulse from a force **F** is given by **F**t where t is the time for which the force acts. When **F** is large and t is small the impulse is denoted by **J**. Impulse is a vector quantity.

10.2   The momentum of a body of mass m travelling with velocity **v** is given by m**v**. Momentum is a vector quantity.

10.3   The impulse–momentum equation is

$$Impulse = final\ momentum - initial\ momentum$$

10.4   The law of conservation of momentum states that when no external forces are acting on a system, the total momentum of the system is constant. Since momentum is a vector quantity this applies to the magnitude of the momentum in any direction.

10.5   Coefficient of restitution $e = \dfrac{\text{speed of separation}}{\text{speed of approach}}$

$$speed\ of\ separation = e \times speed\ of\ approach$$

All the above apply perpendicular to a smooth plane.

10.6   Parallel to a smooth plane the velocity is unchanged.

10.7   The S.I. unit for impulse and momentum is the newton second (Ns).

Frameworks

I have yet to see any problem, however complicated, which, when looked at in the right way, did not become still more complicated.

Paul Anderson

The Millennium Stadium at Cardiff has been built using frameworks made from triangular elements. You will have seen many structures like this.

QUESTION 11.1   Why is the triangle the basic element in so many structures?

This chapter is concerned with the forces in structural frameworks. Not all structures are based on frameworks, but many are because the strength to weight ratio for a framework is usually higher than for a solid structure.

A framework is an arrangement of structural members (rods or cables). Frameworks are almost always made from triangular elements fitted together in two or three dimensions, because a triangle is rigid even if it is freely hinged (or *pin-jointed*) at the corners. If any other shape, such as a quadrilateral, were used, the joints would have to be rigid. This would make both the design of the structure and its successful construction far more difficult.

To analyse the forces in frameworks is quite complex, unless you can make two simplifying assumptions.
i)   All members of the framework are light, so that their masses can be ignored;
ii) All the joints are smooth pin joints. This means that the bars could rotate about the joints without any resistance: there are no moments acting at the joints.

The consequence of these assumptions is that all the internal forces are directed along the rods. Then the only forces that need to be considered are the external forces on the framework, and the tensions or thrusts in the rods of the framework.

QUESTION 11.2 ABCDE is a framework of seven light freely jointed rods, all of the same length, supported at A and D. A weight of 1000 N is hanging from E.

Which of the rods do you think are in tension and which in compression (thrust)?

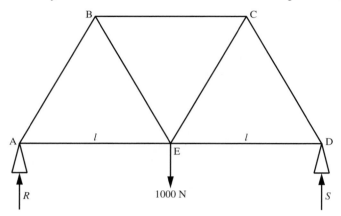

Figure 11.1

Now follow the analysis of this structure in the worked example below and see whether your intuitive ideas were correct.

EXAMPLE 11.1 Calculate the forces in each of the rods in the structure above, stating whether the rod is in tension or compression.

SOLUTION

i) External forces

There are three external forces acting on the framework, R N and S N vertically upwards and 1000 N vertically downwards. Looking at the equilibrium of the structure as a whole:

For vertical equilibrium: $R + S = 1000$ ①

Taking moments about D: $R \times 2l = 1000 \times l$

$\Rightarrow$ $R = 500$

From ① $S = 500.$

Notice that you could have found this result by symmetry of framework and forces.

ii) Internal forces

The next step is to mark the internal forces acting on the joints. In this case they are all unknown and all the rods are assumed to be in tension, as in figure 11.2. If in fact the forces are thrusts they will be found to have negative values.

The tensions (in N) have been called $T_1, T_2, T_3, \ldots T_7$. They could equally have been called $T_{AB}, T_{AE}, \ldots T_{CD}.$

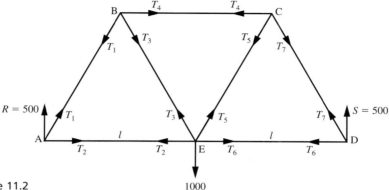

Figure 11.2

Now consider the equilibrium of each joint in turn. It is best to avoid joints where there are more than two unknowns, so start at A or D.

Starting with A (figure 11.3):

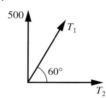

Figure 11.3

Vertical equilibrium: $T_1 \sin 60° + 500 = 0$
$$\Rightarrow \quad T_1 = -577 \text{ (thrust).} \qquad ①$$

Horizontal equilibrium: $T_1 \cos 60° + T_2 = 0$
$$-577 \cos 60° + T_2 = 0$$
$$\Rightarrow \quad T_2 = 289 \text{ (tension).} \qquad ②$$

Moving on to point B (figure 11.4):

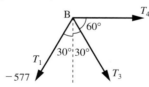

Figure 11.4

Vertical equilibrium: $T_1 \cos 30° + T_3 \cos 30° = 0$
Substitute from ① $-577 \cos 30° + T_3 \cos 30° = 0$
$$\Rightarrow \quad T_3 = 577 \text{ (tension).} \qquad ③$$

Horizontal equilibrium: $T_4 + T_3 \cos 60° - T_1 \cos 60° = 0$
$$\Rightarrow \quad T_4 + 577 \cos 60° + 577 \cos 60° = 0$$
$$\Rightarrow \quad T_4 = 577 \text{ (thrust).} \qquad ④$$

The three remaining tensions can be deduced because both the framework and the external forces are symmetrical about the vertical through E.

$T_7 = T_1 = -577$ (thrust); $T_5 = T_3 = 577$ (tension): $T_6 = T_2 = 289$ (tension).

Alternatively you can find them by considering the equilibrium of the other joints.

Figure 11.5 shows the forces in all of the members.

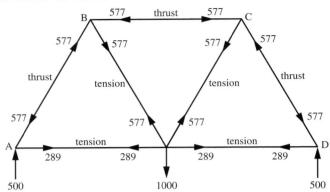

Figure 11.5

EXAMPLE 11.2

Figure 11.6 shows a simple crane. AB is a cable; BC, CD and DB are light freely jointed rods. The framework is freely hinged to its base at D. A load of 5000 N is hanging from C. The rod CD and the cable BA both make angles of 45° with the horizontal and BC is horizontal. Find exact values for

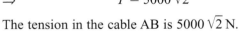

i)   the tension in the cable AB

Figure 11.6

ii) the magnitude of the reaction of the base on the framework at D

iii) the force in each rod, stating whether it is in tension or compression.

SOLUTION

i)   The diagram shows the external forces.

Assume that BD and BC are of length a units.

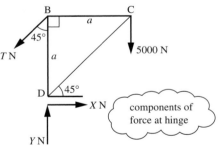

Taking moments about D for the whole framework:

$$T \sin 45° \times a = 5000 \times a$$

$$\Rightarrow \qquad T = 5000\sqrt{2}$$

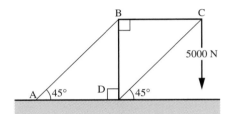

$\sin 45° = \dfrac{1}{\sqrt{2}}$

Figure 11.7

The tension in the cable AB is $5000\sqrt{2}$ N.

ii) Resolving horizontally:

$$X = T \sin 45°$$

$$\Rightarrow \qquad X = 5000$$

Resolving vertically:

$$Y = T \cos 45° + 5000$$

$$\Rightarrow \qquad Y = 10\,000.$$

$\cos 45° = \dfrac{1}{\sqrt{2}}$

The magnitude of the reaction at D is $5000\sqrt{(1^2 + 2^2)} = 5000\sqrt{5}$ N.

iii) The diagram shows all the internal and external forces on the joints.

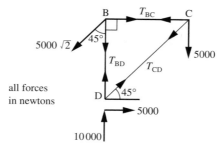

Figure 11.8

all forces
in newtons

For B, vertically: $T_{BD} + 5000 \sqrt{2} \times \cos 45° = 0$
$$T_{BD} = -5000$$

horizontally: $T_{BC} - 5000 \sqrt{2} \times \sin 45° = 0$
$$T_{BC} = 5000.$$

For C, vertically: $T_{CD} \cos 45° - 5000 = 0$
$$T_{CD} = 5000 \sqrt{2}.$$

The stresses in the rods are BD: 5000 N (compression), BC: 5000 N (tension), CD: $5000 \sqrt{2}$ N (compression).

These values are exact. The compression in CD can now be written to any required degree of accuracy.

Chapter 11 Exercises

11.1   The diagram shows a framework made up of three light rods which are freely jointed at A, B and C. A load of 2000 N is applied at the point B.

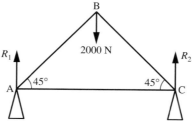

i)   Find the magnitudes of the forces R_1 and R_2.
ii)   Draw a sketch of the framework and mark in tensions T_1, T_2 and T_3 acting in the rods AB, BC and CA respectively.
iii)   If one of these rods is actually in compression, how will the value of the tension show this?
iv)   Write down equations for the horizontal and vertical equilibrium of joint A and solve these equations to find T_1 and T_3.
v)   By considering the equilibrium of joint B, find T_2.
vi)   Show that with the values of T_2 and T_3 which you have found, joint C is also in equilibrium.
vii)   Write down the forces in the three rods, stating whether they are in tension or in compression.

11.2   The diagram shows a framework PQR of light, freely jointed rods in the shape of an equilateral triangle. The framework is freely hinged to the wall at P. A light string connecting Q to S is taut when QS and PR are horizontal, as in the diagram. A weight of 500 N is hanging from R.

i)   By considering the equilibrium of the whole framework PQR, find the tension in the string QS.

ii)   Find the reaction of the wall on the framework at P

    a)   as horizontal and vertical components

    b)   in magnitude and direction form.

iii)   Find the internal forces in the three rods, stating whether they are in tension or in compression.

11.3   The diagram shows a crane supporting a load of 2000 N. The framework is light and freely jointed, and is secured at points A and C.

i)   Find the external forces acting on the framework.

ii)   Find the force in each member of the framework stating whether it is in tension or compression.

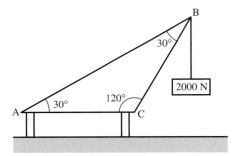

11.4   The light, freely jointed framework shown below is supported at E and C and is carrying loads at A, B and D. Find the magnitudes of the forces in each of the seven rods and state whether they are in tension or in compression.

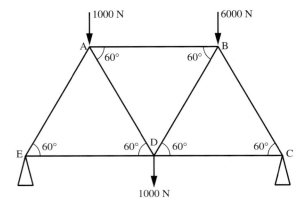

11.5

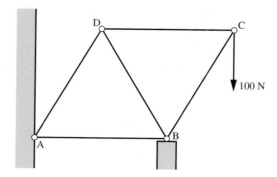

Five light rods AB, BC, CD, DB and DA, each of length 1m, are freely hinged to each other and to a vertical wall at A. The structure rests on a smooth horizontal support at B and has a load of 100 N which acts vertically downwards at C. The rod AB is horizontal and the framework is in a vertical plane.

i) Draw a diagram showing all the external forces acting on the framework and calculate their values.

ii) Draw a diagram showing all the forces acting on the pin-joints, including forces internal to the rods.

iii) Calculate the internal forces in each of the rods, stating whether these are tensions or compressions.

[You may find it helpful to use surds in your answers.]

11.6 The diagram shows a light framework ABCD freely pin-jointed together at A, B and C and freely attached to a vertical wall at A and D. There is a load of 1200 N at C and a vertical force of T N acts at B. The other external forces U, V, X and Y N and essential geometrical information are marked in the diagram. The framework is in equilibrium.

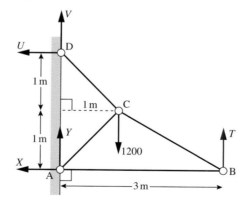

i) Show that $X = -U$ and that $U = \frac{1}{2}(1200 - 3T)$.

ii) By considering the equilibrium at D, show that $U = V$.

iii) Show that $Y = \frac{1}{2}(1200 + T)$ and find expressions in terms of T for the internal forces in each of the rods AB, BC, AC and CD. Show that just one of the rods will change from being in tension to being in thrust as T increases. For what value of T is there no internal force in this rod?

11.7   The diagram shows a vertical framework made from the light, rigid rods AB, BC, CD, DA and BD which are freely pin-jointed at A, B, C and D. ABCD is a parallelogram with AB = CD = BD = 2 m; angle DAB = 30°. The rod AB is horizontal. The framework is freely hinged to a wall at A and rests on a smooth support at B. A vertical load of L N is attached at C.

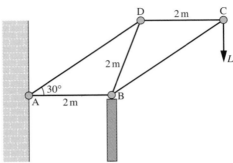

   i)   Show that the support at B exerts a force of $\frac{5}{2}L$ N upwards on the framework and that the wall exerts a force of $\frac{3}{2}L$ N downwards on the framework at A.

   ii) Draw a diagram showing all the forces acting on each of the pin-joints of the framework, including those due to the internal forces in the rods. Calculate the magnitudes of the internal forces in the rods and state for each rod whether it is in tension or compression.

   [You may express your answers in terms of surds where appropriate.]

11.8   A framework consists of three light, rigid rods AB, BC and AC, freely pin-jointed to each other at A, B and C. The framework is freely attached to a vertical wall at A. AB is horizontal and C rests on a smooth plane inclined at 45° to the vertical. AB has length 2 m, AC = BC and C is 2 m vertically below the line AB, as shown in figure (A), so that angle CAB = angle

   CBA = α, where $\sin \alpha = \dfrac{2}{\sqrt{5}}$ and $\cos \alpha = \dfrac{1}{\sqrt{5}}$.

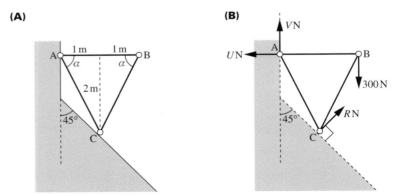

(A)

(B)

   A load of 300 N is attached at B. Figure (B) shows the external forces U N, V N and R N as well as the load at B.

   i)   Write down equations for the equilibrium of the framework in the horizontal and vertical directions.

ii) Show that the anticlockwise moment of R about A is $\dfrac{3R}{\sqrt{2}}$ Nm. Hence show that $R = 200\sqrt{2}$.

iii) Calculate the values of U and V.

iv) By considering the equilibrium of the forces acting at points A and B, or otherwise, calculate the magnitude of the internal force in each of the rods, stating whether it is in tension or compression.

[You may express your answers in terms of $\sqrt{5}$ where appropriate.]

KEY POINTS

Method for solving framework problems

11.1 First calculate the external forces acting on a framework.
- Consider the horizontal and vertical equilibrium of the whole framework.
- Take moments about a suitable point for the whole framework.

11.2 Then calculate the internal forces making these two *assumptions*:
- all the members of the framework are light rods;
- all the joints are smooth so there are no moments acting on them.

11.3 Consider the horizontal and vertical equilibrium of each joint in turn. Make sure that there are no more than two unknown forces each time you move to a new joint.

12 Circular motion

Whirlpools and storms his circling arm invest
With all the might of gravitation blest.

Alexander Pope

These pictures show some objects which move in circular paths. What other examples can you think of?

QUESTION 12.1
What makes objects move in circles?

Why does the moon circle the earth?

What happens to the 'hammer' when the athlete lets it go?

Does the pilot of the plane need to be strapped into his seat at the top of a loop in order not to fall out?

The answers to these questions lie in the nature of circular motion. Even if an object is moving at constant speed in a circle, its velocity keeps changing because its direction of motion keeps changing. Consequently the object is accelerating and so, according to Newton's first law, there must be a force acting on it. The force required to keep an object moving in a circle can be provided in many ways.

Without the earth's gravitational force, the moon would move off at constant speed in a straight line into space. The wire attached to the athlete's hammer provides a tension force which keeps the ball moving in a circle. When the athlete lets go, the ball flies off at a tangent because the tension has disappeared.

Although it would be sensible for the pilot to be strapped in, no upward force is necessary to stop him falling out of the plane because his weight contributes to the force required for motion in a circle.

In this chapter, these effects are explained.

12.1 Notation

To describe circular motion (or indeed any other topic) mathematically you need a suitable notation. It will be helpful in this chapter to use the notation (attributed to Newton) for differentiation with respect to time in which, for

example, $\dfrac{ds}{dt}$ is written as $\dot{s}$, and $\dfrac{d^2\theta}{dt^2}$ as $\ddot{\theta}$.

Figure 12.1 shows a particle P moving round the circumference of a circle of radius r, centre O. At time t, the position vector $\overrightarrow{OP}$ of the particle makes an angle θ (in radians) with the fixed direction $\overrightarrow{OA}$. The arc length AP is denoted by s.

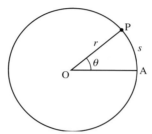

Figure 12.1

12.2 Angular speed

Using this notation,

$$s = r\theta.$$

Differentiating this with respect to time using the product rule gives:

$$\frac{ds}{dt} = r\frac{d\theta}{dt} + \theta\frac{dr}{dt}.$$

Since r is constant for a circle, $\dfrac{dr}{dt} = 0$, so the rate at which the arc length increases is:

$$\frac{ds}{dt} = r\frac{d\theta}{dt} \qquad \text{or} \qquad \dot{s} = r\dot{\theta}. \tag{1}$$

In this equation $\dot{s}$ is the speed at which P is moving round the circle (often denoted by v), and $\dot{\theta}$ is the rate at which the angle θ is increasing, i.e. the rate at which the position vector $\overrightarrow{OP}$ is rotating.

The quantity $\dfrac{d\theta}{dt}$, or $\dot{\theta}$, can be called the *angular velocity* or the *angular speed* of P. In more advanced work, angular velocity is treated as a vector, whose direction is taken to be that of the axis of rotation. In this chapter, $\dfrac{d\theta}{dt}$ is often referred to as angular speed, but is given a sign: positive when θ is increasing (usually anticlockwise) and negative when θ is decreasing (usually clockwise).

Angular speed is often denoted by ω, the Greek letter omega. So the equation $\dot{s} = r\dot{\theta}$ may be written as

$$v = r\omega.$$

Notice that for this equation to hold, θ must be measured in radians, so the angular speed is measured in *radians per second* or rad s^{-1}.

QUESTION 12.2   Angular speeds are often written as multiples of π unless otherwise requested. Why is this?

Figure 12.2 shows a disc rotating about its centre, O, with angular speed ω. The line OP represents any radius.

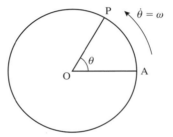

Figure 12.2

Every point on the disc describes a circular path, and all points have the same angular speed. However the *actual* speed of any point depends on its distance from the centre: increasing r in the equation $v = r\omega$ increases v. You will appreciate this if you have ever been at the end of a rotating line of people in a dance or watched a body of marching soldiers wheeling round a corner.

Angular speeds are sometimes measured in revolutions per second or revolutions per minute (rpm) where one revolution is equal to 2π radians. For example, turntables for vinyl records rotate at 45 or $33\frac{1}{3}$ rpm while a computer hard disc might spin at 7200 rpm or more. At cruising speeds, crankshafts in car engines typically rotate at 3000 to 4000 rpm.

EXAMPLE 12.1   A police car drives at 40 mph around a roundabout in a circle of radius 16 m. A second car moves so that it has the same angular speed as the police car but in a circle of radius 12 m. Is the second car breaking the 30 mph speed limit? (Use the approximation 1 mile $= \frac{8}{5}$ km.)

SOLUTION

Converting miles per hour to metres per second gives:

$$40 \, \text{mph} = 40 \times \tfrac{8}{5} \, \text{km h}^{-1}$$

$$= \frac{40 \times 8 \times 1000}{5 \times 3600} \, \text{ms}^{-1}$$

$$= \frac{160}{9} \, \text{ms}^{-1}$$

Using $v = r\omega$,

$$\omega = \frac{160}{9 \times 16} \text{ in rad s}^{-1}$$

$$= \frac{10}{9} \text{ in rad s}^{-1}$$

The speed of the second car is:

$$v = 12\omega$$

$$= \frac{10}{9} \times 12 \text{ in ms}^{-1}$$

$$= \frac{120 \times 5 \times 3600}{9 \times 8 \times 1000} \text{ in mph}$$

$$= 30 \text{ mph}$$

The second car is just on the speed limit.

Notes

1 Notice that working in fractions gives an exact answer.

2 A quicker way to do this question would be to notice that, because the cars have the same angular speed, the actual speeds of the cars are proportional to the radii of the circles in which they are moving. Using this method it is possible to stay in mph. The ratio of the two radii is $\frac{12}{16}$ so the speed of the second car is $\frac{12}{16} \times 40$ mph $= 30$ mph.

12.3 Velocity and acceleration

Velocity and acceleration are both vector quantities. They can be expressed either in magnitude–direction form, or in components. When describing circular motion or other orbits it is most convenient to take components in directions along the radius (*radial* direction) and at right angles to it (*transverse* direction).

For a particle moving round a circle of radius r, the velocity has:

radial component: 0

transverse component: $r\dot{\theta}$ or $r\omega$.

The acceleration of a particle moving round a circle of radius r has:

radial component: $-r\dot{\theta}^2$ or $-r\omega^2$

transverse component: $r\ddot{\theta}$ or $r\dot{\omega}$.

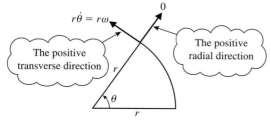

Figure 12.3: Velocity

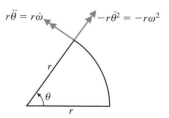

Figure 12.4: Acceleration

The transverse component is just what you would expect, the radius multiplied by the angular acceleration, $\ddot{\theta}$. If the particle has constant angular speed, its angular acceleration is zero and so the transverse component of its acceleration is also zero.

In contrast, the radial component of the acceleration, $-r\omega^2$, is almost certainly not a result you would have expected intuitively. It tells you that a particle travelling in a circle is always accelerating towards the centre of the circle, but without ever getting any closer to the centre. If this seems a strange idea, you may find it helpful to remember that circular motion is not a natural state; left to itself a particle will travel in a straight line. To keep a particle in the unnatural state of circular motion it must be given an acceleration at right angles to its motion, i.e. towards the centre of the circle.

The derivation of these expressions for the acceleration of a particle in circular motion is complicated by the fact that the radial and transverse directions are themselves changing as the particle moves round the circle, in contrast to the fixed x and y directions in the Cartesian system.

12.4 Derivation of expressions

When a particle moves in a circle, it is convenient for the velocity and acceleration to be expressed in the radial and transverse (tangential) directions. For this reason, unit vectors $\hat{\mathbf{r}}$ and $\hat{\boldsymbol{\theta}}$ are defined in these two directions. Unlike the unit vectors $\mathbf{i}$ and $\mathbf{j}$, however, these vectors are not constant. They change in direction as the particle moves round the circle.

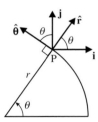

Figure 12.5

Figure 12.5 shows $\hat{\mathbf{r}}$ and $\hat{\boldsymbol{\theta}}$ when the particle P is at a general position. You can see that

$$\hat{\mathbf{r}} = \cos\theta\mathbf{i} + \sin\theta\mathbf{j}$$

and

$$\hat{\boldsymbol{\theta}} = -\sin\theta\mathbf{i} + \cos\theta\mathbf{j}.$$

Differentiating $\hat{\mathbf{r}}$ with respect to time,

$$\frac{d\hat{\mathbf{r}}}{dt} = \frac{d}{dt}(\cos\theta)\mathbf{i} + \frac{d}{dt}(\sin\theta)\mathbf{j}$$

$$= -\sin\theta\frac{d\theta}{dt}\mathbf{i} + \cos\theta\frac{d\theta}{dt}\mathbf{j}$$

$$= (-\sin\theta\mathbf{i} + \cos\theta\mathbf{j})\dot{\theta} \quad \text{where } \dot{\theta} = \frac{d\theta}{dt}.$$

You will recognise that the vector in brackets, $(-\sin\theta\mathbf{i} + \cos\theta\mathbf{j})$, is the same as $\hat{\boldsymbol{\theta}}$, the unit vector in the transverse direction. Consequently,

$$\frac{d\hat{\mathbf{r}}}{dt} = \dot{\theta}\hat{\boldsymbol{\theta}}.$$

Similarly,
$$\frac{d\hat{\boldsymbol{\theta}}}{dt} = -(\cos\theta)\dot{\theta}\mathbf{i} - (\sin\theta)\dot{\theta}\mathbf{j}$$
$$= (-\cos\theta\mathbf{i} + \sin\theta\mathbf{j})\dot{\theta}$$

i.e.
$$\frac{d\hat{\boldsymbol{\theta}}}{dt} = -\dot{\theta}\hat{\mathbf{r}}.$$

These two results, $\dfrac{d\hat{\mathbf{r}}}{dt} = \dot{\theta}\hat{\boldsymbol{\theta}}$ and $\dfrac{d\hat{\boldsymbol{\theta}}}{dt} = -\dot{\theta}\hat{\mathbf{r}}$ are very important, and apply to all motion described in polar co-ordinates, not just circular motion.

In the case of circular motion with radius r, the displacement from the centre at any time is $r\hat{\mathbf{r}}$.

Differentiating this with respect to time, you obtain

$$\text{velocity} = \frac{dr}{dt}\hat{\mathbf{r}} + r\frac{d\hat{\mathbf{r}}}{dt}$$
$$= 0 + r\dot{\theta}\hat{\boldsymbol{\theta}} \quad \text{(since } r \text{ is constant)}$$
$$= r\dot{\theta}\hat{\boldsymbol{\theta}} \quad \text{(i.e. } r\dot{\theta} \text{ in the transverse direction).}$$

Differentiating again you obtain

$$\text{acceleration} = \frac{dr}{dt}\dot{\theta}\hat{\boldsymbol{\theta}} + r\frac{d\dot{\theta}}{dt}\hat{\boldsymbol{\theta}} + r\dot{\theta}\frac{d\hat{\boldsymbol{\theta}}}{dt}$$
$$= 0 + r\ddot{\theta}\hat{\boldsymbol{\theta}} - r\dot{\theta}^2\hat{\mathbf{r}}.$$

Thus the acceleration has two components:

transverse: $\quad r\ddot{\theta}$
radial: $\quad -r\dot{\theta}^2$.

Circular motion with constant speed

In this section, the circular motion is assumed to be uniform and so have no transverse component of acceleration. Later in the chapter (page 267 onwards), situations are considered in which the angular speed varies.

Problems involving circular motion often refer to the actual speed of the object, rather than its angular speed. It is easy to convert the one into the other using the relationship $v = r\omega$. This relationship can also be used to express the magnitude of the acceleration in terms of v and r.

$$\omega = \frac{v}{r}$$

$$a = r\omega^2 = r\left(\frac{v}{r}\right)^2$$

$$\Rightarrow \quad a = \frac{v^2}{r} \text{ towards the centre.}$$

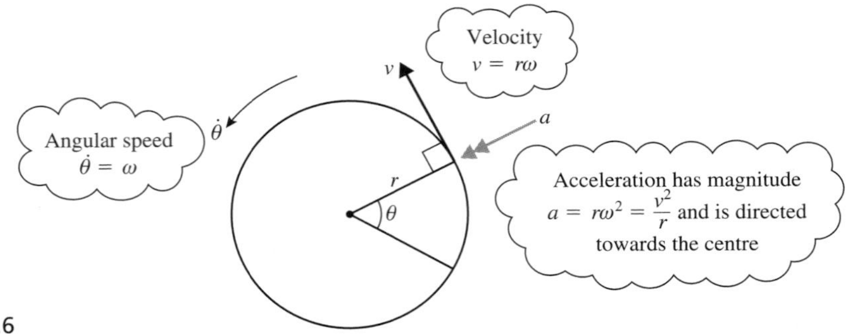

Figure 12.6

EXAMPLE 12.2   A fly is standing on a record at a distance of 8 cm from the centre. If the record is rotating at 45 rpm, find

i)   the angular speed of the fly in radians per second
ii) the speed of the fly in metres per second
iii) the acceleration of the fly.

SOLUTION

i)   One revolution is 2π rad so

$$45 \text{ rpm} = 45 \times 2\pi \text{ rad min}^{-1}$$
$$= \frac{45 \times 2\pi}{60} \text{ rad s}^{-1}$$
$$= \frac{3\pi}{2} \text{ rad s}^{-1}.$$

ii) If the speed of the fly is $v \text{ ms}^{-1}$, v can be found using

$$v = r\omega$$
$$= 0.08 \times \frac{3\pi}{2}$$
$$= 0.377 \ldots$$

So the speed of the fly is 0.38 ms^{-1} (to 2 d.p.).

iii) The acceleration of the fly is given by

$$r\omega^2 = 0.08 \times \left(\frac{3\pi}{2}\right)^2 \text{ ms}^{-2}$$
$$= 1.78 \text{ ms}^{-2}.$$

It is directed towards the centre of the record.

QUESTION 12.3   A wheel of radius r m is rolling in a straight line with forward speed $u \text{ ms}^{-1}$. What are

i)   the speed of the point which is instantaneously in contact with the ground
ii) the angular speed of the wheel
iii) the velocities of the highest point and the point on the edge of the wheel which is level with and behind the axle?

12.5 The forces required for circular motion

Newton's first law of motion states that a body will continue in a state of rest or uniform motion in a straight line unless acted upon by an external force. Any object moving in a circle, such as the police car and the fly in Examples 12.1 and 12.2 must therefore be acted upon by a resultant force in order to produce the required acceleration towards the centre.

A force towards the centre is called a *centripetal* (centre-seeking) force. A resultant centripetal force is necessary for a particle to move in a circular path.

12.6 Examples of circular motion

You are now in a position to use Newton's second law to determine theoretical answers to some of the questions which were posed at the beginning of this chapter. These will, as usual, be obtained using models of the true motion which will be based on simplifying assumptions, for example zero air resistance. Large objects are assumed to be particles concentrated at their centres of mass.

EXAMPLE 12.3

A coin is placed on a rotating turntable. Its centre is 5 cm from the centre of rotation and the coefficient of friction, μ, between the coin and the turntable is 0.5.

i) If the speed of rotation of the turntable is gradually increased, at what angular speed will the coin begin to slide?

ii) What happens next?

SOLUTION

i) Because the speed of the turntable is increased only gradually, it can be assumed that the coin will not slip tangentially.

Figure 12.7 shows the forces acting on the coin, and its acceleration.

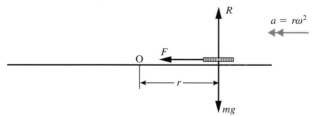

Figure 12.7

The acceleration is towards the centre, O, of the circular path so there must be a frictional force F in that direction.

There is no vertical component of acceleration, so the resultant force acting on the coin has no vertical component.

Therefore
$$R - mg = 0$$
$$R = mg. \qquad \text{①}$$

By Newton's second law towards the centre of the circle:

$$\text{Force } F = ma = mr\omega^2.$$ ②

The coin will not slide so long as $F \leqslant \mu R$.

Substituting from ② and ① this gives

$$mr\omega^2 \leqslant \mu \, mg$$

$$\Rightarrow \quad r\omega^2 \leqslant \mu g.$$

> Notice that the mass, m, has been eliminated at this stage, so that the answer does not depend upon it

Taking g in ms^{-2} as 9.8 and substituting $r = 0.05$ and $\mu = 0.5$

$$\omega^2 \leqslant 98$$

$$\omega \leqslant \sqrt{98}$$

$$\omega \leqslant 9.89 \ldots$$

The coin will move in a circle provided that the angular speed is less than about 10 rad s^{-1} and this speed is independent of the mass of the coin.

ii) When the angular speed increases beyond this, the coin slips to a new position. If the angular speed continues to increase the coin will slip right off the turntable. When it reaches the edge it will fly off in the direction of the tangent.

The conical pendulum

A conical pendulum consists of a small bob tied to one end of a string. The other end of the string is fixed and the bob is made to rotate in a horizontal circle below the fixed point so that the string describes a cone as in figure 12.8.

Figure 12.8

QUESTION 12.4   1 Draw a diagram showing the magnitude and direction of the acceleration of a bob and the forces acting on it.

2 In the case that the radius of the circle remains constant, try to predict the effect on the angular speed when the length of the string is increased or when the mass of the bob is increased. What might happen when the angular speed increases?

3 Draw two circles of equal diameter on horizontal surfaces so that two people can make the bobs of conical pendulums rotate in circles of the same radius.

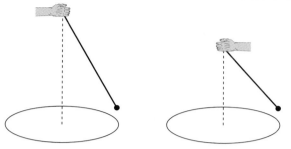

Figure 12.9

  i) Compare pendulums of different lengths with bobs of equal mass.
  ii) Compare pendulums of the same length but with bobs of different masses.

Does the angular speed depend on the length of the pendulum or the mass of the bob?

4 What happens when somebody makes the speed of the bob increase?

5 Can a bob be made to rotate with the string horizontal?level with and behind the axle?

Theoretical model for the conical pendulum

A conical pendulum may be modelled as a particle of mass m attached to a light, inextensible string of length l. The mass is rotating in a horizontal circle with angular speed ω and the string makes an angle α with the downward vertical. The radius of the circle is r and the tension in the string is T, all in consistent units (e.g. S.I. units). The situation is shown in figure 12.10.

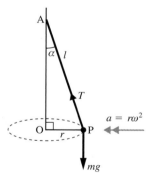

Figure 12.10

The magnitude of the acceleration is $r\omega^2$. The acceleration acts in a horizontal direction towards the centre of the circle. This means that there must be a resultant force acting towards the centre of the circle.

There are two forces acting on this particle, its weight mg and the tension T in the string.

As the acceleration of the particle has no vertical component, the resultant force has no vertical component, so

$$T \cos \alpha - mg = 0.$$ ①

Using Newton's second law towards the centre, O, of the circle

$$T \sin \alpha = ma = mr\omega^2 \qquad \text{②}$$

In triangle AOP $r = l \sin \alpha$.

Substituting for r in ② gives $T \sin \alpha = m(l \sin \alpha)\omega^2$

$$\Rightarrow \qquad T = ml\omega^2.$$

Substituting this in ① gives

$$ml\omega^2 \cos \alpha - mg = 0$$

$$\Rightarrow \qquad l \cos \alpha = \frac{g}{\omega^2}. \qquad \text{③}$$

This equation provides sufficient information to give theoretical answers to the questions posed on page 260.

● When r is kept constant and the length of the string is increased, the length AO $= l \cos \alpha$ increases. Equation ③ indicates that the value of $\frac{g}{\omega^2}$ increases and so the angular speed ω decreases. Conversely, the angular speed increases when the string is shortened.

● The mass of the particle does not appear in equation ③, so it has no effect on the angular speed, ω.

● When the length of the pendulum is unchanged, but the angular speed is increased, $\cos \alpha$ decreases, leading to an increase in the angle α and hence in r.

● If $\alpha \geqslant 90$, $\cos \alpha \leqslant 0$, so $\frac{g}{\omega^2} \leqslant 0$, which is impossible. You can see from figure 12.10 that the tension in the string must have a vertical component to balance the weight of the particle.

EXAMPLE 12.4   The diagram on the right represents one of several arms of a fairground ride, shown on the left. The arms rotate about an axis and riders sit in chairs linked to the arms by chains.

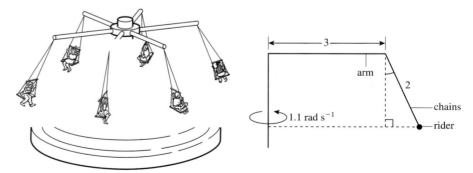

Figure 12.11

The chains are 2 m long and the arms are 3 m long. Find the angle that the chains make with the vertical when the rider rotates at 1.1 rad s^{-1}.

SOLUTION

Let T N be the resultant tension in the chains holding a chair, and m kg the mass of chair and rider.

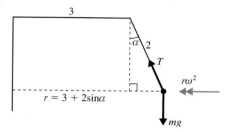

Figure 12.12

If the chains make an angle α with the vertical, the motion is in a horizontal circle with radius given by

$$r = 3 + 2 \sin \alpha.$$

The magnitude of the acceleration is given by

$$r\omega^2 = (3 + 2 \sin \alpha)1.1^2.$$

It is in a horizontal direction towards the centre of the circle. Using Newton's second law in this direction gives

$$\text{Force} = mr\omega^2$$

$$\Rightarrow \quad T \sin \alpha = m(3 + 2 \sin \alpha)1.1^2$$

$$= 1.21m(3 + 2 \sin \alpha). \qquad ①$$

Vertically $\qquad T \cos \alpha - mg = 0$

$$\Rightarrow \quad T = \frac{mg}{\cos \alpha}.$$

Substituting for T in equation ① :

$$\frac{mg}{\cos \alpha} \sin \alpha = 1.21m(3 + 2 \sin \alpha)$$

$$\Rightarrow \quad 9.8 \tan \alpha = 3.63 + 2.42 \sin \alpha.$$

> Since m cancels out at this stage, the angle does not depend on the mass of the rider

This equation cannot be solved directly, but a numerical method will give you the solution 25.5° correct to three significant figures. You might like to solve the equation yourself or check that this solution does in fact satisfy the equation.

Note

Since the answer does not depend on the mass of the rider and chair, when riders of different masses, or even no riders, are on the equipment all the chains should make the same angle with the vertical.

Banked tracks

You may have noticed that when they curve round bends, most roads are banked so that the edge at the outside of the bend is slightly higher than that at the inside. For the same reason the outer rail of a railway track is slightly higher than the inner rail when it goes round a bend. On bobsleigh tracks the bends are almost bowl shaped, with a much greater gradient on the outside.

Figure 12.13 shows a car rounding a bend on a road which is banked so that the cross-section makes an angle α with the horizontal.

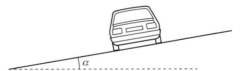

Figure 12.13

In modelling such situations, it is usual to treat the bend as part of a horizontal circle whose radius is large compared to the width of the car. In this case, the radius of the circle is taken to be r metres, and the speed of the car constant at v metres per second.

The car is modelled as a particle which has an acceleration of $\dfrac{v^2}{r}$ ms^{-2} in a horizontal direction towards the centre of the circle. The forces and acceleration are shown in figure 12.14.

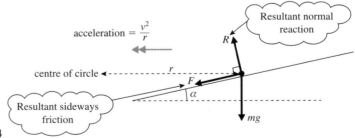

Figure 12.14

The direction of the frictional force F will be up or down the slope depending on whether the car has a tendency to slip sideways towards the inside or outside of the bend.

QUESTION 12.5 Under what conditions do you think each of these will occur?

EXAMPLE 12.5   A car is rounding a bend of radius 100 m which is banked at an angle of 10° to the horizontal. At what speed must the car travel to ensure it has no tendency to slip sideways?

SOLUTION

When there is no tendency to slip there is no frictional force, so in the plane perpendicular to the direction of motion of the car, the forces and acceleration are as shown in figure 12.15. The only horizontal force is provided by the horizontal component of the normal reaction of the road on the car.

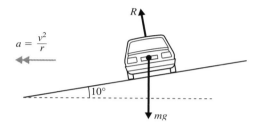

Figure 12.15

Vertically, there is no acceleration so there is no resultant force.

$$R \cos 10° - mg = 0$$

$$\Rightarrow \quad R = \frac{mg}{\cos 10°} \qquad \text{①}$$

By Newton's second law in the horizontal direction towards the centre of the circle

$$R \sin 10° = ma = \frac{mv^2}{r}$$

$$= \frac{mv^2}{100}.$$

Substituting for R from ①

$$\left(\frac{mg}{\cos 10°}\right) \sin 10° = \frac{mv^2}{100}$$

$$\Rightarrow \quad v^2 = 100g \tan 10°$$

$$\Rightarrow \quad v = 13.15 \ldots$$

> The mass, m cancels out at this stage, so the answer does not depend on it

The speed of the car must be about 13.15 ms^{-1} or 30 mph.

There are two important points to notice in this example.

● The speed is the same whatever the mass of the car.
● The example looks at the situation when the car does not tend to slide, and finds the speed at which this is the case. At this speed the car does not depend on friction to keep it from sliding, and indeed it could travel safely round the bend at this speed even in very icy conditions. However, at other speeds there is a tendency to slide, and friction actually helps the car to follow its intended path.

Safe speeds on a bend

What would happen in the previous example if the car travelled either more slowly than $13.15\ ms^{-1}$ or more quickly?

The answer is that there would be a frictional force acting so as to prevent the car from sliding across the road.

There are two possible directions for the frictional force. When the vehicle is stationary or travelling slowly, there is a tendency to slide down the slope and the friction acts up the slope to prevent this. When it is travelling quickly round the bend, the car is more likely to slide up the slope, so the friction acts down the slope.

Fortunately, under most road conditions, the coefficient of friction between tyres and the road is large, typically about 0.8. This means that there is a range of speeds that are safe for negotiating any particular bend.

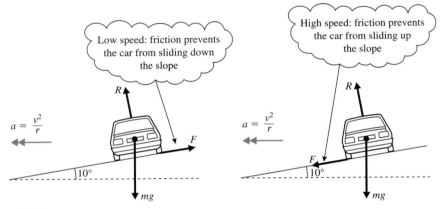

Figure 12.16

QUESTION 12.6 1 Using a particle model for the car, show that it will not slide up or down the slope, provided that

$$\sqrt{rg\frac{(\sin \alpha - \mu \cos \alpha)}{(\cos \alpha + \mu \sin \alpha)}} < v < \sqrt{rg\frac{(\sin \alpha + \mu \cos \alpha)}{(\cos \alpha - \mu \sin \alpha)}}$$

If $r = 100$ and $\alpha = 10°$ (so that $\tan \alpha = 0.176$) the minimum and maximum safe speeds (in mph) for different values of μ are given in the following table.

μ		0	0.1	0.2	0.3	0.4	0.5	0.6	0.7	0.8	0.9	1.0	1.1	1.2
Minimum safe speed	30	19	0	0	0	0	0	0	0	0	0	0	0	
Maximum safe speed	30	37	44	50	55	61	66	70	75	80	84	89	93	

2 Would you regard this bend as safe? How, by changing the values of r and α, could you make it safer?

EXAMPLE 12.6

A bend on a railway track has a radius of 500 m and is to be banked so that a train can negotiate it at 60 mph without the need for a lateral force between its wheels and the rail. The distance between the rails is 1.43 m.

How much higher should the outside rail be than the inside one?

SOLUTION

There is very little friction between the track and the wheels of a train. Any sideways force required is provided by the 'lateral thrust' between the wheels and the rail. The ideal speed for the bend is such that the lateral thrust is zero.

Figure 12.17 shows the forces acting on the train and its acceleration when the track is banked at an angle α to the horizontal.

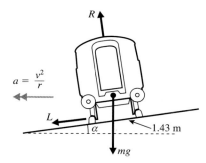

Figure 12.17

When there is no lateral thrust, $L = 0$.

Horizontally: $\qquad\qquad R \sin \alpha = \dfrac{mv^2}{r}.$ ①

Vertically: $\qquad\qquad R \cos \alpha = mg.$ ②

Dividing ① by ② gives $\quad \tan \alpha = \dfrac{v^2}{rg}.$

Using the fact that 60 mph = 26.8 ms⁻¹ this becomes

$$\tan \alpha = 0.147$$
$$\Rightarrow \quad \alpha = 8.4° \text{ (to 2 s.f.)}.$$

The outside rail should be raised by 1.43 sin α metres, i.e. by about 21 cm.

12.7 Circular motion with variable speed

On page 257 you met the general expressions for the acceleration of a particle in circular motion:

● the *transverse component of acceleration* is $r\dot\omega$ or $r\ddot\theta$ along the tangent
● the *radial component of acceleration* is $-r\omega^2$ or $-r\dot\theta^2$ along the radius.

For circular motion with variable speed, the radial component, $-r\dot\theta^2$, is the same as that for circular motion with constant speed. The effect of the varying speed appears in the transverse component, $r\ddot\theta$.

On page 258, the speed was found by differentiating $s = r\theta$ with respect to time to give $\dot s = r\dot\theta$. The transverse component of acceleration for circular motion turns out to be the result of differentiating again: $\ddot s = r\ddot\theta$.

The symbols $\ddot\theta$ and $\dot\omega$ denote the rate of change of the angular velocity, or the *angular acceleration*. This quantity is measured in radians per second squared (rad s^{-2}).

In the next two sections you will be studying two types of circular motion with variable speed:

1 motion with constant angular acceleration
2 unforced motion in a vertical circle.

12.8 Circular motion with constant angular acceleration

When the angular acceleration of a body moving in a circle is constant, it is convenient to use a standard notation

$$\ddot\theta = \alpha.$$

You can find the angular speed and the angular displacement by integrating this with respect to time:

$$\dot\theta = \alpha t + c \quad \text{for some constant } c.$$

It is usual to call the initial angular speed ω_0. In this case $\dot\theta = \omega_0$ when $t = 0$, so $c = \omega_0$.

This gives $\qquad \dot\theta = \omega_0 + \alpha t \quad \text{or} \quad \omega = \omega_0 + \alpha t.$

Integrating again and assuming that $\theta = 0$ when $t = 0$:

$$\theta = \omega_0 t + \tfrac{1}{2}\alpha t^2.$$

These two equations may look familiar to you. They are very much like the equations for motion in a straight line with constant acceleration:

$$v = u + at$$

and $\qquad s = ut + \tfrac{1}{2}at^2.$

There is a direct correspondence between the variables.

Motion in a straight line with constant acceleration	s	u	v	a
Circular motion with constant angular acceleration	θ	ω_0	$\omega(=\dot\theta)$	α

In fact it can be shown that each equation for motion in a straight line with constant acceleration corresponds to an equation for circular motion with constant angular acceleration.

$$v = u + at \longleftrightarrow \omega = \omega_0 + \alpha t$$
$$s = ut + \tfrac{1}{2}at^2 \longleftrightarrow \theta = \omega_0 t + \tfrac{1}{2}\alpha t^2$$
$$v^2 = u^2 + 2as \longleftrightarrow \omega^2 = \omega_0^2 + 2\alpha\theta$$
$$s = \tfrac{1}{2}(u + v)t \longleftrightarrow \theta = \tfrac{1}{2}(\omega_0 + \omega)t$$

EXAMPLE 12.7

Oliver is standing on a playground roundabout at a distance of 2 m from the centre. Imogen pushes the roundabout with constant angular acceleration for 2 s and in this time the angular speed increases from 0.3 rads^{-1} to 1.5 rads^{-1}. Find
i)   the angular acceleration of the roundabout
ii)   the magnitude and direction of the resultant horizontal force acting on Oliver just before Imogen stops pushing the roundabout. (Oliver's mass is 40 kg.)

SOLUTION

i)   Using the standard notation: $\omega_0 = 0.3$, $\omega = 1.5$, $t = 2$. To find α the required equation is:

$$\omega = \omega_0 + \alpha t$$
$$\Rightarrow \qquad 1.5 = 0.3 + 2\alpha$$
$$\Rightarrow \qquad \alpha = 0.6$$

The angular acceleration is 0.6 rads^{-2}.

ii)

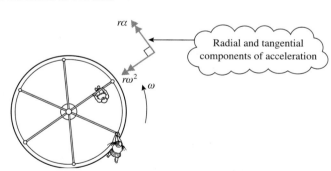

Figure 12.18

Just before Imogen stops pushing, Oliver's acceleration has two horizontal components.

Transverse motion
$$r\dot\theta = r\alpha$$
$$= 2 \times 0.6$$
$$= 1.2 \ (\text{ms}^{-2})$$

Radial motion towards the centre:
$$r\dot\theta^2 = r\omega^2$$
$$= 2 \times 1.5^2$$
$$= 4.5 \ (\text{ms}^{-2})$$

The resultant acceleration has two perpendicular components $1.2\,\text{ms}^{-2}$ and $4.5\,\text{ms}^{-2}$. Its magnitude is $\sqrt{1.2^2 + 4.5^2}$ or $4.66\,\text{ms}^{-2}$ and it makes an angle $\arctan\left(\frac{4.5}{1.2}\right)$ or $75°$ (to the nearest degree) with the transverse direction shown.

By Newton's second law, the resultant horizontal force is given by

$$F = ma$$
$$= 40 \times 4.66\,\text{N}.$$

So the resultant force is $186\,\text{N}$ (to 3 s.f.) at $75°$ to the transverse direction.

12.9 Motion in a vertical circle

Figure 12.19 shows the forces acting on a particle of mass m undergoing free circular motion in a vertical plane. For free motion it is assumed that there is no transverse force.

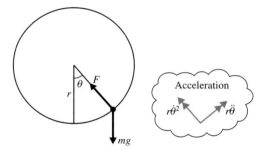

Figure 12.19

For circular motion to take place, there must be a resultant force acting on the particle towards the centre of the circle, so there must be an applied central force, F. When the circle is vertical, the force of gravity also acts in this plane, and is therefore relevant to the motion. When the particle is in the position shown in the diagram, Newton's second law gives the following equations.

Towards the centre $\qquad\qquad F - mg\cos\theta = mr\dot{\theta}^2$ $\qquad\qquad$ ①

Transverse motion $\qquad\qquad -mg\sin\theta = mr\ddot{\theta}$ $\qquad\qquad$ ②

The force F in the first equation might be the tension in a string or the normal reaction from a surface. This force will vary with θ and so equation ① is not helpful in describing how θ varies with time. The second equation, however, does not involve F and may be written as

$$\frac{d^2\theta}{dt^2} = -\frac{g}{r}\sin\theta.$$

This differential equation can be solved, using suitable calculus techniques, to obtain an expression for θ in terms of t.

Using conservation of energy

A different (and at this stage more profitable) approach is to consider the energy of the particle. There is no motion in the radial direction, so F does no work. Gravity is the only other force acting so you can apply the principle of conservation of mechanical energy.

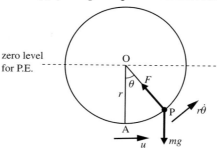

Figure 12.20

Assume that u is the speed of the particle at A, the lowest point of the circle, and that the zero level of gravitational potential energy is through the centre of the circle, O, as shown in figure 12.20.

The total energy at A is $\quad \frac{1}{2}mu^2 - mgr$
$\qquad\qquad\qquad\qquad\quad$ (K.E.) (P.E.)

The total energy at P is $\quad \frac{1}{2}m(r\dot\theta)^2 - mgr \cos \theta$
$\qquad\qquad\qquad\qquad\quad$ (K.E.) $\quad$ (P.E.)

By the principle of conservation of energy

$$\tfrac{1}{2}m(r\dot\theta)^2 - mgr \cos \theta = \tfrac{1}{2}mu^2 - mgr$$

$$\Rightarrow \quad r\dot\theta^2 = \frac{u^2}{r} - 2g(1 - \cos \theta)$$

This tells you the angular speed, $\dot\theta$, of the particle when OP is at any angle θ to OA.

The next two examples show how conservation of energy may be applied to theoretical models of problems involving motion in a vertical circle.

EXAMPLE 12.8

A particle of mass 0.03 kg is attached to the end, P, of a light rod OP of length 0.5 m which is free to rotate in a vertical circle with centre O. The particle is set in motion starting at the lowest point of the circle.

The initial speed of the particle is 2 ms^{-1}.
i) Find the initial kinetic energy of the particle.
ii) Find an expression for the potential energy gained when the rod has turned through an angle θ.
iii) Find the value of θ when the particle first comes to rest.
iv) Find the stress in the rod at this point, stating whether it is a tension or a thrust.
v) Repeat parts i) to iv) using an initial speed of 4 ms^{-1}.
vi) Why is it possible for the first motion (when $v_0 = 2$) to take place if the rod is replaced by a string, but not the second (when $v_0 = 4$)?

SOLUTION

i)
$$\text{Kinetic energy} = \tfrac{1}{2}mv^2$$
$$= \tfrac{1}{2} \times 0.03 \times 2^2$$
$$= 0.06$$

The initial kinetic energy is 0.06 J.

ii) Figure 12.21 shows the position of the particle when the rod has rotated through an angle θ.

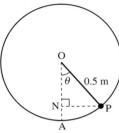

Figure 12.21

It has risen a distance AN where
$$\text{AN} = \text{OA} - \text{ON}$$
$$= 0.5 - 0.5 \cos \theta$$
$$= 0.5(1 - \cos \theta).$$

The gain in potential energy at P is therefore
$$0.03g \times 0.5(1 - \cos \theta) = 0.015g(1 - \cos \theta) \text{ J.}$$

iii) When the particle first comes to rest, the kinetic energy is zero, so by the principle of conservation of energy:
$$0.015g(1 - \cos \theta) = 0.06$$
$$1 - \cos \theta = \frac{0.06}{0.015g}$$
$$= 0.408 \ldots$$
$$\Rightarrow \quad \cos \theta = 0.592$$
$$\theta = 0.938 \text{ rad, i.e. about } 53.7°.$$

iv) The forces acting on the particle and its acceleration are as shown in figure 12.22.

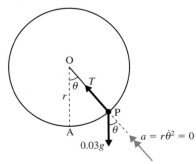

Figure 12.22

The component of the acceleration towards the centre of the circle is $r\dot{\theta}^2$ which equals zero when the angular speed is zero. Resolving towards the centre:

$$T - 0.03g \cos \theta = 0 \qquad \text{①}$$

$$T = 0.174.$$

Since this is positive the stress in the rod is a tension. Its magnitude is 0.174 N.

v) When the initial speed is 4 ms^{-1}, the initial kinetic energy is

$$\tfrac{1}{2} \times 0.03 \times 4^2 = 0.24 \text{ J}.$$

The gain in potential energy at P, as shown in part **ii)**, is

$$0.015g(1 - \cos \theta) \text{ J}.$$

When the particle first comes to rest, the kinetic energy is zero, so by the principle of conservation of energy:

$$0.015g(1 - \cos \theta) = 0.24$$

$$\cos \theta = -0.6327$$

$$\theta = 2.256 \text{ rad, i.e. about } 129°.$$

Now equation ① gives the tension in the rod as

$$T = 0.03g \cos \theta = -0.186.$$

The negative tension means that the stress is in fact a thrust of 0.186 N. Figure 12.23 illustrates the forces acting in this position.

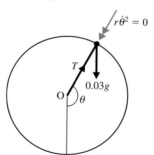

Figure 12.23

vi) A string cannot exert a thrust so, although the rod could be replaced by a string in the first case, it would be impossible in the second. In the absence of any radial thrust the particle will leave its circular path at the point where the tension is zero and before reaching the position where the velocity is zero.

EXAMPLE 12.9

A bead of mass 0.01 kg is threaded on a smooth circular wire of radius 0.6 m and is set in motion with a speed of u ms^{-1} at the bottom of the circle. This just enables the bead to reach the top of the wire.

i) Find the value of u.

ii) What is the direction of the reaction of the wire on the bead when the bead is at the top of the circle?

SOLUTION

i) The initial kinetic energy is

$$\tfrac{1}{2} \times 0.01u^2 = 0.005u^2.$$

The bead will just reach the top if the speed there is zero. If this is the case, its kinetic energy at the top will also be zero.

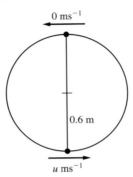

Figure 12.24

It has then risen a height of 2×0.6 m or 1.2 m, so its gain in potential energy is

$$0.01g \times 1.2 = 0.012g$$

By the principle of conservation of energy,

$$\text{loss in K.E.} = \text{gain in P.E.}$$
$$0.005u^2 = 0.012g$$
$$u^2 = 2.4g$$
$$u = \sqrt{2.4g}.$$

The initial speed must be 4.85 ms^{-1} (to 2 d.p.).

ii) The reaction of the wire on the bead could be directed either towards the centre of the circle or away from it. The bead has zero angular speed at the top so the component of its acceleration, and therefore the resultant force towards the centre, is zero. The reaction must be outwards, as shown in figure 12.25, and equal to 0.01g N.

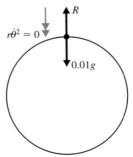

Figure 12.25

QUESTION 12.7 Would this motion be possible if the bead were tied to the end of a string instead of being threaded on a wire?

12.10 The breakdown of circular motion

Activity

⚠ When carrying out this activity, keep well away from other people and breakable objects.

Tie a small object on the end of a piece of strong thread and tie the other end loosely (to minimise friction) round a smooth knitting needle (or a smooth rod with a cork on one end).

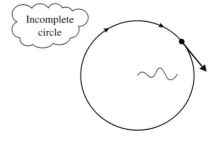

Hold the pointed end of the needle and make the object move in a vertical circle as shown in figure 12.26.

Figure 12.26

Demonstrate these three types of motion.
1 The object travels in complete circles.
2 The object swings like a pendulum.
3 The object rises above the level of the needle but then fails to complete a full circle.

What would happen if the string broke?
What would happen if a rod were used rather than a string?

The three different types of motion mentioned in the activity described above and the case in which the string breaks are illustrated in figure 12.27.

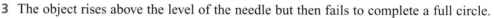

Complete circles

Oscillations

a) Object rotates in complete circles

b) Object oscillates backwards and forwards

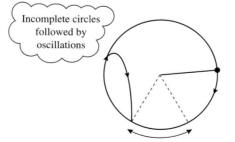

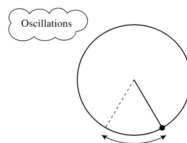

Incomplete circles followed by oscillations

Incomplete circle

c) Object leaves circle at some point and falls inwards

d) String breaks and object starts to move away along a tangent

Figure 12.27

Modelling the breakdown of circular motion

For what reasons might something depart from motion in a circle? For example, under what conditions will a particle attached to a string and moving in a vertical circle fall out of the circle? Under what conditions will a bicycle travelling over a speed bump with circular cross-section leave the road?

A particle on a string

Figure 12.28 shows a particle P of mass m attached to a string of length r, rotating with angular speed ω in a vertical circle, centre O.

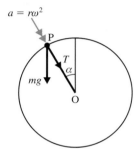

Figure 12.28

There are two forces acting on the particle, its weight, mg, and the tension, T, in the string. The acceleration of the particle is $r\omega^2$ towards the centre of the circle.

Applying Newton's second law towards the centre gives

$$T + mg \cos \alpha = mr\omega^2 \qquad \text{①}$$

where α is the angle shown in the diagram.

While the particle is in circular motion the string is taut, and so $T > 0$. The instant it starts to leave the circle the string goes slack and $T = 0$.

Substituting $T = 0$ in ① gives $\qquad mg \cos \alpha = mr\omega^2$

$$\Rightarrow \quad \cos \alpha = \frac{r\omega^2}{g}.$$

The equation

$$\cos \alpha = \frac{r\omega^2}{g}$$

allows you to find the angle α at which the particle leaves the circle, if it does.

The greatest possible value for $\cos \alpha$ is 1, so if $\dfrac{r\omega^2}{g}$ is greater than 1 throughout the motion, the equation has no solution and this means that the particle never leaves the circle. Thus the condition for the particle to stay in circular motion is that $\omega^2 \geqslant \dfrac{g}{r}$ throughout.

In this example, of a particle on the end of a string, ω varies throughout the motion. As you saw on page 268, the value of ω at any instant is given by the energy equation, which in this case is

$$\tfrac{1}{2}mr^2\omega^2 + mgr(1 + \cos \alpha) = \tfrac{1}{2}mu^2$$

where u is the speed of the particle at the lowest point.

A particle moving on the inside of a vertical circle

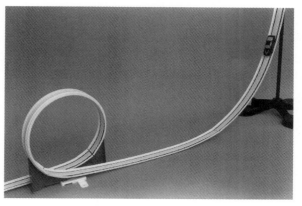

The same analysis applies to a particle sliding around the inside of a smooth circle. The only difference is that in this case the tension, T, is replaced by the normal reaction, R, of the surface on the particle (see figure 12.29). When $R = 0$ the particle leaves the surface.

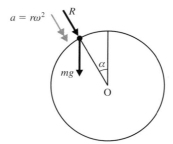

Figure 12.29

QUESTION 12.8 In the toy shown in the photograph above, the track exerts a downwards force on the car as it moves around the top of the circle. What would you feel if you gently touched the top of the track at this instant?

A particle moving on the outside of a vertical circle

The forces acting on a particle moving on the outside of a vertical circle, such as a car going over a hump-backed bridge are the normal reaction, R, acting outwards and the weight of the particle, as shown in figure 12.30.

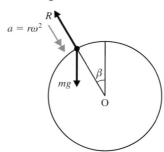

Figure 12.30

Applying Newton's second law towards the centre gives

$$mg \cos \beta - R = mr\omega^2 \qquad \qquad ②$$

where β is the angle shown.

If the normal reaction is zero it means there is no force between the particle and the surface and so the particle is leaving the surface.

Substituting $R = 0$ in ② gives

$$mg \cos \beta = mr\omega^2$$
$$\Rightarrow \qquad \cos \beta = \frac{r\omega^2}{g}$$

QUESTION 12.9   The conditions for the breakdown of circular motion seem to be the same in the cases of a particle on the end of a string and a particle on the outside of a circle. However, everyday experience tells you that circular motion on the end of a string is only possible if the angular speed is large enough whereas a particle will only stay on the outside of a circle if the angular speed is small enough.

How do the conditions $T > 0$ and $R > 0$ explain this difference?

EXAMPLE 12.10   Determine whether it is possible for a particle, P, of mass m kg to be in the position shown in figure 12.31, moving round a vertical circle of radius 0.5 m with an instant angular speed of 4 rads^{-1} when it is
i)   sliding on the outside of a smooth surface
ii) sliding on the inside of a smooth surface
iii) attached to the end of a string OP
iv) threaded on a smooth vertical ring.

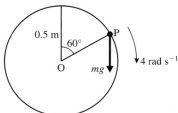

Figure 12.31

SOLUTION

i)   On the outside of a smooth surface:

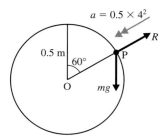

Figure 12.32

The normal reaction R N of the surface on the particle must be acting outwards, so Newton's second law towards the centre gives

$$mg \cos 60° - R = m \times 0.5 \times 4^2$$
$$\Rightarrow \quad R = mg \cos 60° - 8m$$
$$= -3.1m.$$

Whatever the mass, m, this negative value of R is impossible, so the motion is impossible. The particle will already have left the surface.

ii) On the inside of a smooth surface:

The normal reaction of the surface on the particle will now be acting towards the centre and so

$$R + mg \cos 60° = m \times 0.5 \times 4^2$$
$$\Rightarrow \quad R = +3.1m.$$

This is possible.

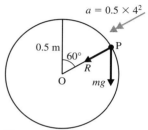

Figure 12.33

iii) Attached to the end of a string:
This example is like part ii) since the tension acts towards the centre, so the motion is possible.

iv) Threaded on a smooth ring:
If the particle is threaded on a ring the normal reaction can act inwards or outwards so the motion can take place whatever the angular speed. This is also the case when a particle is attached to the centre by a light rod. The rod will exert a tension or a thrust as required.

QUESTION 12.10 Which of the situations in the previous example are possible when the instant angular speed is 3 rads^{-1}?

EXAMPLE 12.11 Eddie, a skier of mass m kg, is skiing down a hillside when he reaches a smooth hump in the form of an arc AB of a circle centre O and radius 8 m as shown in figure 12.34. O, A and B lie in a vertical plane and OA and OB make angles of 20° and 40° with the vertical respectively. Eddie's speed at A is 7 ms^{-1}. Determine whether Eddie will lose contact with the ground before reaching the point B.

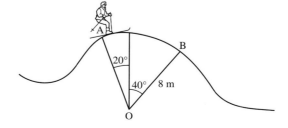

Figure 12.34

SOLUTION

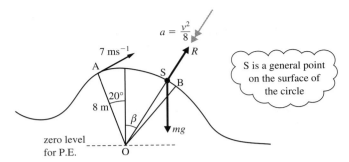

Figure 12.35

Taking the zero level for potential energy to be a horizontal line through O, the initial energy at A is

$$\tfrac{1}{2}m \times 7^2 + mg \times 8 \cos 20°.$$

The energy at any point S is

$$\tfrac{1}{2}mv^2 + mg \times 8 \cos \beta.$$

By the principle of conservation of energy these are equal.

$$\tfrac{1}{2}mv^2 + mg \times 8 \cos \beta = \tfrac{1}{2}m \times 7^2 + mg \times 8 \cos 20°$$

$$\Rightarrow \quad v^2 + 16g \cos \beta = 49 + 147.34$$

$$\Rightarrow \quad v^2 = 196.34 - 16g \cos \beta. \qquad ①$$

Using Newton's second law towards the centre of the circle

$$mg \cos \beta - R = m\frac{v^2}{8}$$

$$\Rightarrow \quad R = m\left(g \cos \beta - \frac{v^2}{8}\right).$$

If Eddie leaves the circle at point S, then $R = 0$

$$\Rightarrow \quad v^2 = 8g \cos \beta.$$

Substituting in ①

$$8g \cos \beta = 196.34 - 16g \cos \beta$$

$$\Rightarrow \quad 24g \cos \beta = 196.34.$$

This gives $\beta \approx 33.4°$, which is less than 40°, so Eddie will lose contact with the ground before he reaches the point B.

Chapter 12 Exercises

12.1 The London Eye observation wheel has a diameter of 135 m and completes one revolution in 30 minutes.

  i) Calculate its angular speed in
    a) rpm   **b)** radians per second.
  ii) Calculate the speed of the point on the circumference where passengers board the moving wheel.

12.2 A lawnmower engine is started by pulling a rope that has been wound round a cylinder of radius 4 cm. Find the angular speed of the cylinder at a moment when the rope is being pulled with a speed of 1.3 ms^{-1}. Give your answers in radians per second, correct to one decimal place.

12.3 The angular speed of an audio CD changes continuously so that a laser can read the data at a constant speed of 12 ms^{-1}. Find the angular speed (in rpm) when the distance of the laser from the centre is
i)   30 mm
ii) 55 mm.

12.4 The minute hand of a clock is 1.2 m long and the hour hand is 0.8 m long.
i)   Find the speeds of the tips of the hands.
ii) Find the ratio of the speeds of the tips of the hands and explain why this is not the same as the ratio of the angular speeds of the hands.

12.5 The position vector of a rider on a helter-skelter is given by
$$\mathbf{r} = 2 \sin t\mathbf{i} + 2 \cos t\mathbf{j} + (8 - \tfrac{1}{2}t)\mathbf{k}$$
where the units are in metres and seconds. The unit vector $\mathbf{k}$ is vertically upwards.
i)   Find an expression for the velocity of the rider at time t.
ii) Find the speed of the rider at time t.
iii) Find the magnitude and direction of the rider's acceleration when $t = \dfrac{\pi}{4}$.

12.6 Two coins are placed on a horizontal turntable. Coin A has mass 15 g and is placed 5 cm from the centre; coin B has mass 10 g and is placed 7.5 cm from the centre. The coefficient of friction between each coin and the turntable is 0.4.
i)   Describe what happens to the coins when the turntable turns at
     a) 6 rads^{-1} b) 8 rads^{-1} c) 10 rads^{-1}.
ii) What would happen if the coins were interchanged?

12.7 A car is travelling at a steady speed of 15 ms^{-1} round a roundabout of radius 20 m.
i)   Criticise this false argument:
     The car is travelling at a steady speed and so its speed is neither increasing nor decreasing and therefore the car has no acceleration.
ii) Calculate the magnitude of the acceleration of the car.
iii) The car has mass 800 kg. Calculate the sideways force on each wheel assuming it to be the same for all four wheels.
iv) Is the assumption in part iii) realistic?

12.8 A fairground ride has seats at 3 m and at 4.5 m from the centre of rotation. Each rider travels in a horizontal circle. Say whether each of the following statements is true, giving your reasons.
i)   Riders in the two positions have the same angular speed at any time.
ii) Riders in the two positions have the same speed at any time.
iii) Riders in the two positions have the same magnitude of acceleration at any time.

12.9 Two spin driers, both of which rotate about a vertical axis, have different specifications as given in the table below.

Model	Rate of rotation	Drum diameter
A	600 rpm	60 cm
B	800 rpm	40 cm

State, with reasons, which model you would expect to be the more effective.

12.10 A satellite of mass M_s is in a circular orbit around the earth, with a radius of r metres. The force of attraction between the earth and the satellite is given by

$$F = \frac{GM_eM_s}{r^2}$$

where $G = 6.67 \times 10^{-11}$ in S.I. units. The mass of the earth M_e is 5.97×10^{24} kg.

i)   Find, in terms of r, expressions for
  a)   the speed of the satellite, $v\,\text{ms}^{-1}$
  b)   the time, T s, it takes to complete one revolution.

ii)   Hence show that, for all satellites, T^2 is proportional to r^3.

A geostationary satellite orbits the earth so that it is always above the same place on the equator.

iii)   How far is it from the centre of the earth?

(The law found in part ii) was discovered experimentally by Johannes Kepler (1571–1630) to hold true for the planets as they orbit the sun, and is commonly known as Kepler's third law.)

12.11 A rotary lawn mower uses a piece of light nylon string with a small metal sphere on the end to cut the grass. The string is 20 cm in length and the mass of the sphere is 30 g.

i)   Find the tension in the string when the sphere is rotating at 2000 rpm assuming the string is horizontal.

ii)   Explain why it is reasonable to assume that the string is horizontal.

iii)   Find the speed of the sphere when the tension in the string is 80 N.

12.12 Experiments carried out by the police accident investigation department suggest that a typical value for a coefficient of friction between the tyres of a car and a road surface is 0.8.

i)   Using this information, find the maximum safe speed on a level circular motorway slip road of radius 50 m.

ii)   How much faster could cars travel if the road were banked at an angle of 5° to the horizontal?

12.13 An astronaut's training includes periods in a centrifuge. This may be modelled as a cage on the end of a rotating arm of length 5 m.

At a certain time, the arm is rotating at 30 rpm.

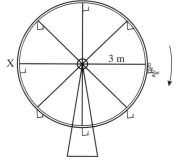

i) Find the angular velocity of the astronaut in radians per second and her speed in metres per second.

ii) Show that under these circumstances the astronaut is subject to an acceleration of magnitude about 5*g*.

At a later stage in the training, the astronaut blacks out when her acceleration is 9*g*.

iii) Find her angular velocity (in rpm) when she blacks out.

The training is criticised on the grounds that, in flight, astronauts are not subject to rotation and the angular speed is too great. An alternative design is considered in which the astronaut is situated in a carriage driven round a circular railway track. The device must be able to simulate accelerations of up to 10*g* and the carriage can be driven at up to 100 ms^{-1}.

iv) What should be the radius of the circular railway track?

12.14 A skater spins at 1 revolution per second with her arms out sideways horizontally. She then takes 2 s to lower her arms and increases her angular speed to 3 revolutions per second. Find

i) the angular acceleration in rad s^{-2} (assumed constant)

ii) the number of complete revolutions she makes during the 2 s while she is lowering her arms.

12.15 The diagram shows the Big Wheel at a fairground. It has radius 3 m. Once it is loaded with passengers it is given a uniform angular acceleration for 20 s then runs at uniform angular speed for 2 minutes. It then slows down at a uniform rate over a further 10 s. During the main part of the ride, the wheel completes 1 revolution every 10 s.

i) Draw a graph showing the angular speed of the wheel against time and state what information is given by the area under the graph.

ii) Find the total angle through which a passenger moves and the distance the passenger travels.

iii) Find the magnitude of the acceleration of a passenger at the top of the ride when it is travelling at maximum speed. Draw a force diagram to show the forces on a passenger of mass 20 kg at the highest point of the ride.

iv) Describe the acceleration vector of a passenger who is at a point X when the wheel is half-way through its acceleration phase.

12.16 As a challenge, a girl is required to swing a bucket of water in a vertical circular arc above her head. If the bucket is not moving fast enough she will get wet. This may be modelled by taking the girl's arm to be 55 cm long and the bucket 35 cm from base to handle. The handle is taken to be rigidly attached to the bucket and is held firmly so that her arm is always in line with the centre of the base. Assume the water behaves as a solid block.

The girl considers three possible depths of water in the bucket: 5 cm, 10 cm and 15 cm.

   i)   For which depth does she need to give the bucket the highest angular speed?

In the event she is forced to have 15 cm of water in the bucket.

   ii) Find the minimum angular speed at the top of the arc for the water to stay in the bucket.

   iii) Deduce the *average* angular acceleration the girl must give the bucket, assuming it is at rest at the lowest point.

One girl, doing this for real, ended with the bucket as well as the water hitting her head.

   iv) What might have happened?

12.17 The diagram illustrates an old road bridge over a river. The road surface follows an approximately circular arc with radius 15 m.

A car is being driven across the bridge and you should model it as a particle.

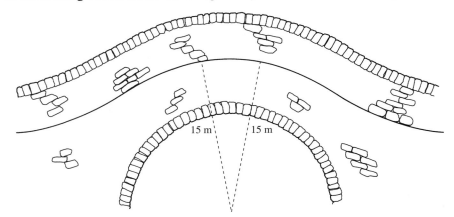

   i)   Calculate the greatest constant speed at which it is possible to drive the car across the bridge without it leaving the road, giving your answer both in ms^{-1} and in mph.

   ii) Comment on the fact that the bridge is old.

   iii) How is it possible to improve the design of the bridge?

12.18 A particle of mass m is attached to one end, A, of a light, inelastic string of length l. The other end, B, of the string is attached to a ceiling so that the particle is free to swing in a vertical plane; the angle between the string and the downward vertical is θ rad. You may assume that the air resistance on the particle is negligible.

Initially $\theta = \dfrac{\pi}{3}$ and the particle is released from rest.

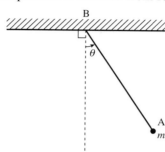

i) Show that the potential energy lost by the particle since leaving its initial position is $\dfrac{mgl}{2}(2\cos\theta - 1)$. Hence find an expression for v^2, where v is the linear speed of the particle, in terms of l, g and θ.

ii) Show that the tension in the string at any point of the motion is $mg(3\cos\theta - 1)$.

iii) Find the greatest tension in the string. What is the position of the particle when the tension in the string is greatest?

Before its release with $\theta = \dfrac{\pi}{3}$, the particle is held in position by means of a second light string inclined at an acute angle, α, to the downward vertical, as shown in the diagram below. The second string is cut to allow the particle to swing.

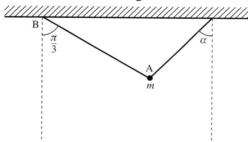

iv) What is the direction of the acceleration of the particle just after the string is cut?

You are given that the tension in the string AB when the particle is being supported is

$$\frac{2mg}{1 + \sqrt{3}\cot\alpha}.$$

v) Calculate the value of α for which the tension in AB remains unchanged when the string is cut.

Personal Tutor

 12.1 Whilst waiting for clearance to land, aircraft are told to fly a fixed pattern in a 'holding area'. For one airport the pattern is a horizontal circuit consisting of two semi-circles connected by straight lines. The plan is shown in the diagram below.

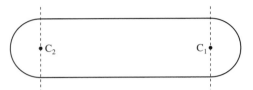

An aircraft flies at a constant speed taking 60 s for each semi-circle.

i) Find the angular speed about C_1 of the aircraft when flying the semi-circle centre C_1.

When flying the semi-circle the aircraft is banked at 26° to the horizontal. The forces acting on the banked aircraft flying in a circle about the point C_1 are shown in the diagram below. The aircraft has mass M kg and a lift force L N.

ii) Copy this diagram and mark in the direction of the acceleration. Write down equations for the vertical and radial components of motion of the aircraft. Hence deduce the radius of the semi-circle and show that the speed of the aircraft is about 91.3 ms^{-1}.

Whilst the aircraft is flying at 1230 m above horizontal ground a wheel falls off.

iii) Neglecting air resistance, show that the wheel travels a horizontal distance of about 1446 m before hitting the ground.

The locus of possible landing points of the wheel is shown in the diagram below.

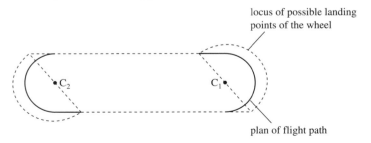

locus of possible landing points of the wheel

plan of flight path

iv) Explain why the curved sections of the locus are semi-circles with C_1 and C_2 as centres and find the common radius of these semi-circles.

12.1 Position, velocity and acceleration of a particle moving on a circle of radius r.

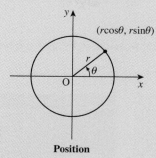
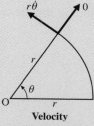
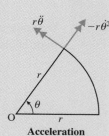

| Position | Velocity | Acceleration |

- position $(r \cos \theta, r \sin \theta)$
- velocity   transverse component: $v = r\dot{\theta} = r\omega$

                                       radial component: 0

         where $\dot{\theta}$ or ω is the angular velocity of the particle.

- acceleration   transverse component: $r\ddot{\theta} = r\dot{\omega}$

                                         radial component: $-r\dot{\theta}^2 = -r\omega^2 = -\dfrac{v^2}{r}$

         where $\ddot{\theta}$ or $\dot{\omega}$ is the angular acceleration of the particle.

12.2 By Newton's second law the forces acting on a particle of mass m in circular motion are equal to

- transverse component $mr\dot{\omega} = mr\ddot{\theta}$

- radial component: $-\dfrac{mv^2}{r} = -mr\omega^2$

- or radial component: $+\dfrac{mv^2}{r} = +mr\omega^2$ towards the centre.

12.3 Problems involving free motion in a vertical circle can be solved using the principle of conservation of energy

$$\text{P.E.} + \text{K.E.} = \text{constant}$$

in conjunction with Newton's second law in the radial direction.

12.4 Circular motion breaks down when the available force towards the centre is $< mr\omega^2$ or $\dfrac{mv^2}{r}$.

12.5 The equations for circular motion with constant angular acceleration α are

- $\omega = \omega_0 + \alpha t$
- $\omega^2 = \omega_0^2 + 2\alpha\theta$
- $\theta = \omega_0 t + \frac{1}{2}\alpha t^2$
- $\theta = \frac{1}{2}(\omega_0 + \omega)t$

where ω_0 is the initial angular speed.

13 Elasticity

The only way of finding the limits of the possible is by going beyond them into the impossible.

Arthur C. Clarke

The pictures show people taking part in the sport of bungee jumping.

This is a dangerous sport which originated in the South Sea islands where creepers were used rather than ropes. In the more modern version, a person jumps off a high bridge or crane to which he is attached by an elastic rope round his ankles.

QUESTION 13.1   If somebody bungee jumping from a bridge wants the excitement of just reaching the surface of the water below, how would you calculate the length of rope required?

The answer to this question clearly depends on the height of the bridge, the mass of the person jumping and the elasticity of the rope. All ropes are elastic to some extent, but it would be extremely dangerous to use an ordinary rope for this sport because the impulse necessary to stop somebody falling would involve a very large tension acting in the rope for a short time and this would provide too great a shock to the system. A bungee is a strong elastic rope, similar to those used to secure loads on cycles, cars or lorries, with the essential property for this sport that it allows the impulse to act over a much longer time so that the rope exerts a smaller force on the jumper.

13.1 Hooke's law

In 1678 Robert Hooke formulated a *Rule or law of nature in every springing body* which, for small extensions relative to the length of the string or spring, can be stated as follows:

> *The tension in an elastic spring or string is proportional to the extension. If a spring is compressed the thrust is proportional to the decrease in length of the spring.*

When a string or spring is described as elastic, it means that it is reasonable to apply the modelling assumption that it obeys Hooke's law. A further assumption, that it is light (i.e. has zero mass), is usual and is made in this book.

There are three ways in which Hooke's law is commonly expressed for a string. Which one you use depends on the extent to which you are interested in the string itself rather than just its overall properties. Denoting the natural length of the string by l_0 and its area of cross-section by A, the different forms are as follows.

- $T = \dfrac{EA}{l_0} x$ In this form E is called *Young's modulus* and is a property of the material out of which the string is formed. This form is commonly used in physics and engineering, subjects in which properties of materials are studied. It is rarely used in mathematics. The S.I. unit for Young's modulus is Nm^{-2}.

- $T = \dfrac{\lambda}{l_0} x$ The constant λ is called the *modulus of elasticity* of the string and will be the same for any string of a given cross-section made out of the same material. Many situations require knowledge of the natural length of a string and this form may well be the most appropriate in such cases. The S.I. unit for the modulus of elasticity is N.

- $T = kx$ In this simplest form, k is called the *stiffness* of the string. It is a property of the string as a whole. You may choose to use this form if neither the natural length nor the cross-sectional area of the string is relevant to the situation. The S.I. unit for stiffness is Nm^{-1}.

Notice that $k = \dfrac{\lambda}{l_0} = \dfrac{EA}{l_0}$

In this book only the forms using the modulus of elasticity and stiffness are used, and these can be applied to springs as well as strings.

EXAMPLE 13.1

A light elastic string of natural length 0.7 m and modulus of elasticity 50 N has one end fixed and a particle of mass 1.4 kg attached to the other. The system hangs vertically in equilibrium. Find the extension of the string.

SOLUTION

The forces acting on the particle are the tension, T N, upwards and the weight, $1.4g$ N, downwards.

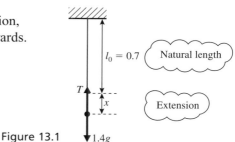

Figure 13.1

Since the particle is in equilibrium

$$T = 1.4g$$

Using Hooke's law:

$$T = \frac{\lambda}{l_0} x$$

$$\Rightarrow \quad 1.4g = \frac{50}{0.7} x$$

$$\Rightarrow \quad x = \frac{0.7 \times 1.4g}{50}$$

$$= 0.192$$

The extension in the string is 0.19 m (correct to 2 s.f.).

EXAMPLE 13.2   The mechanism of a set of kitchen scales consists of a light scale pan supported on a spring. When measuring 1.5 kg of flour, the spring is compressed by 7 mm. Find
i)   the stiffness of the spring
ii)   the mass of the heaviest object that can measured if it is impossible to compress the spring by more than 15 mm.

SOLUTION

i)   The forces on the scale pan with its load of flour are the weight, $1.5g$ N, downwards, and the thrust of the spring, T N, upwards.

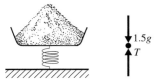

Figure 13.2

Since it is in equilibrium

$$T = 1.5g$$

Applying Hooke's law with stiffness k Nm^{-1}

$$T = k \times 0.007$$
$$\Rightarrow \quad 1.5g = 0.007k$$
$$\Rightarrow \quad k = 2100$$

The stiffness of the spring is 2100 Nm^{-1}.

ii)   Let the mass of the heaviest object be M kg, so the maximum thrust is Mg N. Then Hooke's law for a compression of 15 mm gives:

$$Mg = 2100 \times 0.015$$
$$\Rightarrow \quad M = 3.214 \ldots$$

The mass of the heaviest object that can be measured is 3.21 kg (to 3 s.f.).

Note

These scales would probably be calibrated to a maximum of 3 kg.

13.2 Using Hooke's law with more than one spring or string

Hooke's law allows you to investigate situations involving two or more springs or strings in various configurations.

EXAMPLE 13.3

A particle of mass 0.4 kg is attached to the mid-point of a light elastic string of natural length 1 m and modulus of elasticity λ N. The string is then stretched between a point A at the top of a doorway and a point B which is on the floor 2 m vertically below A.
i)   Find, in terms of λ, the extensions of the two parts of the string.
ii) Calculate their values in the case where $\lambda = 9.8$.
iii) Find the minimum value of λ which will ensure that the lower half of the string is not slack.

SOLUTION

For a question like this it is helpful to draw two diagrams, one showing the relevant natural lengths and extensions, and the other showing the forces acting on the particle.

Since the force of gravity acts downwards on the particle, its equilibrium position will be below the mid-point of AB. This is also shown in the diagram.

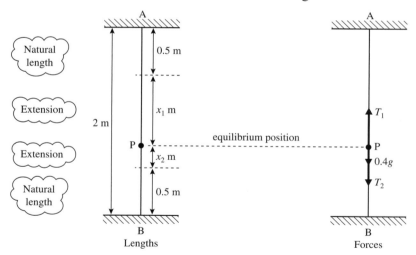

Figure 13.3

i)   The particle is in equilibrium, so the resultant vertical force acting on it is zero.

Therefore $T_1 = T_2 + 0.4g.$ ①

Hooke's law can be applied to each part of the string.

For AP: $T_1 = \dfrac{\lambda}{0.5} x_1$ ②

For BP: $T_2 = \dfrac{\lambda}{0.5} x_2.$ ③

Substituting these expressions in equation ① gives:

$$\frac{\lambda}{0.5}x_1 = \frac{\lambda}{0.5}x_2 + 0.4g$$

$$\Rightarrow \quad \lambda x_1 - \lambda x_2 = 0.5 \times 0.4g$$

$$\Rightarrow \quad x_1 - x_2 = 0.2\frac{g}{\lambda}. \qquad \text{④}$$

> Alternatively, you can use x for x_1 and $(1-x)$ for x_2.

But from the first diagram it can be seen that

$$x_1 + x_2 = 1. \qquad \text{⑤}$$

Adding ④ and ⑤ gives:

$$2x_1 = 1 + 0.2\frac{g}{\lambda} \quad \Rightarrow \quad x_1 = 0.5 + 0.1\frac{g}{\lambda}$$

Similarly, subtracting ④ from ⑤ gives:

$$x_2 = 0.5 - 0.1\frac{g}{\lambda}. \qquad \text{⑥}$$

ii) Since $\lambda = 9.8$ the extensions are 0.6 m and 0.4 m.

iii) The lower part of the string will not become slack providing $x_2 > 0$. It follows from equation ⑥ that:

$$0.5 - 0.1\frac{g}{\lambda} > 0$$

$$\Rightarrow \quad 0.5 > 0.1\frac{g}{\lambda} \quad \Rightarrow \quad \lambda > 0.2g.$$

The minimum value of λ for which the lower part of the string is not slack is 1.96 N, and in this case BP has zero tension.

Historical note

If you search for Robert Hooke (1635–1703) on the internet, you will find that he was a man of many parts. He was one of a talented group of polymaths (which included his rival Newton) who have had an enormous impact on scientific thought and practice. Among other things, he designed and built Robert Boyle's air pump, discovered the red spot on Jupiter and invented the balanced spring mechanism for watches. His work on microscopy led to his becoming the father of microbiology and he was the first to use the term 'cell' with respect to living things. Hooke worked closely with his friend Sir Christopher Wren in the rebuilding of the City of London after the great fire, and was responsible for the realisation of many of his designs including the Royal Greenwich Observatory. Both Hooke and Wren were astronomers and architects and they designed the Monument to the fire with a trapdoor at the top and a laboratory in the basement so that it could be used as an enormous 62 m telescope. Hooke, the great practical man, also used the column for experiments on air pressure and pendulums.

13.3 Work and energy

In order to stretch an elastic spring a force must do work on it. In the case of the muscle exerciser, this force is provided by the muscles working against the tension. When the exerciser is pulled at constant speed, at any given time the force F applied at each end is equal to the tension in the spring; consequently it changes as the spring stretches.

Figure 13.4

Suppose that one end of the spring is stationary and the extension is x as in figure 13.5. By Hooke's law the tension is given by

$$T = kx,$$

and so $$F = kx.$$

The work done by a *constant force F* in moving a distance d in its own direction is given by Fd. To find the work done by a *variable* force the process has to be considered in small stages.

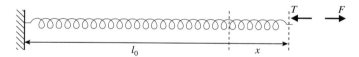

Figure 13.5

Now imagine that the force extends the string a small distance δx. The work done is given by

$$F\delta x = kx\,\delta x.$$

The total work done in stretching the spring many small distances is

$$\sum F\delta x = \sum kx\delta x.$$

In the limit as $\delta x \to 0$, the work done is:

$$\int F\,dx = \int kx\,dx$$
$$= \tfrac{1}{2}kx^2 + c.$$

When the extension $x = 0$, the work done is zero, so $c = 0$.

The total work done in stretching the spring an extension x *from its natural length l_0* is therefore given by:

$$\tfrac{1}{2}kx^2 \quad \text{or} \quad \tfrac{1}{2}\frac{\lambda}{l_0}x^2.$$

The result is the same for the work done in compressing a spring.

Elastic potential energy

The tensions and thrusts in perfectly elastic springs and strings are conservative forces, since any work done against them can be recovered in the form of kinetic energy. A catapult and a jack-in-a-box use this property.

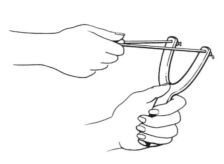

Figure 13.6

The work done in stretching or compressing a string or spring can therefore be regarded as potential energy. It is known as *elastic potential energy*.

The elastic potential energy stored in a spring which is stretched or compressed by an amount x is

$$\tfrac{1}{2}kx^2 \quad \text{or} \quad \tfrac{1}{2}\frac{\lambda}{l_0}x^2.$$

EXAMPLE 13.4

An elastic rope of natural length 0.6 m is extended to a length of 0.8 m. The modulus of elasticity of the rope is 25 N. Find
i)   the elastic potential energy stored in the rope
ii) the further energy required to stretch it to a length of 1.65 m over a car roof-rack.

SOLUTION

i)   The extension of the elastic is $(0.8 - 0.6)\,\text{m} = 0.2\,\text{m}$.

The energy stored in the rope is $\tfrac{1}{2}\dfrac{\lambda}{l_0}x^2$

$$= \frac{25}{2 \times 0.6}(0.2)^2$$

$$= 0.83\text{J (to 2 d.p.)}.$$

ii) The extension of the elastic rope is now $1.65 - 0.6 = 1.05$ m.

The elastic energy stored in the rope is $\dfrac{25}{2 \times 0.6}(1.05)^2$

$$= 22.97 \text{ J}.$$

The extra energy required to stretch the rope is 22.14 J (correct to 2 d.p.).

 In the example above, the string is stretched so that its extension changes from x_1 to x_2 (in this case, from 0.2 m to 1.05 m). The work required to do this is

$$\tfrac{1}{2}kx_2{}^2 - \tfrac{1}{2}kx_1{}^2 = \tfrac{1}{2}k(x_2{}^2 - x_1{}^2) \quad \text{or} \quad \tfrac{1}{2}\dfrac{\lambda}{l_0}(x_2{}^2 - x_1{}^2).$$

You can see by using algebra that this expression is not the same as $\tfrac{1}{2}k(x_2 - x_1)^2$, so it is *not* possible to use the extra extension $(x_2 - x_1)$ directly in the energy expression to calculate the extra energy stored in the string.

EXAMPLE 13.5

A catapult has prongs which are 16 cm apart and the elastic string is 20 cm long. A marble of mass 70 g is placed in the centre of the elastic string and pulled back so that the string is just taut. The marble is then pulled back a further 9 cm and the force required to keep it in this position is 60 N. Find

i)   the stretched length of the string
ii) the tension in the string and its stiffness
iii) the elastic potential energy stored in the string and the speed of the marble when the string regains its natural length, assuming they remain in contact.

SOLUTION

To solve this problem it is necessary to assume that there is no elasticity in the frame of the catapult, and that the motion takes place in a horizontal plane. In addition, any air resistance is ignored.

In figure 13.7, A and B are the ends of the elastic string and M_1 and M_2 are the two positions of the marble (before and after the string is stretched). D is the mid-point of AB.

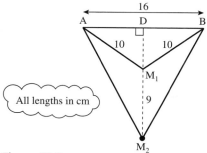

Figure 13.7

i)   Using Pythagoras' theorem in triangle DBM_1 gives

$$DM_1 = \sqrt{10^2 - 8^2} = 6 \text{ cm}.$$

So

$$DM_2 = 9 + 6 = 15 \text{ cm}.$$

Using Pythagoras' theorem in triangle DBM_2 gives

$$BM_2 = \sqrt{15^2 + 8^2} = 17 \text{ cm}.$$

The stretched length of the string is 2×17 cm $= 0.34$ m.

ii) Take the tension in the string to be T N.

Resolving parallel to M_2D

$$2T \cos \alpha = 60.$$

Now $$\cos \alpha = \frac{DM_2}{BM_2} = \frac{0.15}{0.17}.$$

so $$T = 34.$$

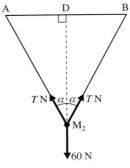

Figure 13.8

The extension of the string is $(0.34 - 0.2)$ m $= 0.14$ m.

By Hooke's law the stiffness k is given by $kx = T$;

$$k = \frac{34}{0.14} = 242.9.$$

The stiffness of the string is 240 Nm^{-1} (to 2 s.f.).

iii) The elastic potential energy stored in the string is

$$\tfrac{1}{2}kx^2 = \tfrac{1}{2} \times 242.9 \times (0.14)^2 = 2.38 \text{ J}.$$

By the principle of conservation of energy, this is equal to the kinetic energy given to the marble. The mass of the marble is 0.07 kg, so

$$\tfrac{1}{2} \times 0.07v^2 = 2.38$$
$$\Rightarrow \qquad v = 8.25 \ldots$$

The speed of the marble is 8.3 ms^{-1}.

13.4 Vertical motion

This chapter began with a bungee jumper undergoing vertical motion at the end of an elastic rope. The next example involves a particle in vertical motion at the end of a spring.

EXAMPLE 13.6   A particle of mass 0.2 kg is attached to the end A of a perfectly elastic spring OA which has natural length 0.5 m and stiffness 40 Nm^{-1}. The spring is suspended from O and the particle is pulled down and released from rest when the length of the spring is 0.7 m. In the subsequent motion the extension of the spring is denoted by x m.
  i)   Write down expressions for the increase in the particle's gravitational potential energy and the decrease in the energy stored in the spring when the extension is x m.
  ii)   Hence find an expression for the speed of the particle in terms of x.
  iii)   Calculate the length of the spring when the particle is at its highest point.

SOLUTION

i)

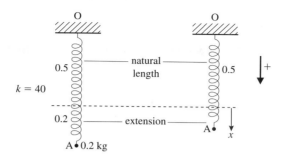

Figure 13.9

The particle has risen a distance $(0.2 - x)$ m.

$$\text{Increase in gravitational P.E.} = mgh$$
$$= 1.96 \times (0.2 - x)$$
$$\text{Stored energy} = \tfrac{1}{2}kx^2$$
$$= 20x^2$$
$$\text{Initial stored energy} = 20 \times 0.2^2$$
$$\text{Decrease in stored energy} = 20\,(0.2^2 - x^2)$$

ii) The initial K.E. is zero.

$$\text{Increase in K.E.} = \tfrac{1}{2} \times 0.2 \times \dot{x}^2 - 0$$
$$= 0.1\dot{x}^2$$

> $\dot{x}$ is the same as v, the speed in the positive direction

Using the law of conservation of mechanical energy,

Increase in K.E. + P.E. = Decrease in stored energy

$$0.1\dot{x}^2 + 1.96\,(0.2 - x) = 20\,(0.2^2 - x^2)$$

and so
$$v = \dot{x} = \sqrt{200\,(0.2^2 - x^2) - 19.6\,(0.2 - x)}.$$

iii) At the highest point, $v = \dot{x} = 0$,

$$200\,(0.2^2 - x^2) - 19.6\,(0.2 - x) = 0$$
$$(0.2 - x)\,[200\,(0.2 + x) - 19.6] = 0$$
$$\Rightarrow \quad x = 0.2 \text{ or } 40 + 200x = 19.6$$
$$\Rightarrow \quad x = 0.2 \text{ or } x = -0.102$$
$$\text{but} \quad x = 0.2 \text{ at the lowest position}$$
$$\text{so} \quad x = -0.102 \text{ at the highest point.}$$

This negative value of x indicates a compression rather than an extension, so at its highest point the spring has length $(0.5 - 0.102)$ m $= 0.398$ m.

Note

For this question it is important that you are dealing with a spring, which still obeys Hooke's law when it contracts, rather than a string which becomes slack.

Chapter 13 Exercises

13.1 In this question take the value of g to be $10\,\text{ms}^{-2}$. The diagram shows a spring of natural length 60 cm which is being compressed under the weight of a block of mass m kg. Smooth supports constrain the block to move only in the vertical direction.

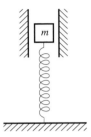

The modulus of elasticity of the spring is 180 N. The system is in equilibrium and the length of the spring is 50 cm. Find
i)   the thrust in the spring
ii)   the value of m
iii) the stiffness of the spring.

More blocks are piled on.
iv) Describe the situation when there are seven blocks in total, all identical to the first one.

13.2 The manufacturer of a sports car specifies the coil spring for the front suspension as a spring of 10 coils with a natural length 0.3 m and a compression 0.1 m when under a load of 4000 N.
i)   Calculate the modulus of elasticity of the spring.
ii)   If the spring were cut into two equal parts, what would be the stiffness of each part?

The weight of a car is 8000 N and half of this weight is taken by two such 10-coil front springs so that each bears a load of 2000 N.
iii) Find the compression of each spring.
iv) Two people each of weight 800 N get into the front of the car. How much further are the springs compressed? (Assume that their weight is carried equally by the front springs.)

13.3 The coach of an impoverished rugby club decides to construct a scrummaging machine as illustrated in the diagrams below. It is to consist of a vertical board, supported in horizontal runners at the top and bottom of each end. The board is held away from the wall by springs, as shown, and the players push the board with their shoulders, against the thrust of the springs.

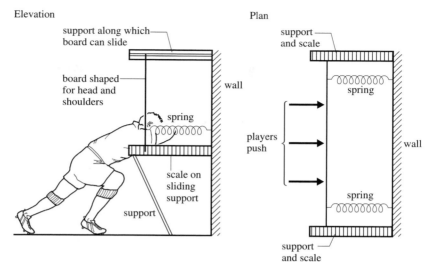

The coach has one spring of length 1.4 m and stiffness 5000 Nm⁻¹, which he cuts into two pieces of equal length.

i) Find the modulus of elasticity of the original spring.
ii) Find the modulus of elasticity and the stiffness of each of the half-length springs.
iii) On one occasion the coach observes that the players compress the springs by 20 cm. What total force do they produce in the forward direction?

13.4 The diagram shows two light springs, AP and BP, connected at P. The ends A and B are secured firmly and the system is in equilibrium.

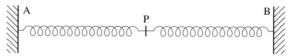

The spring AP has natural length 1 m and modulus of elasticity 16 N.
The spring BP has natural length 1.2 m and modulus of elasticity 30 N.
The distance AB is 2.5 m and the extension of the spring AP is x m.

i) Write down an expression, in terms of x, for the extension of the spring BP.
ii) Find expressions, in terms of x, for the tensions in both springs.
iii) Find the value of x.

13.5 In this question take g to be $10\ \text{ms}^{-2}$. A baby bouncer consists of a light, inextensible harness attached to a spring. It is suspended in a doorway on the end of a chain so that the baby's feet just touch the floor. The baby can then use its feet to bounce up and down. One such harness, with a spring of natural length 20 cm, is set up for Emily (who has mass 10 kg). Before Emily is put into the harness (i.e. when the spring is not extended) the bottom of the harness is 17 cm from the ground. When she sits in the harness with her feet off the ground, the bottom of the harness is 15 cm above the ground.

i) Find the modulus of elasticity of the spring.

The baby bouncer is also used by Charlotte and its height is not adjusted for her. Charlotte's mass is 12 kg (the limit recommended for the bouncer).

ii) What is the extension of the spring when Charlotte sits in the bouncer with her feet off the ground?

iii) Charlotte sits in equilibrium with her feet on the ground. Find an expression for the reaction, R N, between her feet and the ground in terms of the height, h cm, of the harness above the ground.

13.6 Many cafeterias have a device for stacking plates where, for any number of plates in the stack, the top of the top plate is always at the counter level, as shown in the diagram on the left. This is achieved by standing the plates on a metal base which is supported by a number of springs which hang vertically as shown in the diagram on the right. The base is shaped so that when there are no plates on the stack, the top of the base is level with the counter.

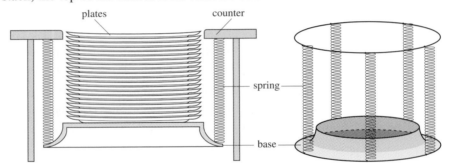

The situation may be modelled by assuming that each plate has a mass of 0.1 kg and raises the vertical height of the stack by 1.5 cm when placed on another plate (or the base). The springs have a natural length of 10 cm and a modulus of 0.8 N. The springs may be assumed to be light and resistances to motion due to friction may be neglected.

i) Suppose that n springs are attached.

a) Show that the extension in each spring when a mass m kg is supported by the system is given by $\dfrac{mg}{8n}$ m.

b) What is the further extension of each spring when one plate is added to the stack?

ii) Show that nine springs are required if the addition of each plate allows the top of that plate to be as near to the counter level as possible but above, not below, the counter level.

iii) Assuming that nine springs are used, approximately how many plates would be above the counter level if there were 40 plates in the stack?

The system can be adjusted so that the top of the plate is at counter level by reducing the natural length of the springs.

iv) When eight springs are used, what should be the natural length of each spring, correct to three decimal places?

13.7   A pinball machine fires small balls of
mass 50 g by means of a spring and a
light plunger. The spring and the ball
move in a horizontal plane.

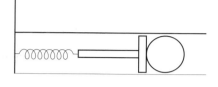

The spring has stiffness 600 Nm^{-1} and is
compressed by 5 cm to fire a ball.
   i)   Find the energy stored in the spring immediately before the ball is fired.
   ii)   Find the speed of the ball when it is fired.

13.8   A simple mathematical model of a railway buffer consists of a horizontal open
coiled spring attached to a fixed point. The stiffness of the spring is 10^5 Nm^{-1}
and its natural length is 2 m.

The buffer is designed to stop a railway truck before the spring is compressed to
half its natural length, otherwise the truck will be damaged.

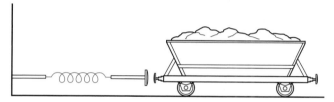

   i)   Find the elastic energy stored in the spring when it is half its natural length.
   ii)   Find the maximum speed at which a truck of mass 2 tonnes can approach the
buffer safely. Neglect any other reasons for loss of energy of the truck.

A truck of mass 2 tonnes approaches the buffer at 5 ms^{-1}.

   iii)   Calculate the minimum length of the spring during the subsequent period of
contact.
   iv)   Find the thrust in the spring and the acceleration of the truck when the spring
is at its minimum length.
   v)   What happens next?

13.9   Two light springs are joined and stretched between two fixed points A and C
which are 2 m apart as shown in the diagram. The spring AB has natural length
0.5 m and modulus of elasticity 10 N. The spring BC has natural length 0.6 m and
modulus of elasticity 6 N. The system is in equilibrium.

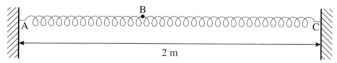

   i)   Explain why the tensions in the two springs are the same.
   ii)   Find the distance AB and the tension in each spring.
   iii)   How much work must be done to stretch the springs from their natural length
to connect them as described above?

A small object of mass 0.012 kg is attached at B and is supported on a smooth
horizontal table. A, B and C lie in a straight horizontal line and the mass is
released from rest at the mid-point of AC.

   iv)   What is the speed of the mass when it passes through the equilibrium
position of the system?

13.10 The end A of a light elastic string AB is fixed to a smooth horizontal table. A small bead of mass 0.05 kg is attached to the end B. Also attached to the end B is another light elastic string BC. The end C is fixed to a point on the table so that the strings lie in a horizontal straight line with the bead resting in equilibrium on the table and with AC equal to 2.3 metres, as shown in the diagram.

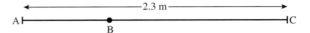

The string AB has natural length 0.4 m and modulus of elasticity 4 N. The string BC has natural length 0.5 m and modulus of elasticity 2 N.
i)   The extension in AB is x_1 m and the extension in BC is x_2 m. Show that $5x_1 = 2x_2$ and calculate x_1 and x_2.

The bead is held at A and released.
ii) Determine whether the string BC goes slack.
iii) Calculate the maximum extension of the string AB.

13.11 A small apple of mass 0.1 kg is attached to one end of an elastic string of natural length 25 cm and modulus of elasticity 5 N. David is asleep under a tree and Sam fixes the free end of the string to the branch of the tree just above David's head. Sam releases the apple level with the branch and it just touches David's head in the subsequent motion. How high above his head is the branch?

13.12 A bungee jump is carried out by a person of mass m kg using an elastic rope which can be taken to obey Hooke's law. It is known that the jump operator does not exceed the total length limit of four times the original length of the rope in any jump. Prove that the tension in the rope is at most $\frac{8}{3}mg$ N.

13.13 A man unloading 25 kg sacks drops each one 2 m, from rest, on to a light horizontal sprung board which breaks the fall. The spring supporting the board is initially 50 cm long and compresses down to a minimum length of 15 cm under the impact of each falling sack which is then removed by another man.

Assuming that there is no energy loss on impact,
i)   show that the potential energy lost by the sack when it reaches its lowest point is about 576 J
ii) show that the stiffness of the spring is 9400 Nm^{-1}
iii) find the maximum deceleration experienced by the falling sack.

13.14 A child suspends a small model aeroplane from the ceiling by means of a light elastic string of modulus 5 N and natural length 0.8 m. The aeroplane is packed with some modelling clay and the total mass is 0.1 kg.
i)   Calculate the length of the string when the aeroplane hangs in equilibrium.
ii) The aeroplane is pulled vertically down until the string is 1.2 m long and then released from rest. Calculate the minimum distance between the aeroplane and the ceiling in the resulting motion.

The child wishes the aeroplane to hang 0.9 m below the ceiling when in equilibrium. The child considers two possibilities. The first possibility is to remove some of the clay.
iii) What mass of clay would need to be removed?

The other possibility is not to remove any clay, but instead to shorten the string to a new natural length.

iv) What would the new natural length need to be?

The child again holds the aeroplane 1.2 m below the ceiling and releases it from rest.

v) Explain, without doing any calculations, whether the aeroplane will get closer to the ceiling than in part **ii)**

    a) if clay has been removed as in part **iii)**

    b) if the string has been shortened as in part **iv)**.

Personal Tutor

 13.1 A light elastic string AB has stiffness k and natural length l_0.

The end A of the string is fixed to a vertical wall, and the end B to a block of mass M which is supported by a smooth horizontal plane. To the other side of the block is attached an inelastic light string which passes over a smooth pulley and is then tied to a particle of mass m at C which hangs freely at all times. The block is initially held against the wall and then is moved gently towards the pulley until it is in equilibrium. The elastic string and the part of the other string between the block and the pulley are horizontal, as shown in the diagram below.

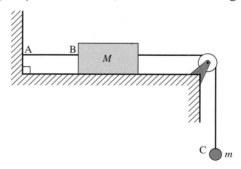

Find expressions for the following in terms of some or all of m, g, k and l_0.

i) The extension in the elastic string.

ii) The potential energy lost by the hanging mass as the system is moved to its equilibrium position.

iii) The energy stored in the elastic string.

iv) Compare your answers to parts **ii)** and **iii)**. Why is energy not conserved?

v) If the string supporting the hanging mass breaks when the system is in equilibrium, show that the block will hit the wall with a speed v given by $v^2 = \dfrac{m^2 g^2}{Mk}$. Find also an expression for the acceleration of the block immediately after the string breaks.

13.1 Hooke's law

The tension T in an elastic string or spring and its extension x are related by:

$$T = kx \quad \text{or} \quad T = \frac{\lambda}{l_0} x$$

where k is the stiffness, λ is the modulus of elasticity and l_0 is the natural length of the string or spring.

13.2 When a spring is compressed, x is negative and the tension becomes a thrust.

13.3 Elastic potential energy

The elastic potential energy stored in a stretched spring or string, or in a compressed spring, is given by E.P.E. where:

$$\text{E.P.E.} = \tfrac{1}{2}kx^2 \quad \text{or} \quad \text{E.P.E.} = \tfrac{1}{2}\frac{\lambda}{l_0}x^2.$$

This is the work done in stretching a spring or string or compressing a spring starting at its natural length.

13.4 The tension or thrust in an elastic string or spring is a conservative force and so the elastic potential energy is recoverable.

13.5 When no frictional or other dissipative forces are involved, elastic potential energy can be used with kinetic energy and gravitational potential energy to form equations using the principle of conservation of energy.

14 Simple harmonic motion

Backwards and forwards half her length
With a short uneasy motion.

Samuel Coleridge Taylor
The Rime of the Ancient Mariner

The guitar, the clock and water ripples all involve *oscillations* or *vibrations* of particles or bodies.

- In the guitar, the strings vibrate in a controllable way, and the instrument transmits them as sound waves (vibrations of air molecules).
- In the clock, the pendulum oscillates in the familiar swinging pattern, and this regular motion is used to operate the clock mechanism.
- Ripples are created when the surface of water (or another liquid) is disturbed. The fluid particles vibrate up and down in a regular wave pattern. The same pattern is visible in ocean waves, or in the wakes of boats.

The remarkable thing about all of these vibrations, and many others that occur in natural and man-made systems, is that they are essentially of the same form. The vibrations don't go on for ever, but over a reasonable interval, you can plot the displacement of a vibrating particle against time for any of these systems and you will obtain a sine wave.

14.1 Oscillating motion

The graph in figure 14.1 shows the displacement of an oscillating particle against time.

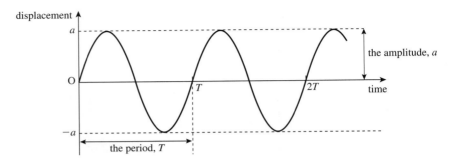

Figure 14.1

From the graph you can see a number of important features of such motion.

● The particle oscillates about a *central position*, O.
● The particle moves between two points with displacements $-a$ and $+a$. The distance a is called the *amplitude* of the motion.
● The motion repeats itself in a cyclic fashion. The number of cycles per second is called the *frequency*, and is usually denoted by ν, the Greek letter 'nu'.
● The motion repeats itself after a time T. The time interval T is called the *period*: it is the time for one complete cycle of the motion.

The frequency and period are reciprocals. For example a period of $\frac{1}{10}$ of a second corresponds to a frequency of 10 cycles per second.

Usually the *period* is used when describing relatively slow mechanical oscillations and the *frequency* for faster oscillations or vibrations.

The S.I. unit for frequency is the hertz (Hz). One hertz is one cycle per second. The unit is named after Heinrich Rudolph Hertz (1857–94) who was the first person to produce electromagnetic waves artificially. You may be familiar with the use of the *megahertz* (MHz) as a measure of the frequency of radio waves. A radio station which is broadcasting on 99.1 megahertz is producing electromagnetic waves with a frequency of 99 100 000 cycles per second.

Historical note

Pythagoras (572 to 497 BC) studied the pitch of notes produced by stretched strings in the first recorded acoustic experiments. He showed that recognisable musical intervals are produced by segments of a stretched string when their lengths are in a simple numerical ratio.

In May 1939, shortly before the outbreak of the second world war, an international conference in London unanimously agreed to adopt a frequency of '440 cycles per second for the note 'A' in the treble clef'. There are then standard relationships between this and the frequencies of other notes. For many years, before electronic music making devices became readily available, an 'A' of this frequency was broadcast regularly on the radio for the use of musicians.

A physical model of an oscillating system

A particle, P, of mass m kg is attached to a light, perfectly elastic spring which has stiffness k Nm^{-1}. The other end, E, of the spring is attached to a fixed point and P moves horizontally.

QUESTION 14.1   How can this be done so that the only horizontal force is the force in the spring?

The spring is its natural length when P is at the fixed point O. The particle is pulled aside to a point A where OA $= a$ m and let go.

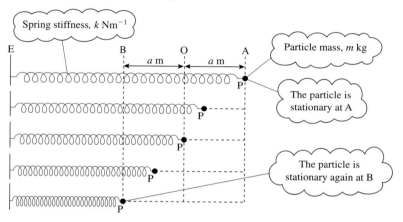

Figure 14.2

QUESTION 14.2   What happens to P as it moves from A to B? Then what happens?

QUESTION 14.3   If you could fix a pen at P and let it draw on a strip of paper moving at a steady rate perpendicular to EA (up the page), what would be the result?

Now think about energy. At A, the spring has elastic energy $\frac{1}{2}ka^2$ and the particle has zero kinetic energy. The particle first comes to rest again at B so the elastic energy at B must also be $\frac{1}{2}ka^2$ and the length OB am. This is the amplitude of the motion.

When the extension of the spring is x m and P has speed v ms^{-1}, the total energy is $\frac{1}{2}mv^2 + \frac{1}{2}kx^2$ so the principle of conservation of energy gives

$$\frac{1}{2}mv^2 + \frac{1}{2}kx^2 = \frac{1}{2}ka^2$$

$$\Rightarrow \quad v^2 = \frac{k}{m}(a^2 - x^2).$$

This gives the speed of the particle in any position, even when x is negative, providing Hooke's law still holds.

QUESTION 14.4   What is its speed at O? What is its maximum speed?

A mathematical model for simple harmonic motion

At the instant the particle, P, is x m to the right of O the forces acting on it are as shown in figure 14.3.

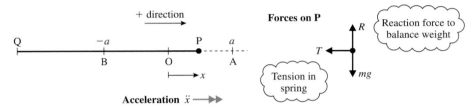

Figure 14.3

Its horizontal equation of motion is $-T = m\ddot{x}$

By Hooke's law, $T = kx$ so $-kx = m\ddot{x}$

$$\ddot{x} = -\frac{k}{m}x \qquad \qquad \textcircled{1}$$

or

$$\frac{d^2x}{dt^2} = -\frac{k}{m}x$$

QUESTION 14.5   What is the effect of the minus sign in this equation?

In equation $\textcircled{1}$, k and m are both positive constants so $\dfrac{k}{m}$ can be written as one positive constant. By convention this is denoted by ω^2 so the equation may be written as

> Writing it as a square ensures that it is always positive.

$$\ddot{x} = -\omega^2 x \qquad \qquad \textcircled{2}$$

The equation for the velocity that you found in the last section can then be written as

$$v^2 = \omega^2(a^2 - x^2) \quad \text{or} \quad \dot{x}^2 = \omega^2(a^2 - x^2). \qquad \qquad \textcircled{3}$$

QUESTION 14.6   What is v when $x = a$, $-a$ and 0?

There are many other similar systems which produce an equation of the same form as equation $\textcircled{2}$. Because the same equation of motion applies to all of them they all have the same type of oscillating motion called simple harmonic motion (SHM).

Simple harmonic motion is defined by the equation $\ddot{x} = -\omega^2 x$. Remember that in this equation, $\ddot{x}$ means acceleration and x means the displacement from the centre. It may be stated in words as

> The acceleration is proportional to the magnitude of the displacement from the centre point of the motion and is directed towards this centre point.

The two equations $v^2 = \omega^2(a^2 - x^2)$ and $\ddot{x} = -\omega^2 x$ tell you a lot about the motion.

From the first equation you can see that

- the motion is symmetrical about the point O where $x = 0$
- x must always lie between $-a$ and a, otherwise v^2 would be negative
- at the extreme points, when $x = \pm a$, the velocity is zero
- the maximum speed is when $x = 0$ and $v = \pm a\omega$.

> This is when the particle passes through O

The second equation tells you that

- $\ddot{x}$ is always directed towards O, so the same must be true of the resultant force
- $\ddot{x}$ and hence the force are zero when $x = 0$, so O is the equilibrium position
- $\ddot{x}$ has a maximum magnitude of $\omega^2 a$ when $x = \pm a$.

These results for $\ddot{x}$ and v in terms of x are shown in the diagram.

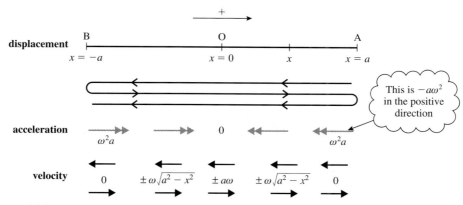

Figure 14.4

The frequency of the motion depends on the value of ω. In the next section on page 310 you will see how to write x, v and $\ddot{x}$ in terms of time. In the meantime you can use the result

$$\omega = 2\pi \times \text{frequency}.$$

EXAMPLE 14.1

When a violin string is playing an 'A' with frequency 880 Hz, a particle on the string oscillates with amplitude of 0.5 mm. Calculate

i)   the time taken for one complete oscillation
ii) ω in terms of π and write down the equation of motion
iii) the maximum speed of the particle and its maximum acceleration
iv) the acceleration and velocity when the particle is 0.25 mm from its central position.

SOLUTION

i) The particle does 880 oscillations per second, so the time for one oscillation is $\frac{1}{880} = 0.001\,13$ s. This is the period of the motion.

ii) $$\omega = 2\pi \times \text{frequency} = 1760\pi$$

The equation of motion is $\quad \ddot{x} = -\omega^2 x$

$$\Rightarrow \quad \ddot{x} = -(1760\pi)^2 x.$$

iii) You can use the equations with mm so long as your units are consistent.

Maximum speed $\qquad a\omega = 0.5 \times 1760\pi \ (\text{mm s}^{-1})$
$$= 2.76 \text{ ms}^{-1} \ (3 \text{ s.f.})$$

Maximum acceleration $\quad a\omega^2 = 0.5 \times (1760\pi)^2 \text{ mm s}^{-1})$
$$= 15\,300 \text{ ms}^{-2} \ (3 \text{ s.f.})$$

iv) When $x = 0.25$, $\qquad \ddot{x} = -(1760\pi)^2 \times 0.25 \text{ mm s}^{-2})$

$$\Rightarrow \quad \text{acceleration} = -7640 \text{ ms}^{-2} \ (3 \text{ s.f.}).$$

The velocity is given by $\quad v^2 = \omega^2(0.5^2 - x^2)$

$$\Rightarrow \quad v = \pm 1760\pi \sqrt{0.25 - 0.0625} \ (\text{mm s}^{-1})$$

$$\Rightarrow \quad \text{velocity} = \pm 2.39 \text{ ms}^{-1}.$$

> The particle can be travelling in either direction

When $x = -0.25$ $\qquad \ddot{x} = +(1760\pi)^2 \times 0.25 \ (\text{mm s}^{-2})$

$$\Rightarrow \quad \text{acceleration} = +7640 \text{ ms}^{-2}$$

The velocity is $\pm 2.39 \text{ ms}^{-1}$ as before.

14.2 Solving the differential equation for SHM

Simple harmonic motion has been defined by the differential equation

$$\ddot{x} = -\omega^2 x.$$

The work which follows shows you how this equation is solved to obtain the general solution in the form $x = a \sin(\omega t + \varepsilon)$.

The acceleration, $\ddot{x} = \dfrac{dv}{dt}$, can be written by using the chain rule as

$$\frac{dv}{dx} \times \frac{dx}{dt} = v\frac{dv}{dx}. \qquad \qquad ①$$

> This is a useful result when acceleration is given in terms of x (or v).

So the SHM equation becomes

$$v\frac{dv}{dx} = -\omega^2 x.$$

Separating the variables gives:

$$\int v \, dv = \int -\omega^2 x \, dx$$

$$\Rightarrow \quad \frac{v^2}{2} = -\frac{\omega^2 x^2}{2} + c.$$

Let the amplitude be a, so that $v = 0$ when $x = a$.

Then $\qquad 0 = -\dfrac{\omega^2 a^2}{2} + c \quad \Rightarrow \quad c = +\dfrac{\omega^2 a^2}{2}$

Therefore $\qquad\qquad\qquad v^2 = \omega^2 \, (a^2 - x^2).$

It follows from this that: $\qquad v = \pm\omega\sqrt{a^2 - x^2} \quad \Rightarrow \quad \dfrac{dx}{dt} = \pm\omega\sqrt{a^2 - x^2}.$

Separating the variables gives:

$$\int \frac{dx}{\pm\sqrt{a^2 - x^2}} = \int \omega \, dt = \omega t + k.$$

For the positive square root in the integral on the left-hand side use the substitution $x = a \sin u$, $dx = a \cos u \, du$.

Then $\qquad\qquad \displaystyle\int \frac{dx}{\sqrt{a^2 - x^2}} = \int \frac{a \cos u \, du}{a \cos u}$

$$= u = \arcsin\left(\frac{x}{a}\right).$$

In this case:

$$\arcsin\left(\frac{x}{a}\right) = \omega t + \varepsilon \text{ where } \varepsilon \text{ is the constant of integration, } k$$

$$\Rightarrow \quad x = a \sin(\omega t + \varepsilon) \qquad\qquad\qquad ②$$

For the negative root use $x = a \cos u$ to give $\arccos\left(\dfrac{x}{a}\right)$ for the left-hand integral.

In this case:

$$\arccos\left(\frac{x}{a}\right) = \omega t + \varepsilon_1 \text{ (now } k = \varepsilon_1)$$

$$\Rightarrow \quad x = a \cos(\omega t + \varepsilon_1) \qquad\qquad\qquad ③$$

Note

The two forms ② and ③ are equivalent.

Simple harmonic motion as a function of time

You have seen how to integrate the SHM equation so that you can write x, $\dot{x}$, and $\ddot{x}$ in terms of the time, but you can also demonstrate how the SHM equation can be satisfied by certain trigonometrical functions as in the next example.

⚠ Remember that all the angles must be in radians when you use calculus.

EXAMPLE 14.2

i) Show that $x = 10 \sin 3t$ satisfies the simple harmonic motion equation
$$\ddot{x} = -9x.$$

ii) Sketch the graph of $x = 10 \sin 3t$ and deduce the amplitude, period and frequency of this motion.

iii) Verify that $\dot{x}^2 = v^2 = \omega^2(a^2 - x^2)$

iv) Show that $x = 25 \sin 3t$ also satisfies $\ddot{x} = -9x$ and comment on this result.

SOLUTION

i) $x = 10 \sin 3t$

Differentiating with respect to t to find the velocity and acceleration:
$$v = \dot{x} = 30 \cos 3t \qquad \qquad ①$$
$$\text{and} \quad \ddot{x} = -90 \sin 3t$$
$$\Rightarrow \quad \ddot{x} = -9x \qquad \qquad ②$$

This is the SHM equation given in the question.

ii) The graph of the function is shown below.

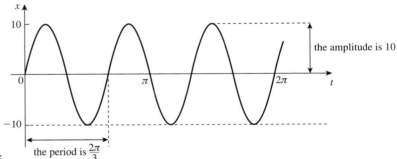

Figure 14.5

the amplitude is 10

the period is $\frac{2\pi}{3}$

The graph shows that the value of x always lies between -10 and 10, so the amplitude of the motion is 10 units.

Each cycle repeats in the time $\dfrac{2\pi}{3}$ so the period of the oscillations, denoted by T, is $\dfrac{2\pi}{3}$. The frequency, f, is $\dfrac{1}{T}$ or $\dfrac{3}{2\pi}$.

iii) From ①
$$\dot{x}^2 = 900 \cos^2 3t$$
$$= 900(1 - \sin^2 3t)$$
$$= 9(100 - 100 \sin^2 3t)$$
$$= \omega^2(a^2 - x^2).$$

$\omega^2 = 9$ from equation ②

iv) Differentiating the function $x = 25 \sin 3t$ twice gives:
$$\dot{x} = 75 \cos 3t$$
$$\ddot{x} = -225 \sin 3t$$
$$= -9(25 \sin 3t)$$
$$\Rightarrow \quad \ddot{x} = -9x \text{ as required}$$

This shows that the amplitude of the oscillation is not determined by the differential equation. In this case it can be 10 or 25, or indeed have any other value.

A general form of SHM

Any motion given by an equation of the form $x = a \sin \omega t$ (where a and ω are positive constants) has a similar displacement–time graph and represents SHM with amplitude a. Differentiating this equation gives:

$$v = \dot{x} = a\omega \cos \omega t$$

and

$$\ddot{x} = -a\omega^2 \sin \omega t = -\omega^2 x.$$

Figure 14.6 shows how the graphs for x, $\dot{x}$ and $\ddot{x}$ are related.

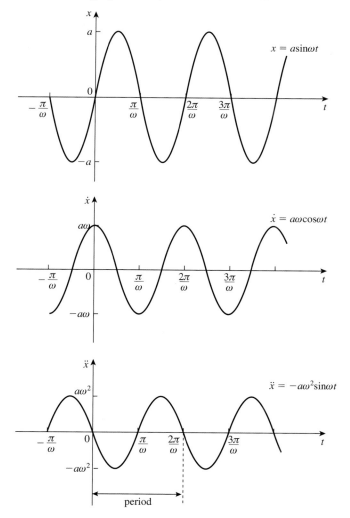

Figure 14.6

The period is shown in figure 14.6, but it can also be found from the equations for x, $\dot{x}$, and $\ddot{x}$. Since $\sin(\omega t + 2\pi) = \sin \omega t$ and $\cos(\omega t + 2\pi) = \cos \omega t$ the values of x, $\dot{x}$, and $\ddot{x}$ remain unchanged when ωt is increased by 2π. This is the smallest increment for which this is true. Also $(\omega t + 2\pi) = \omega\left(t + \dfrac{2\pi}{\omega}\right)$, so whenever you start measuring, all aspects of the motion will be the same after a period of time $T = \dfrac{2\pi}{\omega}$.

In general, $x = a \sin \omega t$ represents SHM with the following properties.

- The amplitude is $a\ (>0)$.
- The period T is $\dfrac{2\pi}{\omega}$.
- The frequency $f = \dfrac{1}{T} = \dfrac{\omega}{2\pi}$.
- $x = 0$ when $t = 0$.

QUESTION 14.7 Show that the equation $x = a \cos \omega t$ also satisfies the definition of SHM. Draw sketches of the graphs of $x = a \sin \omega t$ and $x = a \cos \omega t$ and compare the properties of the motions represented by them.

Phase difference

For $x = a \cos \omega t$, $x = a$ when $t = 0$. Its graph $x = a \cos \omega t$ is a translation of the graph of $x = a \sin \omega t$ along the t axis by a displacement of $-\dfrac{\pi}{2}\left(\text{or } \dfrac{3\pi}{2}, \dfrac{7\pi}{2} \text{ etc.}\right)$.

Both equations describe SHM but their oscillations are at a different stage in the cycle when $t = 0$.

They are said to have a *phase difference* of $-\dfrac{\pi}{2}\left(\text{or } \dfrac{3\pi}{2} \text{ etc.}\right)$

 Angular frequency

In these mathematical models ωt is an angle; ω is sometimes called the angular frequency. It is important to note that, when the model is applied to an oscillating system, ω is a constant which depends on the properties of the system, such as the stiffness of a spring or the length of a pendulum: ω rarely involves a physical angle and, in particular, ω is *not* the angular velocity of the pendulum.

Relationships between acceleration, velocity and displacement

Provided t is measured from an instant when the particle is at the centre and moving in the positive direction, SHM is described by the following equations.

Displacement: $\qquad\qquad\qquad\qquad x = a \sin \omega t$ ①

Velocity: $\qquad\qquad\qquad\qquad v = \dot{x} = a\omega \cos \omega t$ ②

Acceleration: $\qquad\qquad\qquad\qquad \ddot{x} = -a\omega^2 \sin \omega t$ ③

It can be seen from equations ① and ③ that

$$\ddot{x} = -\omega^2 x.$$ ④

Squaring equation ② gives:
$$v^2 = \dot{x}^2 = a^2\omega^2 \cos^2 \omega t$$
$$= \omega^2(a^2 - a^2 \sin^2 \omega t).$$
Therefore
$$v^2 = \omega^2(a^2 - x^2). \qquad ⑤$$

When the velocity is written in terms of t, it is best to denote it by $\dot{x}$ to stress the fact that it is the derivative of x with respect to t.

QUESTION 14.8 You have already shown that $x = a \cos \omega t$ satisfies the equation of motion for SHM. Now show that $\dot{x}$ satisfies the equation $\dot{x}^2 = \omega^2(a^2 - x^2)$.

EXAMPLE 14.3 In a harbour the cycle of tides can be modelled as SHM with a period of 12 hours 30 minutes. On a certain day high water is 10 m above low water.

i) Sketch a graph of the height x m of the water above (or below) the mean level against t, the time in hours since the water was at mean level and rising.

ii) Find a suitable expression to model x in terms of t.

iii) Determine for how long the water is more than 6 m above the low water mark.

iv) Find the rate at which the tide is rising or falling when the water is 6 m above the low water mark.

v) Find the maximum rate at which the tide rises.

SOLUTION

i)

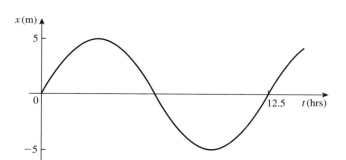

Figure 14.7

ii) First find ω and a.

The period of the tide $T = 12.5$ hours.

Since
$$T = \frac{2\pi}{\omega}$$
$$\omega = \frac{2\pi}{T}$$
$$\omega = 0.503$$

From the graph in part i), the amplitude of the oscillation is 5 m, so $a = 5$.

Since the water has risen x m in t hours after it was at the mean level, a suitable equation is

$$x = a \sin \omega t$$
$$x = 5 \sin (0.503t).$$

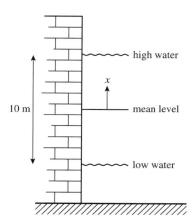

Figure 14.8

iii) Remember that you are working with time in hours, so any values of t will be in hours. The tide is 6 m above low water when $x \geqslant 1$.

Now $x = 1$ when $\quad 5\sin(0.503t) = 1$

$$\Rightarrow \quad \sin(0.503t) = 0.2 = \sin 0.201$$

The first two values of t satisfying this equation are t_1 and t_2 where

$$0.503t_1 = 0.201 \qquad \Rightarrow \qquad t_1 = 0.400$$

and $\qquad\qquad 0.503t_2 = \pi - 0.201 \qquad \Rightarrow \qquad t_2 = 5.846$

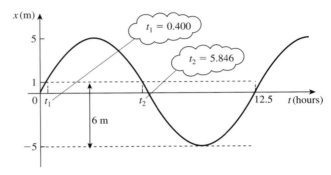

Figure 14.9

The water is 6 m above its lowest level for $(t_2 - t_1)$ hours i.e. about $5\frac{1}{2}$ hours.

iv) The water is 6 m above its lowest level when $x = 1$. The velocity of the oscillation at this point is given by:

$$v^2 = \omega^2(a^2 - x^2)$$

$$\Rightarrow \quad v = \pm 0.503 \sqrt{25 - 1}$$

$$= \pm 2.46.$$

The rate of rise or fall is 2.46 m per hour, or about 4 cm per minute.

v) Since $v^2 = \omega^2(a^2 - x^2)$, the maximum value of v is $a\omega$ when $x = 0$.

The water is rising fastest at the centre of its motion and the rate at which it is rising is:

$$a\omega = 0.503 \times 5$$

$$= 2.515 \text{ m per hour}$$

$$= 4.2 \text{ cm per minute.}$$

Note

A tidal range of 10 m is large, but is realistic for several places around the British coastline. The rate at which the water is rising, 4.2 cm per minute, does not sound high, but if you think about how fast the water would approach your deck chair up a typical beach (with a gradient of say 3°), you will see that it is quite dramatic.

14.3 Alternative forms of the equation for SHM

In the previous section you saw that $x = a \sin \omega t$ and $x = a \cos \omega t$ are solutions of the SHM differential equation $\ddot{x} = -\omega^2 x$. In this section you look at a number of alternative forms of the solution, in each case showing that they satisfy the SHM differential equation.

The forms $x = a \sin(\omega t + \varepsilon)$ and $x = a \cos(\omega t + \varepsilon)$

Since the graph of $x = a \sin \omega t$ passes through the origin, it represents SHM in which the particle is at the centre of its oscillation at time zero. Similarly, $x = a \cos \omega t$ represents SHM in which the particle is initially at maximum positive displacement. There will be times when you need to write an expression for SHM which starts at some other point in the cycle, and this is done by introducing a phase shift of an angle ε (the Greek letter 'epsilon'). This is equivalent to a phase shift in the time of $\dfrac{\varepsilon}{\omega}$ since $\omega t + \varepsilon = \omega\left(t + \dfrac{\varepsilon}{\omega}\right)$.

The equation $x = a \sin \omega t$ then becomes $x = a \sin(\omega t + \varepsilon)$, and the effect of this on the graph is shown in figure 14.10. The effect on $x = a \cos \omega t$ is similar.

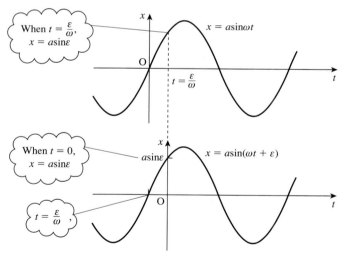

Figure 14.10

EXAMPLE 14.4

The functions f and g are given by

$$f: x = a \sin(\omega t + \varepsilon) \quad \text{and} \quad g: x = a \cos(\omega t + \varepsilon)$$

where ε (epsilon) is a positive constant.

For each of these functions:
i) Differentiate it with respect to time to find v (or $\dot{x}$) and show that $v^2 = a^2(\omega^2 - x^2)$.
ii) Differentiate v with respect to time to find $\ddot{x}$ and show that $\ddot{x} = -\omega^2 x$.

SOLUTION

i) For f:

$$x = a \sin(\omega t + \varepsilon)$$

$$\Rightarrow \qquad v = \dot{x} = a\omega \cos(\omega t + \varepsilon) \qquad \qquad \text{①}$$

Therefore

$$v^2 = a^2\omega^2 \cos^2(\omega t + \varepsilon)$$

$$= \omega^2[a^2 - a^2 \sin^2(\omega t + \varepsilon)]$$

$$\Rightarrow \qquad v^2 = \omega^2(a^2 - x^2)$$

Similarly for g:

$$x = a \cos(\omega t + \varepsilon)$$

$$\Rightarrow \qquad v = \dot{x} = -a\omega \sin(\omega t + \varepsilon) \qquad \qquad \text{②}$$

Therefore

$$v^2 = a^2\omega^2 \sin^2(\omega t + \varepsilon)$$

$$= \omega^2[a^2 - a^2 \cos^2(\omega t + \varepsilon)]$$

$$\Rightarrow \qquad v^2 = \omega^2(a^2 - x^2).$$

ii) Differentiating equation ① gives

$$\ddot{x} = -a\omega^2 \sin(\omega t + \varepsilon)$$

$$\Rightarrow \qquad \ddot{x} = -\omega^2 x. \longleftarrow$$

Similarly differentiating ② gives

$$\ddot{x} = -a\omega^2 \cos(\omega t + \varepsilon)$$

$$\Rightarrow \qquad \ddot{x} = -\omega^2 x. \longleftarrow$$

> These results show that the functions f and g represent SHM

The form $x = A \sin \omega t + B \cos \omega t$

The function $x = A \sin \omega t + B \cos \omega t$ represents the sum of two SHMs, $A \sin \omega t$ and $B \cos \omega t$. These have the same period $\left(\dfrac{2\pi}{\omega}\right)$ and frequency, but different amplitudes (A and B). They are a quarter of a cycle out of phase because

$$\sin \omega t = \cos\left(\omega t - \frac{\pi}{2}\right) = \cos \omega\left(t - \frac{\pi}{2\omega}\right).$$

What is the effect of adding two SHMs in this way?

By differentiating twice with respect to time, you can show that the function obeys the SHM equation $\ddot{x} = -\omega^2 x$.

$$x = A \sin \omega t + B \cos \omega t$$

$$\dot{x} = A\omega \cos \omega t - B\omega \sin \omega t$$

$$\ddot{x} = -A\omega^2 \sin \omega t - B\omega^2 \cos \omega t$$

This expression for $\ddot{x}$ may be written as

$$\ddot{x} = -\omega^2(A \sin \omega t + B \cos \omega t)$$

and so

$$\ddot{x} = -\omega^2 x.$$

This proves that $x = A \sin \omega t + B \cos \omega t$ represents SHM with period $\dfrac{2\pi}{\omega}$. To show that this form is equivalent to $x = a \sin(\omega t + \varepsilon)$, use the following technique.

The function $x = A \sin \omega t + B \cos \omega t$ is rewritten as

$$x = \sqrt{A^2 + B^2}\left(\frac{A}{\sqrt{A^2 + B^2}} \sin \omega t + \frac{B}{\sqrt{A^2 + B^2}} \cos \omega t\right).$$

In the right-angled triangle shown in figure 14.11, for the case when A and B are both positive

$$\cos \varepsilon = \frac{A}{\sqrt{A^2 + B^2}} \text{ and } \sin \varepsilon = \frac{B}{\sqrt{A^2 + B^2}}.$$

The hypotenuse is $\sqrt{A^2 + B^2} = a$.

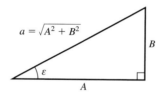

Figure 14.11

Using these results gives

$$x = a(\sin \omega t \cos \varepsilon + \cos \omega t \sin \varepsilon).$$

Using the compound angle formula $\sin(\theta + \phi) = \sin \theta \cos \phi + \cos \theta \sin \phi$, this can be written

$$x = a \sin (\omega t + \varepsilon).$$

Thus the SHM $x = A \sin \omega t + B \cos \omega t$ is equivalent to the SHM $x = a \sin (\omega t + \varepsilon)$, where the amplitude a is $\sqrt{A^2 + B^2}$, and the phase angle ε is given by

$$\sin \varepsilon = \frac{B}{\sqrt{A^2 + B^2}} \text{ and } \cos \varepsilon = \frac{A}{\sqrt{A^2 + B^2}}.$$

This is also equivalent to a cosine form, since for any angle α

$$\sin \alpha = \cos\left(\alpha - \frac{\pi}{2}\right)$$

and so x can be written as

$$x = a \cos\left(\omega t + \varepsilon - \frac{\pi}{2}\right)$$

$$= a \cos (\omega t + \varepsilon') \text{ where } \varepsilon' = \varepsilon - \frac{\pi}{2}.$$

(If ε is an acute angle, ε' is negative.)

The effect of adding together two SHMs of the same period is to create a single SHM, also with the same period but with greater amplitude. This is shown in figure 14.12 for $x = 3 \cos t$ and $x = 4 \sin t$.

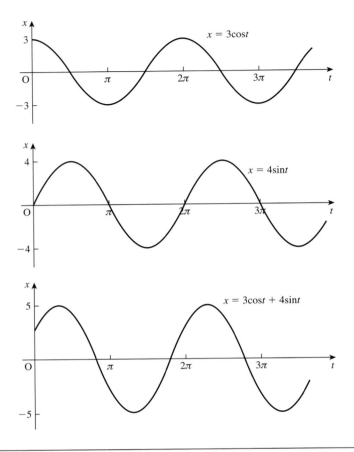

Figure 14.12

QUESTION 14.9   When you hear several musical instruments play notes of the same pitch, the vibrations which reach your ears are a combination of oscillations of different amplitudes. In addition, these may have been set in motion at different stages in the cycle and so be out of phase.

How is it possible for the music to sound pleasant?

The form $x = x_0 + a \sin \omega t$

Although the equation $x = x_0 + a \sin \omega t$ does not satisfy the SHM equation $\ddot{x} = -\omega^2 x$, it does represent SHM about the fixed point x_0 as you can see from figure 14.13.

You may find it helpful to think of this motion in terms of a new variable, z, representing the displacement from the central position. This is given by

$$z = x - x_0$$
$$= a \sin \omega t.$$

The variable z does satisfy the SHM equation $\ddot{z} = -\omega^2 z$ and all the standard SHM results also hold for z.

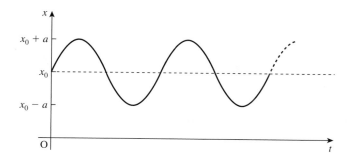

Figure 14.13

Choosing the most appropriate function to model a particular oscillation

You have seen that any particular example of SHM can be described using at least one of a variety of functions.

SHM about $x = 0$ with period $\dfrac{2\pi}{\omega}$, amplitude a	SHM about $x = x_0$ with period $\dfrac{2\pi}{\omega}$, amplitude a
$x = a \sin(\omega t + \varepsilon)$	$x = x_0 + a \sin(\omega t + \varepsilon)$
$x = a \cos(\omega t + \varepsilon)$	$x = x_0 + a \cos(\omega t + \varepsilon)$
$x = A \sin \omega t + B \cos \omega t$	$x = x_0 + A \sin \omega t + B \cos \omega t$
where $\sqrt{A^2 + B^2} = a$	where $\sqrt{A^2 + B^2} = a$

The constants a, A, B and ε are determined by the *initial conditions*, that is the speed, direction and displacement at time zero.

You will find it helpful to sketch the graph of the oscillation. This will show you the initial conditions and help you to choose the most appropriate form.

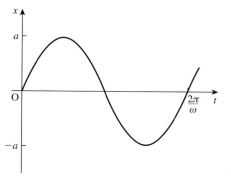

Figure 14.14

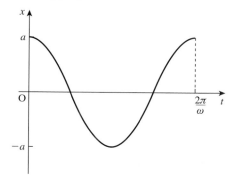

Figure 14.15

In this case when $t = 0$, $x = 0$ and $\dot{x} > 0$; the most appropriate form is
$$x = a \sin \omega t$$

In this case when $t = 0$, $x = a$ and $\dot{x} = 0$; the most appropriate form is
$$x = a \cos \omega t$$

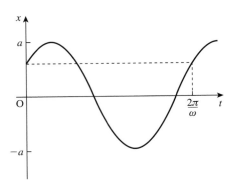

Figure 14.16

When the oscillation starts somewhere between the centre and an extreme, i.e. neither, $x = 0$ nor $\dot{x} = 0$ when $t = 0$, the most appropriate form will be either

$$x = a \sin(\omega t + \varepsilon)$$

or $$x = a \cos(\omega t + \varepsilon)$$

or $$x = A \sin \omega t + B \cos \omega t$$

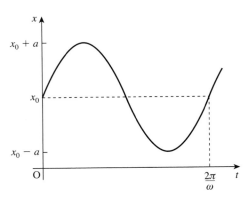

Figure 14.17

When the centre of the oscillations is not at the origin but at a point x_0, the appropriate equation will be one of those above but with x_0 added on. In this case

$$x = x_0 + a \sin \omega t$$

QUESTION 14.10   The graphs below show cases where the initial conditions are different from those covered above. What are the most appropriate forms to model these oscillations?

(i)

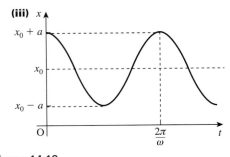

(ii)

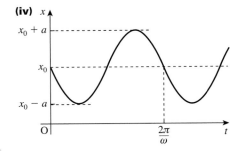

(iii)

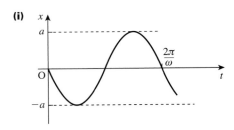

(iv)

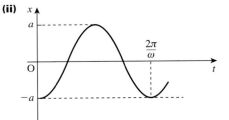

Figure 14.18

EXAMPLE 14.5

A particle is moving with SHM of period π. Initially it is 10 cm from the centre of the motion and moving in the positive direction with a speed of 6 cm s^{-1}. Find an equation to describe the motion.

SOLUTION

The information given is shown in the diagram below.

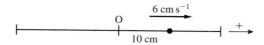

Figure 14.19

The initial speed is positive, so an appropriate equation is

$$x = a \sin(\omega t + \varepsilon),$$

and you need to find the values of a, ω, and ε.

Finding ω
Since the period of the motion is π,

$$\frac{2\pi}{\omega} = \pi \qquad \Rightarrow \qquad \omega = 2.$$

Finding a
Using
$$v^2 = \omega^2(a^2 - x^2)$$
$$6^2 = 2^2(a^2 - 10^2)$$
$$\Rightarrow \qquad a = 109$$

Finding ε
Substituting $t = 0$ in $x = a \sin(\omega t + \varepsilon)$ gives

$$= \sqrt{109} \sin \varepsilon$$
$$\Rightarrow \qquad \varepsilon = 1.28 \text{ rad (see Note below).}$$

So the equation for the motion is

$$x = \sqrt{109} \sin(2t + 1.28).$$

Note

When finding ε you must be careful that you have selected the correct root of the equation. In this case at $t = 0$ the particle has positive displacement and positive velocity (it is on its way out and not on its way back), so $t = 0$ corresponds to an angle between 0 and $\frac{\pi}{2}$.

The next root of the equation $10 = \sqrt{109} \sin \varepsilon$ is $(\pi - 1.28)$.

This lies between $\frac{\pi}{2}$ and π and would be the correct value if the particle were on its way back, with displacement $+10$ and velocity -6.

14.4 SHM as the projection of circular motion

There is a close relationship between circular motion at constant speed and SHM. This can be illustrated by rotating a bob on the end of a string in a horizontal circle with constant angular velocity, thus forming a conical pendulum. If this is done between the light of an overhead projector and the wall, the shadow of the bob on the wall will perform SHM. (For true SHM the rays of light should be parallel and an approximation to this can be achieved if the pendulum is close to the wall and the overhead projector is as far away as possible.)

Assuming the rays of light are parallel, figure 14.20 shows the position, C, of the bob and its shadow, P, at a particular instant. As the bob moves round the circle from A to B, the shadow moves along the straight line from L to N.

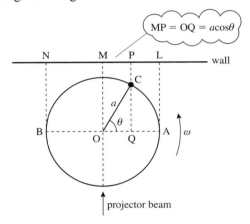

Figure 14.20

Assuming that $t = 0$ when the bob is at A, the angle θ is given by ωt, where ω is the angular speed of the bob. Thus

$$MP = OQ = a \cos \omega t.$$

This is one of the standard forms of SHM.

Note

The projection of uniform circular motion on to a straight line, illustrated here, is the only case where ω actually does represent a physical angular velocity.

EXAMPLE 14.6

An astronomer observes a faint object close to a star. Continued observations show the object apparently moving in a straight line through the star as shown in the diagram

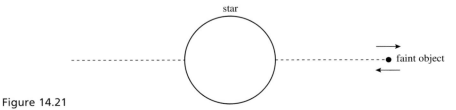

Figure 14.21

The astronomer is able to estimate the apparent distance of the object from the centre of the star and records this at 30-day intervals, resulting in the following table.

Day, t	0	30	60	90	120
Distance (10^{11} m), x	1.0	1.5	1.9	2.0	1.9

The astronomer thinks that the object is a planet moving around the star in a circular orbit and that she is observing it from a point in the plane of the orbit. She decides to model the apparent distance from the centre by the SHM equation:

$$x = 2 \sin(\omega t + \varepsilon).$$

i) Use the values of x for $t = 0$ and $t = 90$ to find values for the constants ω and ε and verify that the other values of x are consistent with this model.

ii) Assuming the model is correct, find

  a) the radius of the orbit

  b) the speed of the planet

  c) the number of earth days the planet takes to go round its star.

SOLUTION

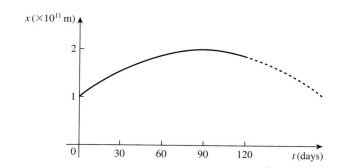

Figure 14.22

i) Using $x = 2 \sin(\omega t + \varepsilon)$:

When $t = 0$, $x = 1$ so $1 = 2 \sin \varepsilon$

$\Rightarrow \quad \varepsilon = \dfrac{\pi}{6}$.

> The equation $\sin \varepsilon = \frac{1}{2}$ has other roots $\left(\text{e.g. } \dfrac{5\pi}{6}\right)$, but the graph shows that $\dfrac{\pi}{6}$ is the one you want.

When $t = 90$, $x = 2$ so $2 = 2 \sin\left(90\omega + \dfrac{\pi}{6}\right)$

$\Rightarrow \quad \sin\left(90\omega + \dfrac{\pi}{6}\right) = 1$

$\Rightarrow \quad 90\omega + \dfrac{\pi}{6} = \dfrac{\pi}{2}$

$\Rightarrow \quad \omega = \dfrac{\pi}{270}$.

So the model is

$$x = 2 \sin\left(\dfrac{\pi t}{270} + \dfrac{\pi}{6}\right)$$

For the other values of t:

t	observed displacement	model's prediction
30	1.5	$2 \sin\left(\dfrac{\pi}{9} + \dfrac{\pi}{6}\right) = 1.53$
60	1.9	$2 \sin\left(\dfrac{2\pi}{9} + \dfrac{\pi}{6}\right) = 1.88$
120	1.9	$2 \sin\left(\dfrac{4\pi}{9} + \dfrac{\pi}{6}\right) = 1.88$

This shows that the model is a very good predictor of the actual position of the object.

ii) **a)** The radius of the orbit is the amplitude a of the motion:

$$\text{radius} = 2 \times 10^{11}\,\text{m}.$$

b) The speed of the planet is

$$a\omega = 2 \times 10^{11} \times \frac{\pi}{270} = 2.3 \times 10^9\,\text{ms}^{-1}.$$

c) The total time for one orbit is the period of the SHM i.e. $\dfrac{2\pi}{\omega}$

$$\text{Number of days} = 2\pi \div \frac{\pi}{270}$$

$$= 540 \text{ days}.$$

Historical note

In 1822, the French mathematician Jean Baptiste Fourier (1768–1830) showed that any function of t can be written as a sum of sines of multiples of t and this is now called a Fourier series. It follows that any vibration which can be written as a function of t can be reproduced by adding simple harmonic vibrations. Fourier accompanied Napoleon to Egypt in 1798 and was made a baron ten years later. He discovered this theorem while working on the flow of heat.

14.5 Oscillating mechanical systems

There are very many mechanical systems which can be modelled using SHM. Two of these are the spring–mass oscillator and the simple pendulum. The motion of the simple pendulum approximates to SHM for small angles as you will see in the next section.

The simple pendulum

A simple pendulum consists of a bob suspended on the end of a light inelastic string as illustrated by the apparatus in figure 14.23.

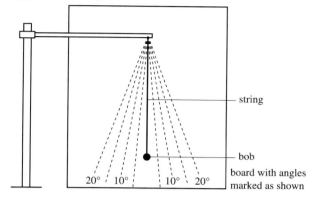

Figure 14.23

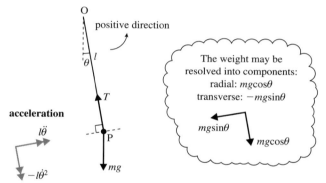

Figure 14.24

The forces acting on the bob are the tension in the string and the force of gravity mg, where m is the mass of the bob as shown in figure 14.24. It swings through a small arc of a circle of radius l where l is the length of the string.

There is no motion in the radial direction. In the transverse direction, the acceleration, $l\ddot{\theta}$, is given by

$$-mg \sin \theta = ml\ddot{\theta}$$

$$\Rightarrow \quad \ddot{\theta} = -\frac{g}{l} \sin \theta.$$

When the angle is measured in radians, $\theta \approx \sin \theta$ for small angles (up to about 0.3 rad for accuracy correct to 2 d.p.). In this case:

$$\ddot{\theta} = -\frac{g}{l} \theta.$$

This is the standard equation for SHM, $\ddot{x} = -\omega^2 x$, with x replaced by θ and ω^2 replaced by $\frac{g}{l}$.

A pendulum is usually set in motion by pulling the bob to one side, say to an angle α, and then releasing it from rest. If this is the case, $\theta = \alpha$ and $\dot{\theta} = 0$ when $t = 0$.

The appropriate form of the SHM equation is

$$\theta = \alpha \cos \sqrt{\frac{g}{l}} \, t.$$

$\theta = \alpha$ and $\dot{\theta} = 0$ when $t = 0$

The period is given by

$$T = \frac{2\pi}{\omega} = 2\pi \sqrt{\frac{l}{g}}.$$

There are several points to note:
- The SHM involves the angular displacement, θ, rather than the linear displacement of the bob.
- SHM is a good model for small values of θ, but the approximation becomes less good with increasing θ.
- The mass of the bob is not present in the SHM equation, so its value does not affect the motion.
- The amplitude is specified only by the initial conditions.
- The period $2\pi \sqrt{\dfrac{l}{g}}$ is not affected by the amplitude of the motion or the mass of the bob. It depends only on g and the length of the string.
- It is important not to confuse the angular velocity, $\dot{\theta}$, of the pendulum with ω, the angular frequency of the SHM.

QUESTION 14.11   Use a calculator to find positive values of θ which differ from $\sin \theta$ by less than

i)   0.05%   ii)   1%   iii) 2%.

EXAMPLE 14.7   A simple pendulum consists of a mass hanging on the end of a light inextensible string. The angle θ radians between the string and the vertical satisfies the differential equation

$$\ddot{\theta} = -\frac{g}{l} \, \theta.$$

i)   Find the length of the pendulum for which the period for small oscillations is 2 s.

The pendulum is released when the string makes an angle of 0.2 rad with the vertical.

ii)   Find an equation for θ in terms of t.
iii)   Find the times when the pendulum is in the equilibrium position.
iv)   Find also the time taken for the pendulum to move from the equilibrium position to a point half-way to the end of the oscillation.

SOLUTION

i)   The equation of motion, $\ddot{\theta} = -\dfrac{g}{l} \, \theta$, is the SHM differential equation with $\omega^2 = \dfrac{g}{l}$.

All the standard results for SHM apply.

The period is 2 s, so $\dfrac{2\pi}{\omega} = 2$ $\Rightarrow$ $\omega = \pi.$

Therefore $\pi^2 = \dfrac{g}{l}$

$$\Rightarrow \quad l = \frac{g}{\pi^2} = \frac{9.8}{(3.14 \ldots)^2}.$$

The length of the pendulum is 0.993 m or about 1 m.

ii) The pendulum is released when $\theta = 0.2$ ($\theta = 0.2$ when $t = 0$). The most appropriate function to use to model the motion is

$$\theta = 0.2 \cos \omega t.$$

You have already seen that $\omega = \pi$.

Therefore $\qquad\qquad\qquad\qquad \theta = 0.2 \cos \pi t.$

iii) The pendulum is in the equilibrium position when $\theta = 0$, i.e. when

$$0.2 \cos \pi t = 0$$

$$\Rightarrow \qquad \cos \pi t = 0$$

$$\Rightarrow \qquad \pi t = \frac{\pi}{2}, \frac{3\pi}{2}, \frac{5\pi}{2} \cdots$$

$$\Rightarrow \qquad t = \tfrac{1}{2}, \tfrac{3}{2}, \tfrac{5}{2}, \ldots$$

iv) The pendulum is first at the equilibrium position, $\theta = 0$, when $t = 0.5$.

It is half-way to the end of the oscillation when $\theta = -0.1$.

The value of t when $\theta = -0.1$ is given by

$$0.2 \cos \pi t = -0.1$$

$$\Rightarrow \qquad \pi t = \frac{2\pi}{3} \qquad\qquad \boxed{\cos \frac{2\pi}{3} = -0.5}$$

$$\Rightarrow \qquad t = \tfrac{2}{3}.$$

The time taken from the centre to the half-way point is $\frac{2}{3} - \frac{1}{2} = \frac{1}{6}$ s.

The graph illustrates the first two swings of the pendulum.

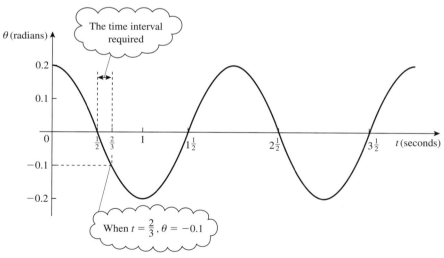

Figure 14.25

⚠ Notice that the time taken to travel from the centre half-way to the end is one third of and not half of the time taken to travel to the end. This is because, for SHM, the velocity is greater near the centre of the motion than near the ends.

The spring–mass oscillator

A spring–mass oscillator is shown in Figure 14.26.

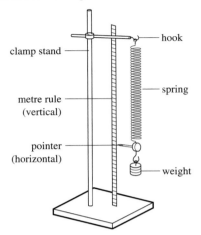

Figure 14.26

The equation of motion of a particle suspended from a spring

Figure 14.27 shows a particle of mass m suspended from a perfectly elastic light spring of natural length l_0, and modulus of elasticity λ. The extension of the spring when the particle is hanging in equilibrium is e.

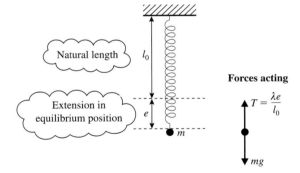

Figure 14.27

To find the equation of motion for the particle, start by finding an expression for e.

The tension T in the spring is given by $T = \dfrac{\lambda e}{l_0}$ and this is equal to the weight of the particle, mg.

Thus
$$\frac{\lambda e}{l_0} = mg. \qquad \text{①}$$

Now look at the situation when the spring has an extension x in the downwards direction from the equilibrium position as in figure 14.28. The total extension is now $e + x$.

The particle is not in equilibrium, and its acceleration in the downwards direction is denoted by $\ddot{x}$.

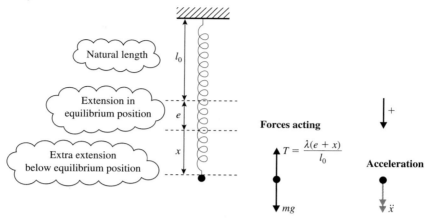

Figure 14.28

There are two forces acting on the particle: the tension in the spring, $T = \dfrac{\lambda(e + x)}{l_0}$, and its weight, mg.

By Newton's second law

$$m\ddot{x} = mg - \frac{\lambda(e + x)}{l_0}$$

$$= mg - \frac{\lambda e}{l_0} - \frac{\lambda x}{l_0}.$$

But from ① $\qquad\qquad \dfrac{\lambda e}{l_0} = mg$

so $\qquad\qquad m\ddot{x} = mg - mg - \dfrac{\lambda x}{l_0}.$

giving the differential equation

$$\ddot{x} = \frac{\lambda}{ml_0} x.$$

This is the standard equation for SHM with

$$\omega^2 = \frac{\lambda}{ml_0}.$$

Its period is $\dfrac{2\pi}{\omega} = 2\pi\sqrt{\dfrac{ml_0}{\lambda}}$.

If k is the stiffness of the spring, $k = \dfrac{\lambda}{l_0}$ so this then becomes $2\pi\sqrt{\dfrac{m}{k}}$.

This means that the period of the oscillations depends only on the mass and the stiffness of the spring.

Note that this model is good only so long as Hooke's law holds, but provided this is so, the motion is an example of true SHM, unlike that of the simple pendulum which only approximates to SHM.

EXAMPLE 14.8

A particle of mass 200 g is attached to a light spring of natural length 40 cm and stiffness 50 Nm^{-1}. The particle is allowed to hang vertically in equilibrium.

i)   Find the extension of the spring in this position.

The spring is now pulled down 3 cm and released from rest.

ii)   Find the length of the spring as a function of time.

SOLUTION

i)   The diagram shows the relevant lengths, and the forces acting on the particle in equilibrium. The extension in the spring is x_0 m.

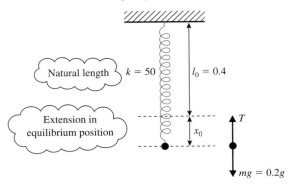

Figure 14.29

By Hooke's law: $T = kx_0 = 50x_0.$

Since the particle is in equilibrium,

$$T = 0.2g$$
$$\Rightarrow \qquad 50x_0 = 0.2g$$
$$x_0 = 0.0392.$$

The extension is 0.0392 m = 3.92 cm.

ii)   Let x m be the displacement from the equilibrium position in the downwards direction.

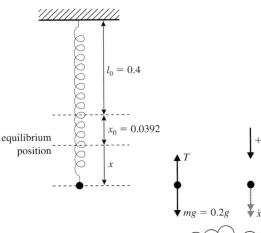

Figure 14.30

The extension of the spring is then $x + x_0$.

You know the value of x_0.

By Hooke's law, $T = 50(x + x_0)$.

Applying Newton's second law in the downwards direction,

$$0.2g - 50(x + x_0) = 0.2\ddot{x}$$

$$\Rightarrow \quad 0.2g - 50x - 50x_0 = 0.2\ddot{x}.$$

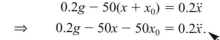

This shows the advantage of measuring x from the equilibrium position. The weight and kx_0 cancel each other out

However, as found in part i), $0.2g = 50x_0$ and so the equation may be simplified to

$$0.2\ddot{x} = -50x$$

$$\Rightarrow \quad \ddot{x} = -250x.$$

This is the SHM equation with $\omega^2 = 250$, and so $\omega = \sqrt{250}$.

Since the particle starts at the end of the oscillation with $x = 0.03$, the appropriate equation is of the form $x = a \cos \omega t$, with $a = 0.03$ and $\omega = \sqrt{250}$.

So, at time t, $\quad x = 0.03 \cos \sqrt{250}t$

and the total length is $0.4 + 0.0392 + 0.03 \cos \sqrt{250}t$

$$= 0.4392 + 0.03 \cos \sqrt{250}t.$$

QUESTION 14.12 Would the period of the oscillations in the above example be the same on the moon?

EXAMPLE 14.9

Two springs have the same modulus of elasticity, 30 N, but are of natural lengths 0.4 m and 0.6 m. An object of mass 0.5 kg is attached to one end of each spring and the other ends are attached to two points which are 1.2 m apart on a smooth horizontal table. Find the period of small oscillations of this system.

SOLUTION

The vertical forces acting on the object, its weight and the normal reaction of the table, have no effect on the motion: they balance each other.

The first step is to find the equilibrium position of the object. The diagram shows the relevant lengths and the horizontal forces acting when the object is in equilibrium. The extensions in the springs are e_1 m and e_2 m and the tension, T N is the same on both sides.

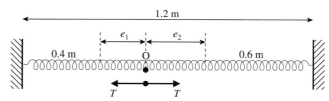

Figure 14.31

Hooke's law applied to each spring gives

$$T = \frac{30}{0.4} \times e_1 = 75e_1$$

and

$$T = \frac{30}{0.6} \times e_2 = 50e_2$$

$$\Rightarrow \quad 75e_1 = 50e_2$$

$$\Rightarrow \quad e_2 = 1.5e_1.$$

The total length is 1.2 m, so $e_1 + e_2 = 0.2$,

$$\Rightarrow \quad 2.5e_1 = 0.2.$$

Hence

$$e_1 = 0.08 \text{ and } e_2 = 0.12.$$

The next step is to find the equation of motion, so it is necessary to consider a general position for the object, and this is given by the displacement x m from the equilibrium position. The direction towards the right is taken to be positive.

The tensions in the springs are now different.

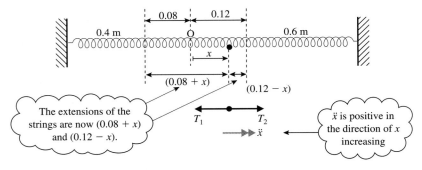

Figure 14.32

Hooke's law gives

$$T_1 = \frac{30}{0.4}(0.08 + x) = 75(0.08 + x)$$

$$T_2 = \frac{30}{0.6}(0.12 - x) = 50(0.12 - x)$$

Newton's second law can now be applied giving

$$T_2 - T_1 = m\ddot{x}$$

$$\Rightarrow \quad 50(0.12 - x) - 75(0.08 + x) = 0.5\ddot{x}$$

$$\Rightarrow \quad -125x = 0.5\ddot{x}$$

$$\Rightarrow \quad \ddot{x} = -250x.$$

This is the equation of motion for SHM with $\omega = \sqrt{250}$.

The period is $\dfrac{2\pi}{\omega}$ or 0.40 s (to 2 d.p.).

Historical note

The story of the quest for a means to measure longitude shows why scientists at the time of Hooke were interested in both astronomy and accurate time keeping. The problem was solved by John Harrison whose increasingly accurate clocks are preserved at the old Greenwich Observatory. You can find out more about Hooke and Harrison at the website of the National Maritime Museum: *www.nmm.ac.uk*.

Chapter 14 Exercises

In exercises **14.1–14.4** *assume that time is measured in seconds.*

14.1 Musical notes which are an octave apart have frequencies in the ratio 1:2. The note A above middle C has a frequency of 440 Hz. On a full-size keyboard there are 4 'A's below it and 3 above it (the range on the keyboard is just over 7 octaves).

    i) Work out the values of ω corresponding to the frequencies of these seven 'A's.

    ii) Find the maximum speed of points on piano strings which are vibrating with amplitude 1 mm to produce the highest and lowest of these notes.

14.2 A piston in an engine oscillates with a period of 0.03 s and an amplitude of 0.35 m. Modelling the oscillations as SHM,

    i) draw a sketch graph to illustrate the oscillations

    ii) calculate the frequency of the motion and hence find ω

    iii) calculate the maximum speed of the piston.

14.3 Air sickness might be caused by the rhythmic vibrations of an aircraft. It has been observed that about 50% of the passengers of an aircraft suffer air sickness when it bounces up and down with a frequency of about 0.3 Hz and a maximum acceleration of 4 ms^{-2}. Assuming SHM find

    i) ω and hence the amplitude of the motion

    ii) the greatest vertical speed during this motion.

14.4 A jig-saw operates at 3000 strokes per minute with the tip of the blade moving 17 mm from the top to the bottom of a stroke. (One stroke is a complete cycle.) Assuming that the motion is simple harmonic, find

    i) the maximum speed of the blade

    ii) the maximum acceleration of the blade

    iii) the speed of the blade when it is 6 mm from the central position.

14.5 A particle is performing SHM described by the equation

$$x = a \sin \omega t$$

Its maximum speed is 10 cm s^{-1} and its period is $\dfrac{\pi}{2}$ s. Lengths are in cm.

    i) Find the values of ω and a.

    ii) Differentiate to find an expression for the velocity $\dot{x}$ and show that $\dot{x}$ is first zero when t is one quarter of the period.

    iii) Find the velocity when t is

        a) 0.2   **b)** 1.

        In which direction is the particle travelling in each case?

    iv) How far does the particle travel after leaving the central position in

        a) 0.2 s   **b)** 1 s?

14.6 The top of a piston has a motion which is modelled as simple harmonic with a period of 0.1 s and an amplitude 0.2 m about a mean position A.

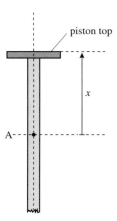

piston top

A

x

i) Show that x, the displacement of the top of the piston from A after t s, is given by the equation $x = 0.2 \sin (20\pi t)$, given that $x = 0$ when $t = 0$.

Where appropriate your answers to the following questions should be expressed in terms of π.

ii) What is the greatest piston speed?

iii) What is the greatest magnitude of the acceleration of the piston?

iv) For what fraction of the period is $x > 0.1$?

14.7 After a boat passes, a duck floating on a river bobs up and down for a short time with SHM period 1 s and amplitude 15 cm. When there is no disturbance, the depth of the water under the duck is 0.5 m.

i) What are the greatest and least depths during the motion?

ii) Assuming the water rises initially, write down an expression for the depth t seconds after the motion starts.

iii) Calculate the first time that the depth is
   a) 0.55 m   b) 0.4 m.

14.8 At a certain location, in a year of 365 days, the latest sunrise is at 8.05 am on 1 January and the earliest is at 3.45 am on 1 July. Sunrise times over a year are to be modelled by a simple harmonic variation of time about a mean time.

i) Determine the mean time in minutes after midnight.

ii) Find the amplitude of the sunrise variation in minutes.

iii) Sketch a graph of sunrise times for a year against day of the year measured from 1 January as day 1.

iv) Write down the angular frequency ω of this variation.

v) Derive an expression for the time of sunrise in minutes after midnight in terms of the day, D, of the year.

vi) Use the result in part v) to find the time of sunrise in hours and minutes (am) at the end of April (day 120 of the year).

14.9 An engineer seeks to model the oscillatory motion of a suspension system as simple harmonic motion. She observes the motion of a point P which moves vertically about a mean position O. The displacement above O is x metres. At time t seconds the displacement of P above O is x metres. When $t = 0$, P is at its highest point.

The engineer records the distance between the highest and lowest points reached by P to be 0.1 m and the time for 20 oscillations to be 6.4 seconds.

i) Write down the amplitude and the period of the motion.

ii) Calculate the velocity and acceleration of the point P when it is first 0.02 m below O, specifying the direction in each case.

iii) Sketch a graph of x against t for the first oscillation.

iv) Write down an expression for x in terms of t. Hence, or otherwise, calculate the time at which P first reaches the point 0.02 m below O while travelling upwards.

14.10 A SHM is described by the equation

$$x = 2 \cos \frac{\pi}{4}t + \sqrt{5} \sin \frac{\pi}{4}t.$$

i)   Plot the graph of x against t for $0 \leqslant t \leqslant 20$.

The motion may be described by an equation of the form

$$x = a \sin (\omega_1 t + \varepsilon).$$

ii) Determine the values of a, ω_1 and ε.

Another way of describing the motion is given by the equation

$$x = b \cos (\omega_2 t + \delta).$$

iii) State which of b, ω_2 and δ are equal, respectively, to a, ω_1 and ε.

iv) State the values of b, ω_2 and δ.

14.11 As a result of storms in different places, two swell wave patterns, both running in the same direction, occur at the same time over a stretch of open sea. Their heights above the mean sea level can be modelled as follows.

Wave pattern A: $h_A = 1.5 \sin \frac{\pi}{15}t$ (t is in s, h is in m)

Wave pattern B: $h_B = 2 \cos \frac{\pi}{15}t$

i)   Plot the two wave patterns on the same piece of graph paper taking values of t from 0 to 45 at 5 s intervals.

ii) The overall height of the water, h, is given by

$$h = h_A + h_B$$

Plot the values of h for $0 \leqslant t \leqslant 45$ on the same graph that you used for h_A and h_B.

iii) Show algebraically that the effect of the two wave patterns is that of a single wave pattern described by

$$h = a \sin (\omega t + \varepsilon)$$

and state the values of a, ω and ε.

14.12 The centre of the London Eye observation wheel is approximately 70 m above the ground (about as high as Big Ben which is on the opposite side of the Thames). The wheel's diameter is 135 m and it takes 30 minutes to make one revolution. Assuming that it turns continuously at a steady speed, find the angular velocity of the wheel.

i)   Write down a suitable equation for the height above the ground of a passenger t minutes after boarding at the lowest level.

ii) Estimate for how long during a flight of one revolution passengers are higher than the highest point on the Houses of Parliament (102 m).

14.13 The height, h m, of water above the bottom of a harbour varies with the tide.

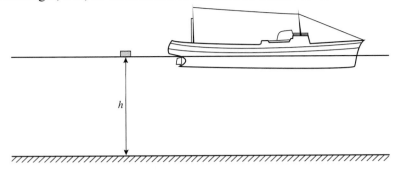

The vertical motion of a piece of wood floating on the surface may be modelled by the equation

$$h = h_0 + a \sin(\omega t + \varepsilon)$$

where t is the time in hours since midnight.

i)   Describe the vertical motion of the piece of wood in terms of h_0, a and ω.

The heights at certain times on one day are given by:

t (hours)	2	5	8	11	12
y (metres)	10	6	2	6	8

ii)   Find the values of h_0, a, ω and ε.
iii) What are the vertical speed and acceleration of the piece of wood when $t = 8.6$?
iv)   What is the greatest speed at which the water level rises?

In exercises **14.14–14.16** *take g as 9.8 ms*$^{-2}$ *(unless otherwise stated).*

14.14 A pendulum has a bob of mass 0.25 kg. The period of the pendulum is 2s. The amplitude of its swing is 3°.

i)   Find the length of the pendulum.
ii) State the effect (if any) on the period of the pendulum of
   a)   making the mass of the bob 0.75 kg
   b)   doubling the length of the pendulum
   c)   halving the amplitude of its swing
   d)   moving it to the moon where the acceleration due to gravity is one sixth of that on the earth.

14.15 Zeb, who should be modelled as a particle of mass m kg, sits on a bouncy toy which behaves like a vertical elastic spring of stiffness $k\,\mathrm{Nm}^{-1}$.

i)   The spring is compressed a length x_0 m when Zeb sits still in the equilibrium position, E. Show that

$$kx_0 = mg.$$

Zeb sets the toy moving so that it performs simple harmonic oscillations.

ii)   Draw diagrams similar to those on page 120 showing relevant lengths, forces and Zeb's acceleration when he is x m below E. Include the natural length, l, of the spring. Write down Zeb's equation of motion.
iii) Obtain the equation for the simple harmonic motion.
iv)   What is the maximum amplitude if Zeb stays fully in contact with the toy?

14.16 A weighing machine is being designed. It consists of a square platform of mass 2.5 kg supported by a number of identical springs each of stiffness 25000 Nm^{-1}, which are attached to a fixed horizontal base as shown in the diagram.

Throughout this question assume that the platform remains horizontal.

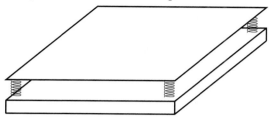

Initially the designer uses four springs and the system is in equilibrium.

i) Calculate the compression in each spring before any object is placed on the platform.

A child of mass 30 kg is standing on the platform, which is at rest.

ii) Calculate the compression in each of the four springs.

iii) Calculate the minimum number of additional springs required to reduce the compression to less than 0.002 m.

The 30 kg child is standing on the platform supported by four springs as in the original design. The child's father lifts her off quickly, allowing the platform to oscillate freely in a vertical direction.

iv) The displacement of the platform *below* the equilibrium position at time t seconds is y metres. Write down the equation of motion for the platform. Hence show that the platform performs simple harmonic motion of period $\frac{1}{100}\pi$ s. Calculate the maximum speed of the platform.

Personal Tutor

 14.1 A thin horizontal plate is being driven by a mechanism so that its vertical motion is simple harmonic. It moves through two complete oscillations per second and the distance from the lowest to the highest point of the motion is 0.5 m. At the instant from which time is measured the plate is moving upwards through the centre of its motion. At a time t s later the plate is x m above this central level, as shown in the diagram.

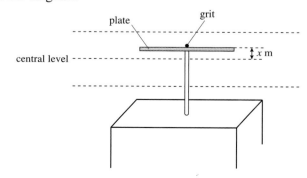

i) Show that the acceleration, $\ddot{x}$, of the plate is given by
$$\ddot{x} = -16\pi^2 x.$$

ii) Find an expression for x at time t.

iii) Find an expression for v^2 in terms of x, where $v\,\mathrm{ms}^{-1}$ is the speed of the plate.

When the plate is at its lowest point, it picks up a small piece of grit of mass m kg.

iv) Draw a diagram showing the forces acting on the piece of grit while it is in contact with the plate, and derive the equation of motion for the piece of grit. Explain why the grit leaves the plate when $\ddot{x} = -g$.

v) How far does the piece of grit travel while in contact with the plate? What is its speed when it leaves the plate?

vi) For how long is the piece of grit in contact with the plate?

KEY POINTS

14.1 Motion for which $\ddot{x} = -\omega^2 x$ is simple harmonic.

14.2 The graph illustrates the amplitude and period of simple harmonic motion.

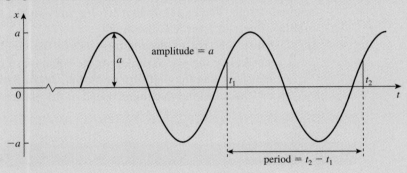

14.3 The speed at displacement x is given by $v^2 = \omega^2(a^2 - x^2)$.

14.4 The period is
$$T = \frac{2\pi}{\omega}.$$

14.5 The frequency is
$$f = \frac{1}{T} = \frac{\omega}{2\pi}.$$

14.6 Different ways of expressing SHM

Initial conditions	Displacement x	Velocity $v = \dot{x}$	Acceleration $\ddot{x}$
$x = 0$ and $\dot{x} > 0$	$a \sin \omega t$	$a\omega \cos \omega t$	$-a\omega^2 \sin \omega t$
$x = a$ and $\dot{x} = 0$	$a \cos \omega t$	$-a\omega \sin \omega t$	$-a\omega^2 \cos \omega t$
Otherwise i)	$a \cos(\omega t + \varepsilon)$	$-a\omega \sin(\omega t + \varepsilon)$	$-a\omega^2 \cos(\omega t + \varepsilon)$
or ii)	$a \sin(\omega t + \varepsilon)$	$a\omega \cos(\omega t + \varepsilon)$	$-a\omega^2 \sin(\omega t + \varepsilon)$
or iii)	$A \cos \omega t + B \sin \omega t$	$-A\omega \sin \omega t + B\omega \cos \omega t$	$-\omega^2 (A \cos \omega t + B \sin \omega t)$
where $a = \sqrt{A^2 + B^2}$			

For all these functions $x = 0$ in the position of equilibrium. If x takes another value, x_0, in the equilibrium position, then $\ddot{x} = -\omega^2(x - x_0)$ and x_0 must be added to functions in the displacement column.

14.7 For mechanical systems oscillating with simple harmonic motion, $\ddot{x} = -\omega^2 x$,
- the centre of motion is always a position of equilibrium
- for a spring–mass oscillator, x is not the extension of the spring
- for a simple pendulum x is the angular displacement, θ, which is small
- ω is not the angular velocity of a simple pendulum.

15 Damped and forced oscillations

One cannot escape the feeling that these mathematical formulae have an independent existence and an intelligence of their own, that they are wiser than we are, wiser even than their discoverers, that we get more out of them than was originally put into them.

Heinrich Hertz

QUESTION 15.1   Cleo Laine can break a wine glass by singing at it.
   i)   How is it possible?
   ii) Can you do it?

All structures have natural frequencies of vibration. If an external agent causes them to vibrate at or near one of these frequencies, large oscillations are seen to build up. This phenomenon is called *resonance*. Over the years there have been several well-known instances of resonance occurring with dramatic and destructive effect.

- In 1963 a cooling tower at the Ferrybridge power station near Leeds collapsed. It had been caused to vibrate violently by vortices formed when the wind blew past a row of towers in front of it.
- In 1940 a wind of moderate speed caused large oscillations in the Tacoma Narrows bridge in Washington, USA. The bridge eventually collapsed into the river below.
- In 1831 a column of soldiers marching over the Broughton suspension bridge, near Manchester, set up a forced vibration which had a frequency close to one of the natural frequencies of the bridge. The bridge collapsed. Since this disaster, it has become normal for troops to break step when crossing a bridge.

● In biblical times, after camping outside the walled town of Jericho for seven days, Joshua gave orders which brought the siege to an end in a curious way:

So they blew trumpets, and when the army heard the trumpet sound, they raised a great shout, and down came the walls.

Could this, too, have been the effect of forced vibrations?

These instances show how important it is for engineers to be able to predict the natural frequencies of any structures they are designing, and the frequencies of any possible forcing agents. Nor is this a matter for civil engineers alone: engines which run at particular frequencies cause ships and aircraft to shake themselves to pieces.

15.1 Improving the SHM model

Simple harmonic motion has constant amplitude and goes on for ever. For many real oscillating systems, SHM is not a very good model: usually the amplitude of the oscillations gradually decreases, and the motion dies away.

When you find that a model is unsatisfactory, you need to look again at your assumptions. Real oscillating systems are almost always *damped*: that is they are affected to some degree by the resistive forces of friction and/or air resistance. They perform *damped oscillations*.

In many systems the damping force is proportional to the speed of the object. This is often represented on a diagram by a device called a linear dashpot, as shown in figure 15.1.

Figure 15.1

A dashpot exerts a force on the system which is proportional to the rate at which it is being extended or compressed, and which acts in the direction opposite to that of the motion. This is illustrated in figure 15.2.

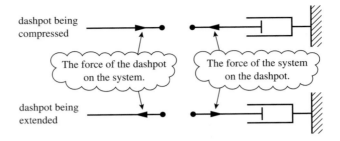

dashpot being compressed

The force of the dashpot on the system.

The force of the system on the dashpot.

dashpot being extended

Figure 15.2

The force R that the dashpot exerts on the system at time t is given by

$$R = r\frac{dL}{dt}$$

where the constant of proportionality r is called the *dashpot constant* (or the *damping constant*). The amount of travel still left in the dashpot (see figure 15.3) is denoted by L.

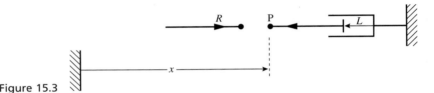

Figure 15.3

It is important to be clear in your mind about the direction of the force R and the signs involved. Look at the point P on the moving part of the dashpot.

- When P is moving from right to left, L is increasing; $\dfrac{dL}{dt}$ is positive and the force is in the same direction as that marked for R in figure 15.3. The dashpot is opposing the right to left motion.

- When P is moving from left to right, L is decreasing, $\dfrac{dL}{dt}$ is negative and the force is in the opposite direction to that marked for R in figure 15.3. The dashpot is now opposing the left to right motion.

Thus the sign of the dashpot force looks after itself as the motion changes.

However you will not usually be interested in the quantity L so much as the distance of the point P from some fixed point of the system. This distance is shown as x in figure 15.3. All the systems that you will meet in this book are set up so that as x increases, L decreases, and vice versa, so

$$\frac{dx}{dt} = -\frac{dL}{dt}.$$

Consequently the force that the dashpot exerts on the system is given by

$$R = -r\frac{dx}{dt}$$

in the direction of increasing x.

Note

Here we are using a dashpot as a symbol to represent damping effects but a dashpot is actually a real device that is often used in man-made systems. It is used when a certain (predictable) amount of damping, over and above that provided by friction and air resistance, is desired. For example, a car's shock-absorber includes a linear dashpot, as shown in the photograph.

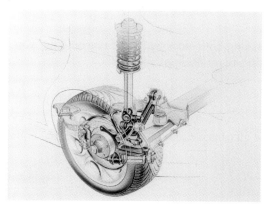

Such a dashpot consists of a cylinder containing a viscous liquid, often oil. When the cylinder is compressed or extended a disc moves along the cylinder, and is opposed by a force which is proportional to the speed of compression or extension. The constant of proportionality depends on the viscosity of the liquid.

EXAMPLE 15.1

A simple oscillating system is being modelled as a damped spring-mass oscillator, in which an object of mass 2 kg is attached to fixed points by a spring of natural length 0.5 m, stiffness $20\,\text{Nm}^{-1}$ and by a dashpot of constant $12\,\text{Nm}^{-1}$ s. The spring-mass-dashpot system lies on a smooth horizontal surface, as shown in the figure 15.4.

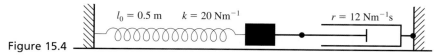

Figure 15.4

i)   Formulate the differential equation of motion for this system.

The system is released from rest when the spring length is 0.6 m.

ii) Find the particular solution of the differential equation that models this situation.

SOLUTION

i)   Figure 15.5 shows the spring-mass-dashpot system at some general time t (seconds), when the extension of the spring is x. The horizontal forces are the tension in the spring, T, and the damping force R.

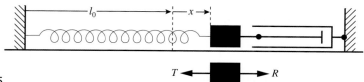

Figure 15.5

The tension in the spring is $T = kx = 20x$.

The dashpot force is $R = -r\dfrac{dx}{dt} = -12\dfrac{dx}{dt}$.

Applying Newton's second law $F = ma$ at any instant gives

This side is *ma*.

$$2\frac{d^2x}{dt^2} = -12\frac{dx}{dt} - 20x.$$

This side is F where $F = R - T$.

Dividing both sides by 2, and rearranging, you obtain the equation of motion of the spring-mass-dashpot system:

$$\frac{d^2x}{dt^2} + 6\frac{dx}{dt} + 10x = 0.$$

Notice that you again have a linear homogeneous second order equation with constant coefficients.

ii) The auxiliary equation for the differential equation is

$$\lambda^2 + 6\lambda + 10 = 0$$

$$\Rightarrow \quad \lambda = \frac{-6 \pm \sqrt{6^2 - 40}}{2} = -3 \pm j.$$

The general solution of the differential equation is $x = ae^{-3t}\sin(t + \varepsilon)$.

At the start of the motion, the length of the spring is 0.6 m and the object is at rest, so the initial conditions are $x = 0.1$ and $\dfrac{dx}{dt} = 0$ when $t = 0$.

When $t = 0$, $x = 0.1$ $\quad\Rightarrow\quad a\sin\varepsilon = 0.1$. ①

By differentiating the general solution you obtain

$$\frac{dx}{dt} = -3e^{-3t}\,a\,\sin(t + \varepsilon) + ae^{-3t}\cos(t + \varepsilon).$$

When $t = 0$, $\dfrac{dx}{dt} = 0$ $\quad\Rightarrow\quad 0 = -3a\sin\varepsilon + a\cos\varepsilon$

$$\Rightarrow \quad \tan\varepsilon = \tfrac{1}{3}.$$

From ① we see that $\sin\varepsilon = \dfrac{0.1}{a}$, which is positive, and so ε must be an angle in the first quadrant.

$$\Rightarrow \quad \varepsilon = 0.322 \text{ (radians) and } a = \frac{0.1}{\sin 0.322} = 0.316.$$

The particular solution in this case is

$$x = 0.316e^{-3t}\sin(t + 0.322).$$

The initial amplitude of the motion is 0.316 m (to 3 decimal places) and the period in seconds is $\dfrac{2\pi}{1} = 2\pi$.

The amplitude decays exponentially. In this case the oscillation decays very quickly.

Figure 15.6 shows the graph of a typical damped oscillation. There are many real situations where the oscillations decrease gradually in amplitude like this. Oscillations of this type are often called *lightly damped* or *underdamped*.

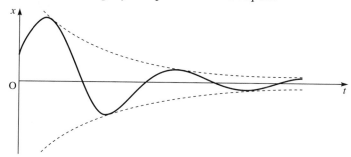

Figure 15.6

15.2 The general equation for damped oscillations

The differential equation of motion in the spring-mass-dashpot system above is an example of the general differential equation of a linearly damped system:

$$\frac{d^2y}{dt^2} + \alpha \frac{dy}{dt} + \omega^2 y = 0,$$

where α and ω are positive constants. For a spring-mass-dashpot system, $\alpha = \dfrac{r}{m}$ where r is the dashpot constant, and $\omega = \sqrt{\dfrac{k}{m}}$ where k is the stiffness of the spring. The quantity $\dfrac{\omega}{2\pi}$ is called the *natural frequency* of the system: it is the same whether the system is damped or undamped.

The solution of this differential equation can give several different types of motion, depending on the relative sizes of the parameters α and ω. The auxiliary equation is

$$\lambda^2 + \alpha\lambda + \omega^2 = 0.$$

This has two solutions,

$$\lambda_1 = \frac{-\alpha + \alpha^2 - 4\omega^2}{2} \quad \text{and} \quad \lambda_2 = \frac{-\alpha - \alpha^2 - 4\omega^2}{2}.$$

The *discriminant*, $\alpha^2 - 4\omega^2$, determines the nature of the solution. There are three possibilities, as follows.

● *Overdamping*: $\alpha^2 - 4\omega^2$, is positive, and the system does not oscillate (figure 15.7).

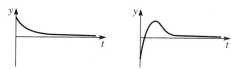

Figure 15.7

● *Underdamping*: $\alpha^2 - 4\omega^2$ is negative, and oscillations occur (figure 15.8).

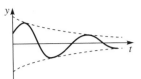

Figure 15.8

● *Critical damping*: $\alpha^2 - 4\omega^2 = 0$ (figure 15.9).

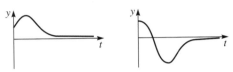

Figure 15.9

Critical damping is the borderline between overdamping and underdamping. It is not obvious in a physical situation when damping is critical, since the pattern of motion for critical damping can be very similar to that in the overdamped case.

QUESTION 15.2   In many systems damping is desirable or even necessary. For each of the following systems, decide whether damping is desirable and if so whether overdamping, underdamping or critical damping is preferable.
i)   A clock pendulum   ii)   A car suspension system
iii) A set of kitchen scales   iv) A robotic welding arm.

15.3 Modelling forced vibrations: the undamped case

In order to understand the mathematics of forced oscillations, including resonance, we look at the simplest suitable case, that of an object hanging on a light, perfectly elastic spring, without damping. (The case in which both forcing and damping occur is considered later in this chapter, on page 354.)

In figure 15.10 the top end A of the spring is forced to vibrate so that its displacement at time t is $y = a \sin \Omega t$. (This can be achieved experimentally, to a reasonable approximation, by attaching the supporting string over a pulley to a rotating cam, as shown in figure 15.14 on page 354.)

If the natural length of the spring is l_0, the stiffness of the spring is k and the object has mass m, then in equilibrium, point A coincides with O, the object is at rest and the extension of the spring is $e = \dfrac{mg}{k}$. At a general time t during the forced motion, the extension of the spring below the equilibrium position is denoted by x (figure 15.11).

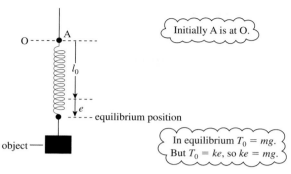

Figure 15.10

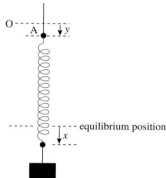

Figure 15.11

There are two forces acting on the object, the force of gravity mg and the tension T. The acceleration of the object is $\dfrac{\mathrm{d}^2x}{\mathrm{d}t^2}$. Applying Newton's second law gives $m\dfrac{\mathrm{d}^2x}{\mathrm{d}t^2} = mg - T$. The extension of the spring is $(e + x - y)$, so the tension in the spring is

$$T = k(e + x - y).$$

The equation of motion is therefore

$$m\frac{\mathrm{d}^2x}{\mathrm{d}t^2} = mg - k(e + x - y).$$

Expanding the right-hand side and recalling that $mg - ke = 0$ (figure 15.10), this becomes

$$m\frac{\mathrm{d}^2x}{\mathrm{d}t^2} + kx = ky.$$

Dividing both sides by m and putting $\omega^2 = \dfrac{k}{m}$,

$$\frac{\mathrm{d}^2x}{\mathrm{d}t^2} + \omega^2x = \omega^2y \qquad \text{or} \qquad \frac{\mathrm{d}^2x}{\mathrm{d}t^2} + \omega^2x = \omega^2\mathrm{f}(t) \qquad \qquad \text{①}$$

In the system we have described, $y = \mathrm{f}(t) = a \sin \Omega t$. This is *forced harmonic motion*, and $\dfrac{\Omega}{2\pi}$ is called the forcing frequency. The differential equation of motion may be written

$$\frac{\mathrm{d}^2x}{\mathrm{d}t^2} + \omega^2x = a\omega^2 \sin \Omega t. \qquad \qquad \text{②}$$

The complementary function is given by

$$A \sin \omega t + B \cos \omega t.$$

For the particular integral, try $x = p \sin \Omega t + q \cos \Omega t$. This gives

$$\frac{dx}{dt} = p\Omega \cos \Omega t - q\Omega \sin \Omega t$$

and

$$\frac{d^2x}{dt^2} = -p\Omega^2 \sin \Omega t - q\Omega^2 \cos \Omega t.$$

> Normally you would use l and m in the trial function, but in this example p and q are used instead because l and m are representing length and mass.

Substituting these in the differential equation of motion gives

$$-p\Omega^2 \sin \Omega t - q\Omega^2 \cos \Omega t + p\omega^2 \sin \Omega t + q\omega^2 \cos \Omega t \equiv a\omega^2 \sin \Omega t.$$

Equating coefficients:

$\sin \Omega t$: $\quad -p\Omega^2 + p\omega^2 = a\omega^2.$

$\cos \Omega t$: $\quad -q\Omega^2 + q\omega^2 = 0.$

Assuming $\Omega \neq \omega$, this gives $p = \dfrac{a\omega^2}{\omega^2 - \Omega^2}$ and $q = 0$.

The particular integral is therefore $\dfrac{a\omega^2}{\omega^2 - \Omega^2} \sin \Omega t.$

The general solution of the differential equation ② is therefore

$$x = A \sin \omega t + B \cos \omega t + \frac{a\omega^2}{\omega^2 - \Omega^2} \sin \Omega t \qquad (\Omega \neq \omega). \qquad\qquad ③$$

QUESTION 15.3 What is the physical significance of the terms

i) $A \sin \omega t + B \cos \omega t$

ii) $\dfrac{a\omega^2}{\omega^2 - \Omega^2} \sin \Omega t$

in the general solution above?

What would happen to the solution if Ω were increased steadily from zero towards ω?

Describing forced oscillations

You should have made the following deductions.

- The terms $A \sin \omega t + B \cos \omega t$ represent the natural or free oscillations of the system, as they would occur if the cam were not rotating. The natural frequency of the system is $\dfrac{\omega}{2\pi}$.

- The term $\dfrac{a\omega^2}{\omega^2 - \Omega 2} \sin \Omega t$ represents the oscillations caused by the rotating cam.

● As the value of Ω approaches that of ω, the quantity $\omega^2 - \Omega^2$ in the denominator tends to zero: the forced oscillations increase in amplitude. Consequently a small input amplitude a leads to a much larger output amplitude $\dfrac{a\omega^2}{\omega^2 - \Omega 2}$. This effect is known as *resonance*.

● This solution is only valid in cases where Ω does not actually equal ω.

Over the last few pages we have set up and solved a differential equation to model a simple case of forced oscillations. This has given us a mathematical explanation for the phenomenon of resonance which we described earlier in this chapter.

The case when $\Omega = \omega$

Resonance occurs when the frequency of the driving function is the same as the natural frequency of the system. Looking back at the differential equation for the general spring-mass system (equation ② on page 349), this occurs when $\Omega = \omega$ and so the differential equation becomes

$$\frac{d^2x}{dt^2} + \omega^2 x = a\omega^2 \sin \omega t.$$

Since it is unaffected by the function on the right-hand side of the equation, the complementary function is still $A \sin \omega t + B \cos \omega t$. To obtain the particular integral, given the function on the right-hand side, you would normally try $x = p \sin \omega t + q \cos \omega t$ but this is included in the complementary function. So you multiply the usual trial function by the independent variable.

In this case try $x = t(p \sin \omega t + q \cos \omega t)$.

Differentiating this gives

$$\frac{dx}{dt} = (p \sin \omega t + q \cos \omega t) + t(p\omega \cos \omega t - q\omega \sin \omega t)$$

and

$$\frac{d^2x}{dt^2} = p\omega \cos \omega t - q\omega \sin \omega t + p\omega \cos \omega t - q\omega \sin \omega t + t(-p\omega^2 \sin \omega t - q\omega^2 \cos \omega t)$$
$$= 2p\omega \cos \omega t - 2q\omega \sin \omega t - p\omega^2 t \sin \omega t - q\omega^2 t \cos \omega t.$$

Substituting these into the differential equation gives

$$2p\omega \cos \omega t - 2q\omega \sin \omega t - p\omega^2 t \sin \omega t - q\omega^2 t \cos \omega t + p\omega^2 t \sin \omega t$$
$$+ q\omega^2 t \cos \omega t \equiv a\omega^2 \sin \omega t.$$

Equating coefficients:

sin ωt: $\qquad\qquad -q\omega - p\omega^2 t + p\omega^2 t = a\omega^2$

$\Rightarrow \qquad\qquad\qquad q = -\tfrac{1}{2}a\omega.$

cos ωt: $\qquad\qquad 2p\omega - q\omega^2 t + q\omega^2 t = 0$

$\Rightarrow \qquad\qquad\qquad p = 0.$

So a particular integral is $-\dfrac{a\omega t}{2}\cos \omega t$ and the general solution in the case when $\Omega = \omega$ is given by

This is called the forcing term.

$$x = A \sin \omega t + B \cos \omega t - \frac{a\omega t}{2} \cos \omega t.$$

Note that as t increases the forcing term dominates the solution. It represents an oscillation whose amplitude is proportional to t and so grows linearly with time. This is a mathematical description of resonance. It occurs when the forcing frequency is identical to the natural frequency of the system.

Drawing graphs of forced oscillations

To illustrate the general results we have just established, we take a particular set of values for the variables involved and specify the initial conditions.

The general solution of $\dfrac{d^2x}{dt^2} + \omega^2 x = a\omega^2 \sin \Omega t$ for $\Omega \neq \omega$ is given by

$$x = A \sin \omega t + B \cos \omega t + \frac{a\omega^2}{\omega^2 - \Omega^2} \sin \Omega t$$

with $\omega = \sqrt{\dfrac{k}{m}}$.

We take the following values:

the stiffness of the spring:	$k = 20 \text{ Nm}^{-1}$
the mass of the object:	$m = 0.2 \text{ kg}$
the amplitude of the forcing motion:	$a = 0.02 \text{ m}$
acceleration due to gravity:	$g = 10 \text{ ms}^{-2}$,

and we assume that initially the object is stationary at the equilibrium position so that

when $t = 0$, $x = 0$ and $\dfrac{dx}{dt} = 0$.

Activities

15.1 Show that under these circumstances the object's displacement, x, is given by

$$x = \frac{1}{(100 - \Omega 2)} (2 \sin \Omega t - 0.2\Omega \sin 10t) \qquad (\Omega \neq 10).$$

The graphs in figure 15.12 show the variation of x with t for various values of Ω.

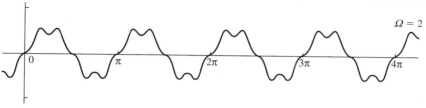

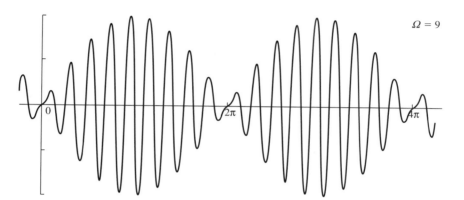

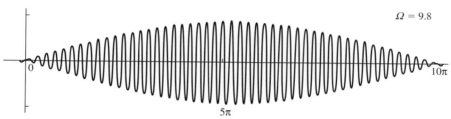

Figure 15.12

15.2 In the case when $\Omega = \omega$, the general solution was found to be

$$x = A \sin \omega t + B \cos \omega t - \frac{a \omega t}{2} \cos \omega t.$$

Show that with the values given on page 350, this result may be written as

$$x = 0.01 \sin 10t - 0.1t \cos 10t$$

Figure 15.13 shows the graph of this solution.

Experiment 15.2 will allow you to see how the theory we have just developed matches what happens in practice.

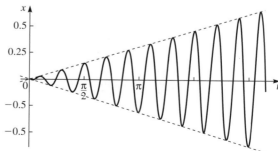

Figure 15.13

15.4 Damped forced oscillations

Earlier in this chapter you saw the effect on simple harmonic motion of introducing a linear damping device called a dashpot. The oscillations, if they occurred, decayed to zero. If the damping constant r was large compared with the stiffness of the spring and the mass of the object, oscillations did not occur at all.

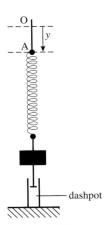

You have also seen the effect of forcing an undamped system. Most real systems do have an element of damping, so in this section we explore the effect of including a linear dashpot in the system. To make comparison easy, we shall look again at the spring-mass oscillator with the values given on page 349.

Figure 15.14

Figure 15.14 shows the spring-mass system which, as before, is forced to oscillate. A linear dashpot has been added below the object.

As before, the spring has stiffness 20 Nm^{-1}, the object has mass 0.2 kg and the amplitude of the forcing oscillation is 2 cm. The dashpot constant is 1 Nm^{-1}s and g is taken to be 10 ms^{-2}. The object starts from rest in its equilibrium position.

The first step is to formulate a differential equation to model this system.

Figure 15.15 shows the system and the forces acting, first in equilibrium and then at some general time t during the motion.

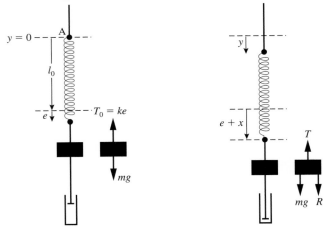

Figure 15.15

In equilibrium, the extension of the spring is $e = \dfrac{mg}{k} = \dfrac{2}{20} = 0.1$. At the general time t the object is displaced a distance x below the equilibrium level. The acceleration of the object is $\dfrac{d^2x}{dt^2}$.

The net force on the object in the direction of positive x (i.e. downwards) is $mg + R - T$.

Applying Newton's second law at any instant t gives

$$m\frac{d^2x}{dt^2} = mg + R - T.$$

The extension of the spring is $e + x - y$ so $T = 20(0.1 + x - y)$.

Let the length of the dashpot at time t be L, and its length in equilibrium be L_0.

Then $L = L_0 - x$, and $\dfrac{dL}{dt} = -\dfrac{dx}{dt}$.

The dashpot force R is given by $r\dfrac{dL}{dt}$, where r is the dashpot constant. So in this case,

$$R = 1\frac{dL}{dt} = -\frac{dx}{dt}.$$

The equation of motion becomes

This is mg.

$$0.2\frac{d^2x}{dt^2} = 2 - \frac{dx}{dt} - 20(0.1 + x - y)$$

$$\Rightarrow \quad \frac{d^2x}{dt^2} + 5\frac{dx}{dt} + 100x = 100y.$$

If the displacement of the forcing point is $y = 0.02 \sin \Omega t$, the differential equation modelling the system is

$$\frac{d^2y}{dt^2} + 5\frac{dx}{dt} + 100x = 2 \sin \Omega t.$$

The next stage is to solve the differential equation, which is (as before) a non-homogeneous linear equation with constant coefficients. Its auxiliary equation is

$$\lambda^2 + 5\lambda + 100 = 0$$

whose roots are

$$\lambda = \frac{-5 \pm \sqrt{375}j}{2}$$

$$\Rightarrow \quad \lambda = -2.5 \pm 9.68j.$$

The complementary function is therefore $e^{-2.5t}(A \sin 9.68t + B \cos 9.68t)$.

For the particular integral, try $x = p \sin \Omega t + q \cos \Omega t$.

Differentiating this gives

$$\frac{dx}{dt} = p\Omega \cos \Omega t - q\Omega \sin \Omega t$$

and $\quad \dfrac{d^2x}{dt^2} = -p\Omega^2 \sin \Omega t - q\Omega^2 \cos \Omega t.$

Substituting these in the differential equation for the system, gives

$-p\Omega^2 \sin \Omega t - q\Omega^2 \cos \Omega t + 5p\Omega \cos \Omega t - 5q\Omega \sin \Omega t + 100p \sin \Omega t + 100q \cos \Omega t$
$\equiv 2 \sin \Omega t.$

Equating coefficients:

$\sin\Omega t$:　　$-p\Omega^2 - 5q\Omega + 100p = 2.$

$\cos\Omega t$:　　$-q\Omega^2 + 5p\Omega + 100q = 0.$

Solving these equations for p and q gives

$$p = \frac{2(100 - \Omega^2)}{(100 - \Omega^2)^2 + 25\Omega^2}$$

and

$$q = -\frac{10\Omega}{(100 - \Omega^2)^2 + 25\Omega^2}.$$

The particular integral is therefore

$$\frac{2(100 - \Omega^2)}{(100 - \Omega^2)^2 + 25\Omega^2}\sin\Omega t - \frac{10\Omega}{(100 - \Omega^2)^2 + 25\Omega^2}\cos\Omega t.$$

The general solution is the sum of the complementary function and the particular integral:

$$x = e^{-2.5t}(A\sin 9.68t + B\cos 9.68t) + \frac{2(100 - \Omega^2)}{(100 - \Omega^2)^2 + 25\Omega^2}\sin\Omega t$$
$$- \frac{10\Omega}{(100 - \Omega^2)^2 + 25\Omega^2}\cos\Omega t.$$

As t increases, the natural damped oscillations, given by the complementary function, decay because of the $e^{-2.5t}$ term, leaving

$$x = \frac{2(100 - \Omega^2)}{(100 - \Omega^2)^2 + 25\Omega^2}\sin\Omega t - \frac{10\Omega}{(100 - \Omega^2)^2 + 25\Omega^2}\cos\Omega t.$$

This is called the *steady state* solution. It is the particular integral of the differential equation. It describes the oscillations that occur after the unforced oscillations have died away. Figure 15.16 shows graphs of this steady state solution for two values of Ω.

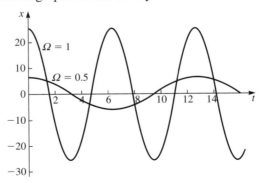

Figure 15.16

Remember that in the undamped case the value of Ω for resonance was calculated by setting the denominator to zero in the particular integral. Catastrophic resonance does not occur in the damped case, because the denominator of each part of the particular integral is always greater than zero. However, the amplitude of the forced vibrations does still depend on the value of Ω.

The amplitude of the steady state oscillations is the square root of the sum of the squares of the coefficients of cos Ωt and sin Ωt in the steady state solution, i.e.

$$\sqrt{\left(\frac{2(100 - \Omega^2)}{(100 - \Omega^2)^2 + 25\Omega^2}\right)^2 + \left(\frac{10\Omega}{(100 - \Omega^2)^2 + 25\Omega^2}\right)^2}.$$

This can be simplified to $\dfrac{2}{\sqrt{(100 - \Omega^2)^2 + 25\Omega^2}}$.

This result shows how the amplitude of the steady state solution depends on Ω. Figure 15.17 shows a graph of this amplitude against Ω.

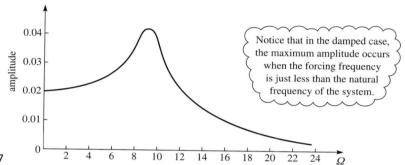

Figure 15.17

When Ω and ω are close in value to each other, then it follows that the forcing frequency $\dfrac{\Omega}{2\pi}$ and the natural frequency $\dfrac{\omega}{2\pi}$ are close in value. When this is the case, the steady state oscillations become large compared with the input amplitude a. The motion of the system at these relatively large amplitudes is still called resonance, though the amplitude of the vibrations does not increase without limit as it does in the undamped case.

You have now seen the effect of linear damping on the system. What would be the effect of varying the damping constant, r? To predict this, look at the differential equation for the same damped spring–mass system ($m = 0.2$ and $k = 20$), but this time use a general damping constant r. The equation becomes

$$0.2\frac{\mathrm{d}^2x}{\mathrm{d}t^2} + r\frac{\mathrm{d}x}{\mathrm{d}t} + 20x = 0.4 \sin \Omega t.$$

We know that the complementary function decays and that the steady state oscillations are given by the particular integral. In this case it is

$$x = \frac{0.4(20 - 0.2\Omega^2)}{(20 - 0.2\Omega^2)^2 + r^2\Omega^2} \sin \Omega t - \frac{0.4r\Omega}{(20 - 0.2\Omega^2)^2 + r^2\Omega^2} \cos \Omega t.$$

The amplitude of the steady state forced oscillation is

$$\frac{2}{\sqrt{(100 - \Omega^2)2 + 25r^2\Omega^2}}.$$

Figure 15.18 shows graphs of the steady state amplitude against Ω for different values of r.

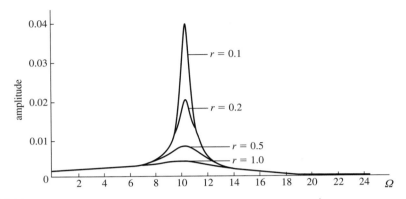

Figure 15.18

The graphs show that as r decreases (i.e. the amount of damping is reduced), the amplitude at the resonant frequency increases. In each case resonance occurs when Ω is very near in value to ω (in this case 10). In any real system there is always some damping but, as you can see, if the damping constant is small the resonance can still be damaging.

Activity

15.3   The previous example involved a particular case of damped forced motion in which the various parameters of the system were given particular values.

   i)   Show that the differential equation modelling the general case may be written

$$\frac{\mathrm{d}^2x}{\mathrm{d}t^2} + \alpha\frac{\mathrm{d}x}{\mathrm{d}t} + \omega^2x = a\omega^2 \sin \Omega t.$$

   ii)   Find   **a)** the general solution;

   b) the particular solution corresponding to

$$x = 0 \text{ and } \frac{\mathrm{d}x}{\mathrm{d}t} = 0 \text{ at } t = 0;$$

   c) the amplitude of the steady state oscillations.

Chapter 15 Exercises

15.1   An electrical circuit consists of a 0.2 henry inductor, a 1 ohm resistor and a 0.8 farad capacitor in series. The charge q coulombs on the capacitor is modelled by the differential equation

$$0.2\frac{\mathrm{d}^2q}{\mathrm{d}t^2} + \frac{\mathrm{d}q}{\mathrm{d}t} + \frac{1}{0.8}q = 0.$$

Initially q is 2 and $\dfrac{\mathrm{d}q}{\mathrm{d}t}$ (the current in ampères) is 4.

   i)   Find an equation for the charge as a function of time.

ii) Sketch the graphs of charge and current against time. Describe how the charge and current change.

iii) What is the charge on the capacitor and the current in the circuit after a long period of time?

15.2 The temperature of a chemical undergoing a reaction is modelled by the differential equation

$$2\frac{d^2T}{dt^2} + \frac{dT}{dt} = 0$$

where T is the temperature in °C and t is the time in minutes.

For a particular experiment, the temperature is initially 50°C, and it is 45°C one minute later.

i) Find an expression for the temperature T at any time.

ii) What will the temperature be after two minutes?

iii) Sketch a graph of T against t.

iv) What is the steady state temperature?

15.3 The angular displacement from its equilibrium position of a swing door fitted with a damping device is modelled by the differential equation

$$\frac{d^2\theta}{dt^2} + 4\frac{d\theta}{dt} + 5\theta = 0.$$

The door starts from rest at an angle of $\frac{\pi}{4}$ from its equilibrium position.

i) Find the general solution of the differential equation.

ii) Find the particular solution for the given initial conditions.

iii) Sketch a graph of the particular solution, and hence describe the motion of the door.

iv) What does your model predict as t becomes large?

15.4 The diagram shows a block of mass 10 kg being dragged across a horizontal surface by means of an elastic spring AB of unstretched length 0.5 m. The end A is being pulled with constant speed 5 ms⁻¹. The block is attached to the end B of the spring.

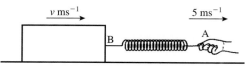

The block is subject to a resistance force given by $4v$ N, where v is the speed of the block in ms⁻¹. The only other horizontal force acting on the block is the tension in the spring given by $T = k(y - 0.5)$ N, where y is the length of the spring in metres and k is a constant (the stiffness of the spring).

i) By considering the horizontal forces acting on the block, write down the equation of motion of the block in terms of $\frac{dv}{dt}$, v, y and k. Use the fact that $\frac{dy}{dt} = 5 - v$ to eliminate v, and show that $10\frac{d^2y}{dt^2} + 4\frac{dy}{dt} + ky = 0.5k + 20$.

ii) You are given that $k = 20$. Find the general solution of the differential equation.

At the beginning of the motion, the spring is unstretched and the block is at rest.

iii) Find an expression for y at time $t > 0$.

iv) What are the limiting values of v and y after a long period of time? Explain briefly how these values would be affected if a different spring were used with a lower value of k.

15.5 An electrical circuit consists of a 1 henry inductor and a 10^{-4} F capacitor in series with a sinusoidal power source.

The charge q, in coulombs (C), stored in the capacitor is given by the differential equation

$$\frac{d^2q}{dt^2} + 10\,000q = 1000\sin\Omega t$$

where t is the time in seconds. Initially the charge q and current $\frac{dq}{dt}$ in the circuit are both zero.

i) Find the particular solution for the charge q.

ii) State the value of Ω for which resonance occurs. Find the particular solution for $q(t)$ in this case. Calculate the time at which the charge first exceeds $10\,000$ C.

iii) Use a graphics calculator or computer to graph the solution in part ii).

15.6 A sphere of mass m and radius a is falling vertically through a liquid which produces a linear resistance force. The motion of the sphere is modelled by the differential equation

$$m\frac{d^2x}{dt^2} + ra\frac{dx}{dt} = mg$$

where x is the distance fallen in t seconds and r is a constant. The sphere is released from rest so that $x = 0$ and $\frac{dx}{dt} = 0$ when $t = 0$.

i) Find the solution for x and hence the velocity of the sphere as a function of time.

ii) Draw a graph of the velocity against time, and describe the motion of the sphere.

15.7 During the design of a car, part of the suspension system is tested by subjecting it to violent displacements. One such test is modelled by the differential equation

$$\ddot{x} + 2k\dot{x} + x = 1$$

where x is displacement, and initially $x = 0$ and $\dot{x} = 0$. The parameter $k\ (> 0)$ is known as the damping coefficient and can be varied during the tests.

For optimum road holding a 'hard' suspension is desirable and it is believed that to achieve this the damping should be critical.

i) Find the value of k for critical damping.

ii) Determine x as a function of time t in this case.

For a more comfortable ride a 'soft' suspension is proposed in which $k = 0.6$.

iii) Determine x as a function of time t for the 'soft' suspension.

iv) Find the maximum displacement of the 'soft' suspension.

15.8 The current in an electrical circuit consisting of an inductor, resistor and capacitor in series with an alternating power source, is described by the equation

$$\frac{\mathrm{d}^2 I}{\mathrm{d}t^2} + 25\frac{\mathrm{d}I}{\mathrm{d}t} + 100I = -170 \sin 20t$$

where I is the current in amperes and t is the time in seconds after the power source is switched on.

i) Find the general solution.

When $t = 0$, $\dfrac{\mathrm{d}I}{\mathrm{d}t} = I = 0$.

ii) Find the solution.

The exponentially decaying terms in the solution describe what is known as the transient current. The non-decaying terms describe the steady state current.

iii) Write down an expression for the steady state current for the solution in part **ii)**. Why would this expression remain unchanged if the initial conditions were different?

iv) Express the steady state current in the form $R \sin (20t + \alpha)$, where R and α are to be determined. Verify that, after only 1 second, the magnitude of the transient current is close to 1% of the steady state amplitude, R.

Personal Tutor

15.1 A hydrometer is an instrument used to measure the densities of liquids. It works on the principle that a floating body experiences an upthrust equal in magnitude to the weight of liquid displaced by that part of the body which is submerged.

A particular hydrometer consists of a light uniform cylindrical rod of radius a and length b with a small mass m countersunk into one end. In a liquid the hydrometer floats with its axis vertical and the amount by which it protrudes above the surface gives a measure of the density of the liquid.

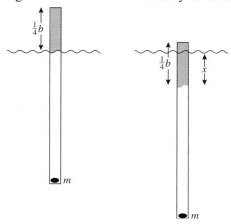

Such a hydrometer floats in equilibrium in a liquid of density ρ with one quarter of its length above the surface.

i) Find an expression for the density ρ in terms of a, b and m.

The hydrometer is pushed vertically downwards and then released. The resistance to motion when the speed is v is given by $2mkv$ where k is a constant. At time t the displacement of the equilibrium level below the surface is denoted by x. The differential equation governing the motion is $m\dfrac{d^2x}{dt^2} = -2mk\dfrac{dx}{dt} - \dfrac{4mg}{3b}x$.

ii) Explain the physical significance of each of the three terms.

iii) Show that for the hydrometer to perform damped simple harmonic motion
$$\frac{4g}{3b} > k^2.$$

Initially the hydrometer is pushed vertically downwards until just submerged and then released from rest.

iv) Denoting $\dfrac{4g}{3b} - k^2$ by ω^2, solve the differential equation governing the motion

and express the solution as an equation for x in terms of t, b, k and ω.

v) Describe the motion of the hydrometer by sketching the graph of x against t.

Four complete oscillations after release, one eighth of the hydrometer is above the surface.

(vi) Show that the periodic time T of the oscillations is given by $T = \dfrac{\ln 2}{4k}$.

15.1 Linear damping is provided by a linear dashpot.

15.2 Motion for which the differential equation is

$$\frac{d^2x}{dt^2} + \alpha\frac{dx}{dt} + \omega^2 x = 0 \quad \text{(for } \alpha > 0\text{)}$$

is called *damped harmonic motion*. A spring-mass-dashpot system performs damped harmonic motion.

15.3 The following table shows the features of damped harmonic motion.

$\alpha^2 - 4\omega^2$	Solution	Type of damping
$\alpha = 0$	$x = A \sin \omega t + B \cos \omega t$ or $x = a \sin (\omega t + \varepsilon)$	no damping
positive $(\alpha > 2\omega)$	$x - Ae^{\lambda_1 t} + Be^{\lambda_2 t}$	overdamping
zero $(\alpha = 2\omega)$	$x = (A + Bt)e^{-\frac{\alpha t}{2}}$	critical damping
negative $(\alpha < 2\omega)$	$x = ae^{-\frac{\alpha t}{2}} \sin (pt + \varepsilon)$, where $p = \frac{1}{2}\sqrt{4\omega^2 - \alpha^2}$	underdamping

15.4 Motion for which the differential equation is $\dfrac{d^2x}{dt^2} + \omega^2 x = \omega^2 f(t)$ is called

undamped forced harmonic motion; f(t) is the forcing term. For a sinusoidal forcing term, $f(t) = a \sin \Omega t$ and the general solution of the differential equation is

$$x = A \sin \omega t + B \cos \omega t + \frac{a\omega^2}{\omega^2 - \Omega^2} \sin \Omega t \qquad (\omega \neq \Omega).$$

15.5 The quantity $\dfrac{\Omega}{2\pi}$ is called the forcing frequency, and $\dfrac{\omega}{2\pi}$ is the natural frequency

of the system. When $\Omega = \omega$ resonance occurs. In this case $f(t) = a \sin \omega t$ and the general solution is

$$x = A \sin \omega t + B \cos \omega t - \frac{a\omega t}{2} \cos \omega t.$$

15.6 With linear damping, the differential equation model becomes

$$\frac{d^2x}{dt^2} + \alpha \frac{dx}{dt} + \omega^2 x = a\omega^2 \sin \Omega t \qquad (a > 0).$$

15.7 The complementary funciton of this differential equation decays to zero. The particular integral remains and is called the stead state solution. If $f(t) = a \sin \Omega t$ then the steady state solution is

$$x = \frac{(\omega^2 - \Omega^2)a\omega^2}{(\omega^2 - \Omega^2)^2 + \Omega^2 a^2} \sin \Omega t - \frac{a\alpha\Omega\omega^2}{(\omega^2 - \Omega^2)^2 + \Omega^2\alpha^2} \cos \Omega t.$$

15.8 The amplitude of this steady state solution is

$$\frac{a\omega^2}{\sqrt{(\omega^2 - \Omega^2)^2 + \Omega^2\alpha^2}}.$$

15.9 When Ω takes the *w*alue which maximises the amlitude of the stedy state

oscillaitons, $\dfrac{\Omega}{2\pi}$ is called the resonant frequency.

16 Dimensional analysis

But whatever his weight in pounds, shillings and ounces,
He always seems bigger because of his bounces.

The House at Pooh Corner, A.A. Milne

What makes the above rhyme about Tigger sound so ridiculous?

Look at the units in the first line of the rhyme. Pounds can be units of either money or mass, shillings are units of money and ounces are units of mass. You can see at once that not only is there a mixture of units, but also of the underlying quantities which they are measuring.

In this chapter you will look at the quantities which you have met in mechanics and classify them according to their *dimensions*. The dimensions of a quantity are closely related to the units in which it is measured.

All of the quantities you have met so far can be described in terms of three fundamental quantities: Mass, Length and Time. Take, for example, the quantity *area*; there are many familiar ways of finding area, depending on the shape involved (see figure 16.1).

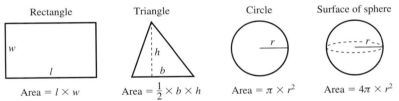

Figure 16.1

All of these formulae have essentially the same form:

$$\text{area} = \text{number} \times \text{length} \times \text{length}.$$

In the case of the rectangle the 'number' is 1 and the two 'lengths' are the length and breadth of the shape. For the circle and the sphere the two 'lengths' are the same, namely the radius, and the 'numbers' are multiples of π. However, the structure of the formula is the same for all four shapes, and indeed for any other area formula you can think of (for example, the area of a trapezium or the surface area of a cone).

The dimensions of the quantity Area can therefore be summarised using the formula

$$[\text{Area}] = L^2.$$

The square brackets [ ] mean the *dimensions of*, and the letter L represents the quantity Length. So the whole statement reads:

The dimensions of area = Length squared.

There are two important points to notice about this.

1 The numbers at the start of the formulae (in the four cases above they are $1, \frac{1}{2}, \pi$ and 4π) do not feature in the dimensions of the quantity because they are dimensionless.
2 The units in which a quantity is measured are derived directly from its dimensions. In the S.I. system the unit of length is 1 metre. Since the dimensions of area are L^2, it follows that the S.I. unit for area is 1 metre squared, usually written m^2.

Note

The area under a curve is often found by calculus methods, using the formula
$$\text{area} = \int y \, dx.$$
In this case both y and the infinitesimal quantity dx represent lengths and so the integral has dimensions L^2, as must be the case since it represents an area.

QUESTION 16.1   How many formulae can you find for the volumes of objects?

Show that all your formulae for volume involve multiplying three lengths together so that
$$[\text{Volume}] = L^3.$$

16.1 The dimensions of further quantities

So far the quantities discussed, area and volume, have only involved lengths. What are the dimensions of speed (or velocity), and of acceleration?

To find the dimensions of any quantity, start by writing down a simple formula which you might use to calculate it. For example, to calculate speed you might use

$$\text{Speed} = \frac{\text{Distance}}{\text{Time}}.$$

The dimensions follow immediately from this. A distance is clearly a length, so

$$[\text{Speed}] = \frac{[\text{Distance}]}{[\text{Time}]} = \frac{L}{T} = LT^{-1}.$$

It follows that the S.I. unit for speed is 1 metre per second, written ms^{-1}. Notice that a second fundamental dimension, T (Time), is now involved.

Similarly
$$\text{Acceleration} = \frac{\text{Change in velocity}}{\text{Time taken}}$$

and so
$$[\text{Acceleration}] = \frac{[\text{Speed}]}{[\text{Time}]}$$
$$= \frac{LT^{-1}}{T}$$
$$= LT^{-2}.$$

The corresponding S.I. unit is therefore $1 \ ms^{-2}$.

Exactly the same procedure allows you to find the dimensions of force.

$$\text{Force} = \text{Mass} \times \text{Acceleration}$$
$$[\text{Force}] = [\text{Mass}] \times [\text{Acceleration}]$$
$$= M \times LT^{-2}$$
$$= MLT^{-2}.$$

Therefore the S.I. unit for force is 1 kilogram metre per second squared. Because force is an important concept and this unit would be such a mouthful, it is given a name of its own, the newton (N).

$$1\,\text{N} = 1\,\text{kg}\,\text{ms}^{-2}.$$

Notice that now all three of the basic dimensions, M, L and T are being used.

EXAMPLE 16.1

i) Find the dimensions of
    a) kinetic energy
    b) gravitational potential energy
    c) work.

ii) Comment on the significance of your answers.

SOLUTION

i) **a)**
$$\text{Kinetic energy} = \tfrac{1}{2}mv^2$$
$$[\text{K.E.}] = \text{M} \times (\text{LT}^{-1})^2$$
$$= \text{ML}^2\text{T}^{-2}$$

g is an acceleration so it has dimensions LT^{-2}

    b) Gravitational potential energy $= mgh$
$$[\text{P.E.}] = \text{M} \times \text{LT}^{-2} \times \text{L}$$
$$= \text{ML}^2\text{T}^{-2}$$

    c)
$$\text{Work} = \text{Force} \times \text{Distance}$$
$$= \text{MLT}^{-2} \times \text{L}$$
$$= \text{ML}^2\text{T}^{-2}$$

ii) All three have the same dimensions because they are examples of the same underlying quantity, energy. The S.I. unit for this could be written as $1\,\text{kg}\,\text{m}^2\text{s}^{-2}$ but is actually given the special name of 1 joule, J.

Dimensionless quantities

Some quantities have no dimensions. For example, angles are dimensionless. The angle θ in figure 16.2 is defined as

$$\theta\,(\text{radians}) = \frac{\text{arc length}}{\text{radius}}.$$

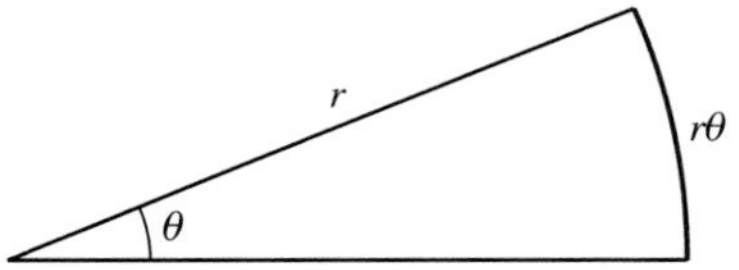

Figure 16.2

Consequently $[\theta] = \dfrac{\text{L}}{\text{L}}$

and θ is dimensionless.

Notice that although in this case the formula used gave θ in radians, the same result would have been obtained for θ in degrees, using

$$\theta = 360 \times \frac{\text{arc length}}{2\pi \times \text{radius}}.$$

The dimensions of a quantity cannot be altered by changing the units in which it is measured.

All numbers are dimensionless, including the trigonometrical ratios, for example $\sin\theta$, $\cos\theta$, $\tan\theta$, and irrational numbers such as π and e. This is the reason why you ignore any numbers in a formula when considering its dimensions; numbers, being dimensionless, cannot be included in this analysis.

16.2 Other systems of units

The units used in this chapter, and indeed in all the chapters in this book, are almost entirely S.I. units. There are other self-consistent sets of units, such as the 'cgs' system (centimetre, gram, second) or the Imperial system (foot, pound, second). Knowing the dimensions of a quantity allows you to find the appropriate unit within any system.

The set of dimensions may also be extended to include quantities that arise in electricity and magnetism. For instance Q represents the dimension of electrical charge.

Change of units

It is helpful to know the dimensions of a quantity when you are changing the units in which it is to be measured, as in the following example.

EXAMPLE 16.2

The energy of a body is 5000 foot poundals (imperial units based on feet, pounds and seconds). Write this in S.I. units, i.e. joules.
(To 3 s.f., 1 pound (lb) = 0.453 kg and 1 foot = 0.305 m.)

SOLUTION

The dimensions of energy are $ML^2T^{-2} = \dfrac{ML^2}{T^2}$

So
$$1 \text{ foot poundal} = \frac{1 \text{ lb} \times (1 \text{ foot})^2}{(1 \text{ second})^2}$$
$$= \frac{0.453 \text{ kg} \times (0.305 \text{ m})^2}{(1 \text{ second})^2}$$
$$= 0.0421 \text{ J}$$

and
$$5000 \text{ foot poundals} = 5000 \times 0.0421 \text{ J}$$
$$= 211 \text{ J (to 3 s.f.)}.$$

On the next page is a table which contains many standard quantities, along with each of their formulae, their dimensions and their S.I. units. You may find it useful to make a copy of this table for reference.

Quantity	Formula	Dimensions	S.I. unit
Area	$l \times w$	L^2	m^2
Volume	$l \times w \times h$	L^3	m^3
Speed	$\dfrac{d}{t}$	LT^{-1}	ms^{-1}
Acceleration	$\dfrac{v}{t}$	LT^{-2}	ms^{-2}
g	acceleration	LT^{-2}	ms^{-2}
Force	$F = ma$	MLT	N, newton
Weight	mg	MLT	N, newton
Kinetic energy	$\frac{1}{2}mv^2$	ML^2T^{-2}	J, joule
Gravitational potential energy	mgh	ML^2T^{-2}	J, joule
Work	Fs	ML^2T^{-2}	J, joule
Power	Fv	ML^2T^{-3}	W, watt
Impulse	Ft	MLT^{-1}	Ns
Momentum	mV	MLT^{-1}	Ns
Pressure	$\dfrac{\text{Force}}{\text{Area}}$	$ML^{-1}T^{-2}$	Nm^{-2}, pascal
Density	$\dfrac{m}{V}$	ML^{-2}	$kg\ m^{-3}$
Moment	Fd	ML^2T^{-2}	Nm
Angle	$s = r\theta$	Dimensionless	(radian)
Angular velocity	$\dfrac{\text{Angle}}{\text{Time}}$	T^{-1}	$rad\ s^{-1}$
Angular acceleration	$\dfrac{\text{Angular velocity}}{\text{Time}}$	T^{-2}	$rad\ s^{-2}$
Modulus of elasticity	$T = \dfrac{\lambda x}{l_0}$	MLT^{-2}	N
Stiffness	$T = kx$	MT^{-2}	Nm^{-1}
Gravitational constant, G	$F = \dfrac{Gm_1m_2}{d^2}$	$M^{-1}L^3T^{-2}$	$Nm^2\,kg^{-2}$
Period	Time interval	T	s
Frequency	$\dfrac{1}{\text{Period}}$	T^{-1}	Hz, hertz
Coefficient of restitution	$\dfrac{V\text{ separation}}{V\text{ approach}}$	Dimensionless	—
Coefficient of friction	$F = \mu R$	Dimensionless	—

16.3 Dimensional consistency

In any equation or formula, all the terms must have the same dimensions. If that is the case, the equation is said to be *dimensionally consistent*. If not, for example if force and area are being added or subtracted, or if a mass is being equated to a velocity, the equation is said to be dimensionally inconsistent.

Any statement that is dimensionally inconsistent must be wrong. In the rhyme at the start of this chapter, the phrase '*his weight in pounds, shillings and ounces*' is nonsense because it is dimensionally inconsistent.

EXAMPLE 16.3

Show that the formula

$$s = ut + \tfrac{1}{2}at^2$$

is dimensionally consistent.

SOLUTION

There are three terms, with dimensions as follows:

$$[s] = L$$
$$[ut] = LT^{-1} \times T = L$$
$$[\tfrac{1}{2}at^2] = LT^{-2} \times T^2 = L.$$

All three terms have the same dimensions, and so the formula is dimensionally consistent.

The fact that a formula is dimensionally consistent does not mean it is necessarily right, but if it is dimensionally inconsistent it is certainly wrong.

Good mathematicians develop the habit of automatically checking the dimensional consistency of anything they write. A consequence of this is that they tend to leave everything written in symbols until the final calculation. You know, for example, that g has dimensions LT^{-2} but if it is replaced by an approximate value, e.g. 10, its dimensions are no longer evident: $h = \tfrac{1}{2}gt^2$ is dimensionally consistent but $h = 5t^2$ seems not to be.

EXAMPLE 16.4

A mathematician writes down the equation

$$\tfrac{1}{2}m_1v_1{}^2 = m_2gh + m_2v_2.$$

Show that it must be wrong.

SOLUTION

Checking for dimensional consistency:

$$[\tfrac{1}{2}m_1v_1{}^2] = M \times (LT^{-1})^2 = ML^2\,T^{-2}$$
$$[m_2gh] = M \times LT^{-2} \times L = ML^2T^{-2}$$
$$[m_2v_2] = M \times LT^{-1} = MLT^{-1}.$$

The third term is dimensionally different from the other two and so the equation is incorrect.

16.4 Finding the form of a relationship

It is sometimes possible to determine the form of a relationship just by looking at the dimensions of the quantities likely to be involved, as in the following example.

A pendulum consists of a light string of length l with a bob of mass m attached to the end. You would expect that the period, t, might depend in some way on the variables l and m, and also on the value of g.

This can be expressed in the form

$$t = kl^\alpha \, m^\beta \, g^\gamma$$

where k is dimensionless and the powers α, β and γ are to be found.

Writing down the dimensions of each side gives

$$T = L^\alpha \times M^\beta \times (LT^{-2})^\gamma$$
$$T = L^{\alpha+\gamma}M^\beta \, T^{-2\gamma}.$$

The left-hand side of the equation, T, may be written as $L^0M^0T^1$ and so

$$L^0M^0T^1 = L^{\alpha+\gamma}M^\beta \, T^{-2\gamma}.$$

Equating the powers of L, M and T gives:

$$\begin{aligned} \text{L:} \quad & 0 = \alpha + \gamma \\ \text{M:} \quad & 0 = \beta \\ \text{T:} \quad & 1 = -2\gamma. \end{aligned}$$

Solving these gives $\alpha = \frac{1}{2}$, $\beta = 0$ and $\gamma = -\frac{1}{2}$ and so the relationship is

$$t = k\sqrt{\frac{l}{g}}.$$

You may remember that you derived this formula in the chapter on simple harmonic motion on page 328. You found it in the form

$$t = 2\pi\sqrt{\frac{l}{g}}.$$

That result was obtained by writing down the equation of motion of the pendulum bob and comparing it with the equation for simple harmonic motion. In this method all you had to do was to think about the dimensions, a very beautiful piece of mathematics. Of course the dimensions method does not tell you that the value of k is 2π, but it does provide the correct form of the relationship.

Notice that the mass of the pendulum bob, m, does not feature in the formula for the period. This is found to be the case in practice: a heavier bob makes no difference to the period.

What about the angle of swing of the pendulum, θ? Since angles are dimensionless it cannot feature in this argument. You may like to think of its effect as being included within the dimensionless k. In fact the value of k is approximately constant (equal to 2π) for small values of θ, but it does vary with θ for larger swings.

In the next example the work is taken a step further. First the form of a relationship is proposed, then data are used to determine the constant involved. The resulting formula is used to predict the outcome of a future experiment. Notice that the units look after themselves providing you are consistent with their use; in this case they happen to be in the cgs system.

EXAMPLE 16.5

In an experiment, small spheres are dropped into a container of liquid which is sufficiently deep for them to attain terminal velocity. For any sphere, the terminal velocity, V, is thought to depend on its radius, r, its weight mg and the viscosity of the liquid, η (the Greek letter 'eta'). (Viscosity is a measure of the 'stickiness' of a liquid and has dimensions $ML^{-1}T^{-1}$.)

i)   Write down a formula for V as the product of unknown powers of r, mg, η, and a dimensionless constant k.

ii) Find the powers of r, mg and η and write out the formula for V with these values substituted in.

In a particular liquid it is found that a sphere of mass 0.02 g and radius 0.1 cm has terminal velocity 5 cm s^{-1}.

iii) Find the value of $\dfrac{k}{\eta}$ taking g to be 1000 cm s^{-2}.

iv) Find the terminal velocity of a sphere of mass 0.03 g and radius 0.2 cm.

SOLUTION

i) $V = kr^{\alpha}(mg)^{\beta}\eta^{\gamma}$.

ii) Taking dimensions of both sides of the equation gives

$$LT^{-1} = L^{\alpha} \times (MLT^{-2})^{\beta} \times (ML^{-1}T^{-1})^{\gamma}$$
$$M^{0}L^{1}T^{-1} = M^{\beta+\gamma}L^{\alpha+\beta-\gamma}T^{-2\beta-\gamma}$$

Equating powers: M: $0 = \beta + \gamma$

L: $1 = \alpha + \beta - \gamma$

T: $-1 = -2\beta - \gamma$

Solving these gives $\alpha = -1$, $\beta = 1$ and $\gamma = -1$

and so the formula is $V = \dfrac{kmg}{r\eta}.$ ①

iii) Substituting $m = 0.02$, $r = 0.1$, $g = 1000$ and $V = 5$ gives

$$5 = \frac{k \times 0.02 \times 1000}{0.1 \times \eta}$$

$$\frac{k}{\eta} = 0.025 \text{ (in cgs units)}.$$

iv) Substituting $m = 0.03$, $r = 0.2$, $g = 1000$ and $\dfrac{k}{\eta} = 0.025$ in ①

gives $$V = \frac{0.025 \times 0.03 \times 1000}{0.2}$$

$$= 3.75.$$

The terminal velocity of this sphere is 3.75 cm s^{-1}.

The method of dimensions

This method for finding the form of a proposed formula is sometimes called *the method of dimensions*; at other times it is simply referred to as *dimensional analysis*. There are a number of points which you should realise when using it.

1 In mechanics, relationships are based on three fundamental quantities Mass, Length and Time and so the right-hand side of the formula can involve no more than three independent quantities. Otherwise you will end up with three equations to find four or more unknown powers. (The last example actually involved four quantities, m, g, r and η, but the first two of these were tied together as the weight, mg, and were not, therefore, independent.)

2 The method requires you to make modelling assumptions about which quantities are going to be important and which can be ignored. You must always be prepared to review these assumptions.

3 This method can only be used when a quantity can be written as a product of powers of other quantities. There are many situations which cannot be modelled by this type of formula and for these situations this method is not appropriate. For example, the formula $s = ut + \frac{1}{2}at^2$ has two separate terms which are added together on the right-hand side. The form of this relationship could not be predicted using the method of dimensions.

Chapter 16 Exercises

16.1 A scientist thinks that the speed, v, of a wave travelling along the surface of an ocean depends on the depth of the ocean, h, the density of the water in the ocean, r (the Greek letter 'rho'), and g.

i) Write down the dimensions of each of the quantities v, h, r and g.

The scientist expresses his idea using the formula

$$v = kh^\alpha \rho^\beta g^\gamma \qquad \text{where } k \text{ is dimensionless.}$$

ii) Use dimensional analysis to find the values of α, β and γ and write out the formula with these values substituted in.

iii) Do waves travel faster in deep water or shallow water?

iv) Do waves travel faster in winter (when the water density is greater) or in summer?

16.2 A student who has not yet learnt about the energy of rotating bodies observes a symmetrical flywheel of mass m and radius r rotating in a horizontal plane about a vertical axis through its middle with angular speed ω. The student decides that the rotational energy, E, of the flywheel can only depend on a product of powers of these three quantities and a dimensionless constant. (Note: the dimensions of energy are ML^2T^{-2}.

ii) Express the student's opinion in the form of a formula for E.

iii) Use dimensional analysis to find the powers of m, r and ω in the formula.

The student conducts some experiments which show that when $m = 20$ kg, $r = 0.5$ m and $\omega = 4$ rad s^{-1}, $E = 40$ J.

iv) Given that the student's experiments are accurate, find the complete formula for E.

v) A similar flywheel, mounted in the same way, has mass 50 kg and radius 1.2 m. Find its energy when it is rotating at 12 rad s^{-1}.

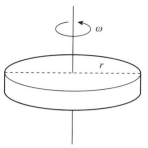

16.3 The magnitude of the force of gravitational attraction, F, between two objects of mass m_1 and m_2 at a distance d apart is given by

$$F = \frac{Gm_1m_2}{d^2}$$

where G is the universal constant of gravitation.

i) Find the dimensions of G.

An astronomer proposes a model in which the lifetime, t, of a star depends on a product of powers of its mass, m, its initial radius r_0, G and a dimensionless constant.

ii) Use the method of dimensions to find the resulting formula for t.

Observation shows that the larger the initial radius the longer the lifetime of the star, but that the larger the mass the shorter the lifetime of the star.

iii) Is the model consistent with these observations?

iv) Show that the model can be expressed more simply if the initial density, ρ_0, of the star is used as one of the variables.

16.4 In the early seventeenth century Mersenne (1588–1648) conducted experiments with long lengths of rope and so obtained the law for the frequency of transverse vibrations of strings.

Assuming that the frequency depends on products of powers of T, the tension in the rope, l, the length of the rope, and m, the mass per unit length of the rope,

i) find, by dimensional analysis, the form of the relationship.

A rope of length 24 m and mass 0.5 kg m^{-1} under tension of 72 N is found to vibrate with a frequency of $\frac{1}{4}$ of a cycle per second.

ii) State the exact relationship between the frequency, T, l and m.

iii) Find the frequency of vibration of a string of length 20 cm and mass 0.005 g cm^{-1} under a tension of 8×10^5 dynes. (The dyne is the cgs unit of force: 1 dyne is the force required to give a mass of 1 g an acceleration of 1 cm s^{-2}.)

16.5 For a wave travelling with speed c, the wavelength is denoted by λ and the frequency, which is the number of complete oscillations per second, by f.

i) Verify that the relation $c = \lambda f$ is dimensionally consistent.

For a sound wave travelling in either air or water, the frequency, f, depends on the wavelength, λ, the density, ρ, and the pressure, P. [Pressure is force per unit area.]

ii) State the dimensions of P and ρ.

iii) Assuming a relation of the form $f = K\lambda^\alpha P^\beta \rho^\gamma$, find the values of α, β and γ given that K is a dimensionless constant with a different value for air and water. How can the value of K be found?

The speed of propagation, c, of a sound wave is known to be given by

$$c = \sqrt{\frac{AP}{\rho}} \quad \text{where } A = 1.4 \text{ for air.}$$

iv) By using this expression for c and the equation in part i), find the value of K for sound waves in air.

v) It is observed that the density of water is much greater than the density of air and that the speed of sound in water is much greater than in air. What does this imply about the value of A for water? Explain your answer.

16.6 An experiment consists of stretching a length of wire and fixing its ends. The wire is then plucked to produce transverse vibrations. The frequency of these vibrations, f, is modelled by the equation

$$f = kP^\alpha l^\beta z^\gamma$$

where P is the tension in the wire, l is the length of the stretched wire, z is the mass per unit length of the stretched wire and k is a dimensionless constant.

i) Given that the dimensions of frequency are T^{-1}, use dimensional analysis to show that $\alpha = \frac{1}{2}$, $\beta = -1$ and $\gamma = -\frac{1}{2}$.

Two pieces of wire, A and B, used in the experiment both have modulus of elasticity 500 N.

ii) Piece A has natural length 0.9 m and stretched length 0.945 m. Calculate the tension in the wire.

iii) Piece B is stretched to a length of 0.81 m. The tension in the wire is 40N. Calculate the natural length of the wire.

iv) *In their unstretched state*, the two wires had the same mass per unit length. In the experiment, wires A and B produce frequencies f_A and f_B respectively. Calculate the ratio $f_A : f_B$.

16.7 An artillery officer is conducting a series of experiments to find the maximum range, R, of a gun which fires shells from ground level across level ground.

i) Show, by considering the shell as a projectile with initial velocity u at an angle to the horizontal and ignoring air resistance, that

$$R = \frac{u^2}{g} \sin 2\alpha$$

and that this has a maximum value, R_M, when α is 45°.

ii) Show that the expression for R in part i) is dimensionally consistent.

The officer finds that the shells fall some distance short of their predicted range. Deciding that this must be due to air resistance, he proposes a model in which the magnitude of the force of resistance, F, is proportional to the speed of the shell, v:

$$F = cv.$$

iii) Show that the constant c is not dimensionless and write down its dimensions.

The officer then suggests that a better model for the range of the shell, when fired at 45° to the horizontal, is given by:

$$R = R_M(1 - \varepsilon)$$

where ε (epsilon) is a number less than 1 representing the proportion of the range which is lost.

iv) State the dimensions of ε.

The officer believes that ε depends on a product of powers of the speed of projection u, the constant c, g and the mass m of the shell.

v) Use dimensional analysis to find an expression for ε in which the power of u is 1. Express the improved model as a formula for R.

It is suggested that a more accurate and more general model still would have the form

$$R = R_M(1 + a_1\varepsilon + a_2\varepsilon^2 + a_3\varepsilon^3 + ...)$$

where a_1, a_2, a_3, ... are dimensionless coefficients, taking into account (among other things) the angle of projection. (Clearly, a_1 would be negative.)

vi) Show that such a model is dimensionally consistent.

16.8 A simple pendulum consists of a particle of mass m attached to the end of a light inextensible string of length l. The particle is held with the string taut and slightly displaced from the vertical. It is then released and the period of the resulting small oscillations is believed to be of the form $k_1 m^\alpha l^\beta g^\gamma$ where k_1 is a dimensionless constant.

i) Use dimensional analysis to determine α, β and γ.

The pendulum is now set in motion as a conical pendulum so that the particle moves in a horizontal circle with constant angular velocity ω. The string makes a constant angle θ with the vertical throughout the motion as shown in the diagram.

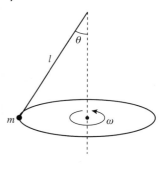

ii) For a particular value of θ, the period for one revolution of a conical pendulum is believed to depend only on m, l and g. Explain why the period will be of the form $k_2 \sqrt{\dfrac{l}{g}}$ where k_2 is a dimensionless constant.

An expression for k_2 is to be found from the equations of motion for the conical pendulum.

iii) Find an expression for ω^2 in terms of l, g and θ. For what range of values of ω^2 is the motion possible?

iv) Find an expression for k_2 in terms of θ. Given that for the simple pendulum in part **i)** $k_1 = 2\pi$, which motion has the longer period? Explain your reasoning.

16.9 i) State the dimensions of density and angular velocity. Show that the dimensions of power are ML^2T^{-3}.

The power P required to drive a propeller of diameter d at an angular velocity ω in a fluid of density r is believed to obey a law of the form

$$P = k_1 d^\alpha \omega^\beta \rho^\gamma$$

where k_1 is a dimensionless constant.

ii) Determine the values of α, β and γ.

It is suggested that the viscosity, η, of the fluid should be included in the equation. The formula

$$P = k_2 d^\alpha \omega^\beta \rho^\gamma \left(\frac{\omega \rho d^2}{\eta} \right)^\delta \qquad \text{①}$$

is proposed. In this formula, k_2 is a dimensionless constant; α, β and γ have the same values as in part ii); δ is unknown.

iii) The quantity $\dfrac{\omega \rho d^2}{\eta}$ is dimensionless. Find the dimensions of η.

Explain briefly whether it is possible to find the value of δ from equation ①.

Equation ① can be used to scale results from a model propeller to a full-size propeller. For the model, $d = 0.6$, $\rho = 1000$ and $\eta = 0.001$, all measured in S.I. units. For the full-size propeller, $d = 1.8$, $\rho = 850$ and $\eta = 0.012$, again all measured in S.I. units.

To enable ① to be used, $\dfrac{\omega \rho d^2}{\eta}$ must have the same value for both the model and the full-size propeller.

iv) If the full-size propeller is to be run at $25\,\text{rad s}^{-1}$, calculate the angular velocity of the model propeller.

At this speed, the model requires 3 kW of power.

v) Calculate the power required by the full-size propeller.

KEY POINTS

16.1 Any quantity in mechanics may be expressed in terms of the three fundamental dimensions:

mass, M; length, L; time, T.

16.2 The unit for any quantity is derived from the three fundamental dimensions.

16.3 Numbers are dimensionless.

16.4 All formulae and equations must be dimensionally consistent.

16.5 Using dimensional analysis you can sometimes find the form of a relationship as the product of powers of the quantities involved, and a dimensionless constant.

17 Use of vectors

This grand book – the Universe... is written in the language of mathematics

Galileo Galilei (1623)

Vectors are important in modelling and solving two- and three-dimensional problems in mechanics. You have already used vector methods. This chapter will help you to review and consolidate your knowledge of vectors with particular reference to their application in mechanics.

17.1 Vector basics

A vector is a quantity with both magnitude (also known as length, size or modulus) and direction. In mechanics, vectors are used to represent quantities such as displacement, velocity, acceleration, force and momentum.

A vector is usually written $\overrightarrow{PQ}$ or a and is typeset as $\overrightarrow{PQ}$ or as bold **PQ** or **a**.

In figure 17.1, $\overrightarrow{PQ} = \overrightarrow{QA}$ because their lengths and directions are the same.

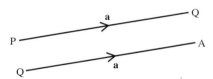

Figure 17.1

Vectors can be added: $\mathbf{a} + \mathbf{b} = \mathbf{c}$

and subtracted: $\mathbf{a} - \mathbf{b} = \mathbf{d}$

and multiplied by a scalar: λa has magnitude $\lambda \times$ magnitude of a.

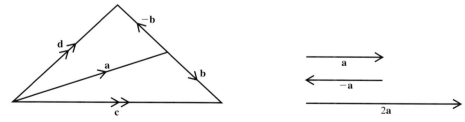

Figure 17.2

The length of the vector **a** is denoted $|\mathbf{a}|$ or a. When $a = 1$, **a** is a *unit vector*. A unit vector is denoted by having a hat $\hat{\mathbf{a}}$.

The *position vector* of a point A relative to a point O is the vector $\overrightarrow{OA}$. Normally, O is the origin of Cartesian axes and the position vector of a point is usually denoted by the corresponding lower-case letter, thus $\overrightarrow{OA} = \mathbf{a}$, $\overrightarrow{OB} = \mathbf{b}$, and so on. Then $\overrightarrow{AB} = \mathbf{b} - \mathbf{a}$, where **b** and **a** are the position vectors of A and B with respect to an assumed origin (figure 17.3) which is not necessarily drawn on the diagram.

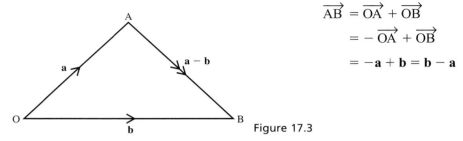

$$\overrightarrow{AB} = \overrightarrow{OA} + \overrightarrow{OB}$$
$$= -\overrightarrow{OA} + \overrightarrow{OB}$$
$$= -\mathbf{a} + \mathbf{b} = \mathbf{b} - \mathbf{a}$$

Figure 17.3

Cartesian components

Vectors can also be expressed in Cartesian component form, with respect to some appropriate origin and axes. In three dimensions:

$$\overrightarrow{OA} = \begin{pmatrix} x \\ y \\ z \end{pmatrix} = x\mathbf{i} + y\mathbf{i} + z\mathbf{k}$$

where a, y, z are the displacements represented by the vector in the x, y and z directions and **i**, **j** and **k** are the unit vectors in the directions of the co-ordinate axes.

Vector addition, subtraction and multiplication by a scalar are easy when vectors are given in Cartesian component form, as you simply apply the operation to each component.

$$\text{Given } \overrightarrow{OA} = \begin{pmatrix} a_1 \\ a_2 \\ a_3 \end{pmatrix} \quad \overrightarrow{OB} = \begin{pmatrix} b_1 \\ b_2 \\ b_3 \end{pmatrix} \quad \text{then} \quad 2\overrightarrow{OA} + \overrightarrow{OB} = \begin{pmatrix} 2a_1 + b_1 \\ 2a_2 + b_2 \\ 2a_3 + b_3 \end{pmatrix}$$

17.2 Vector equation of a line

A line is specified by giving its direction and a point through which it passes. The vector equation of a line is

$$\mathbf{r} = \mathbf{a} + \lambda\mathbf{u}$$

or in component form $\mathbf{r} = (a_1 + \lambda u_1)\mathbf{i} + a_2 + \lambda u_2)\mathbf{j} + (a_3 + \lambda u_3)\mathbf{k}$

$$\text{or} \begin{pmatrix} x \\ y \\ z \end{pmatrix} = \begin{pmatrix} a_1 \\ a_2 \\ a_3 \end{pmatrix} + \lambda \begin{pmatrix} u_1 \\ u_2 \\ u_3 \end{pmatrix} = \begin{pmatrix} a_1 + \lambda u_1 \\ a_2 + \lambda u_2 \\ a_3 + \lambda u_3 \end{pmatrix}.$$

The line passes through the point A with position vector **a** and it is in the direction of the vector **u** (figure 17.4). λ is a scalar variable. The vector **r** is the position vector of the point R with co-ordinates (x, y, z). As λ varies, R always lies on the line through A in the direction of **u**. Since A can be chosen as *any* point on the line and since any multiple of **u** (e.g. $3\mathbf{u}$ or $-10\mathbf{u}$) can be used instead of **u**, the same line has many different forms of equation.

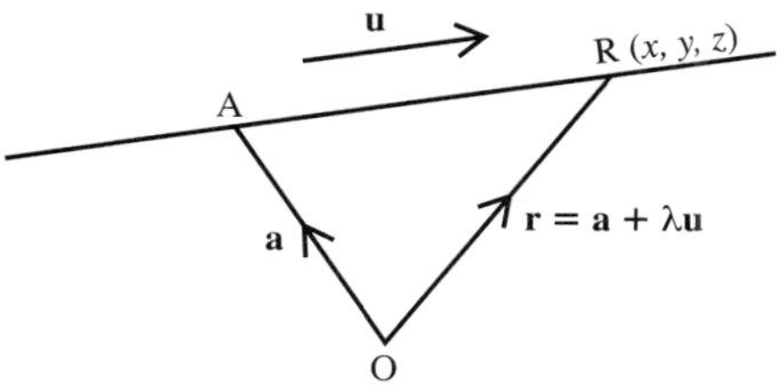

Figure 17.4

Particle moving with constant velocity

Earlier, you saw that a particle moving with a constant velocity **v** passing through a point with position vector $\mathbf{r}_0$ will have at a time t later a position

$$\mathbf{r} = \mathbf{r}_0 + t\mathbf{v}.$$

This is the equation of a line through $\mathbf{r}_0$ in the direction of **v**. In other words, it is the vector equation of the path of the particle, just as you would expect.

Line of action of a force

Although a force is often given as a vector, direction and magnitude alone are not enough full to determine the effect of a force. For example, given that the force $\mathbf{F} = \mathbf{i} + 2\mathbf{j}$ is acting on a body, you have sufficient information to resolve **F** in any direction, but not sufficient to find its *moment* about any point. For this, you need to know a point on the *line of action* of the force. Suppose the force is acting at the point P with position vector $2\mathbf{i}$ (figure 17.5). Then the equation of the line of action of **F** is

$$\mathbf{r} = 2\mathbf{i} + \lambda(\mathbf{i} + 2\mathbf{j}).$$

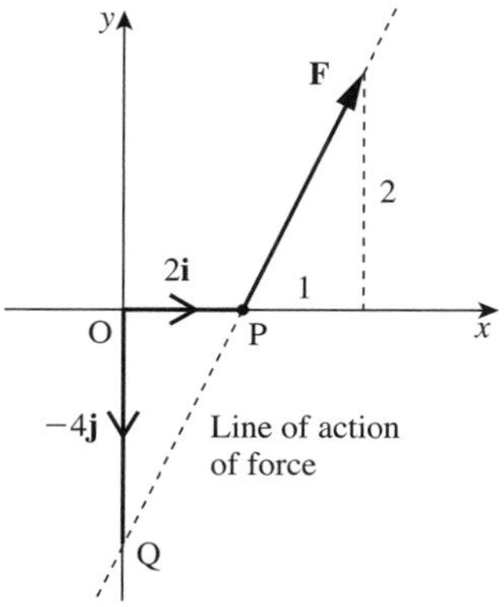

Figure 17.5

Notice that you may just as well have been given any other point on the line of action of **F**. The point Q with position vector $-4\mathbf{j}$ is also on the line of action of the force (put $\lambda = -2$ in the equation). This means that the effect of this force will be exactly the same whether it is acting at $2\mathbf{i}$ or $-4\mathbf{j}$.

The example illustrates the use of the vector equation of a line in describing a force.

EXAMPLE 17.1

In the xy plane, a force $\mathbf{F} = F_x\mathbf{i} + F_y\mathbf{j}$ of magnitude 10 acts along the line

$$\mathbf{r} = \mathbf{i} + \mathbf{j} + \lambda(3\mathbf{i} + 4\mathbf{j})$$

in the direction of increasing λ.
i)   Find F_x and F_y.
ii) **F** acts through the point A$(a, 0)$. Find a.
iii) Deduce the moment of **F** about O.

SOLUTION

i) **F** is in the same direction as $3\mathbf{i} + 4\mathbf{j}$.

$$|3\mathbf{i} + 4\mathbf{j}| = \sqrt{(3^2 + 4^2)} = 5$$

So the unit vector in the direction of **F** is $\frac{1}{5}(3\mathbf{i} + 4\mathbf{j})$. Also $|\mathbf{F}| = 10$, hence

$$\mathbf{F} = \tfrac{10}{5}(3\mathbf{i} + 4\mathbf{j}) = 6\mathbf{i} + 8\mathbf{j}$$
$$\Rightarrow \quad F_x = 6 \quad \text{and} \quad F_y = 8$$

ii)
$$\mathbf{r} = \mathbf{i} + \mathbf{j} + \lambda(3\mathbf{i} + 4\mathbf{j})$$
$$= (1 + 3\lambda)\mathbf{i} + (1 + 4\lambda)\mathbf{j}$$
$$\Rightarrow \quad x = 1 + 3\lambda \quad \text{and} \quad y = 1 + 4\lambda$$

where (a, y) is a point on the line.

When $y = 0$, $\lambda = -\tfrac{1}{4} \Rightarrow x = \tfrac{1}{4}$; hence $a = \tfrac{1}{4}$.

iii) To find the moment about O, regard the force as acting at A and resolved into its two components F_x and F_y. F_x does not contribute, so the result is

$$aF_y = \tfrac{1}{4} \times 8 = 2 \ \text{ anticlockwise}$$

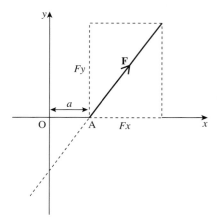

17.3 Use of vectors to describe motion

Two and three-dimensional motion is often described using *vector* notation.

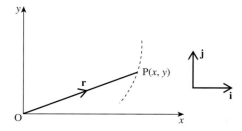

Figure 17.6

Figure 17.6 shows a particle P moving in the *xy* plane. Its position at any time *t* can be given by its *x* and *y* co-ordinates. These co-ordinates can also be regarded as the components of the position vector of P relative to the origin. Denote this vector OP by **r**. Then

$$\mathbf{r} = x\mathbf{i} + y\mathbf{j}$$

where **i** and **j** are, as usual, the unit vectors in the *x* and *y* directions.

In the *x* direction, the speed of the particle is $\dot{x}$ (i.e. d*x*/d*t*) and in the *y* direction $\dot{y}$, so the velocity of the particle, denoted by $\dot{\mathbf{r}}$, is the vector

$$\mathbf{v} = \dot{\mathbf{r}} = \dot{x}\mathbf{i} + \dot{y}\mathbf{j}.$$

Similarly, the acceleration of P can be found by differentiating the components again:

$$\frac{\mathrm{d}\mathbf{v}}{\mathrm{d}t} = \ddot{x}\mathbf{i} + \ddot{y}\mathbf{j}.$$

The following examples demonstrate the use of vectors to represent velocity and acceleration.

EXAMPLE 17.2

A particle moves in the *xy* plane so that its position vector with respect to the origin at a given time *t* is

$$\mathbf{r} = 3t^2\mathbf{i} + (4t - 6)\mathbf{j}$$

Find the vector expressions for the velocity and acceleration of the particle at time *t*.

SOLUTION

Differentiate one for velocity

$$\mathbf{v} = \frac{\mathrm{d}\mathbf{r}}{\mathrm{d}t} = 6t\mathbf{i} + 4\mathbf{j}$$

*Notice that the velocity in the **j** direction is constant*

and again for acceleration

$$\frac{\mathrm{d}\mathbf{v}}{\mathrm{d}t} = 6\mathbf{i}.$$

*... and that acceleration is zero in the **j** direction*

EXAMPLE 17.3

The acceleration of a particle at time *t* is given by

$$\mathbf{a} = \cos t\mathbf{i} + \sin t\mathbf{j}.$$

At time $t = 0$, its position relative to the origin is $-\mathbf{i}$ and its velocity is $-\mathbf{j}$. Prove that the particle always lies on the circle $x^2 + y^2 = 1$.

SOLUTION

The velocity and then the position of the body are obtained by successive integration of

$$\mathbf{a} = \cos t\mathbf{i} + \sin t\mathbf{j}.$$

The velocity is therefore

$$\mathbf{v} = (\sin t + c_1)\mathbf{i} + (-\cos t + c_2)\mathbf{j}$$

where c_1 and c_2 are the constants of integration.

When $t = 0$, $\mathbf{v} = c_1\mathbf{i} + (-1 + c_2)\mathbf{j}$. But the velocity is $-\mathbf{j}$ when $t = 0$, so c_1 and c_2 are both zero. Hence

$$\mathbf{v} = \sin t\mathbf{i} - \cos t\mathbf{j}.$$

Integrating again gives

$$\mathbf{r} = (-\cos t + k_1)\mathbf{i} + (-\sin t + k_2)\mathbf{j}.$$

Again, the constants of integration k_1 and k_2 are zero, since $\mathbf{r}$ is $-\mathbf{i}$ when t is zero. So

$$\mathbf{r} = -\cos t\mathbf{i} - \sin t\mathbf{j}.$$

The co-ordinates of the particle at time t are given by $x = -\cos t$, $y = -\sin t$. This always lies on the circle $x^2 + y^2 = 1$ (since $\sin^2 t + \cos^2 t = 1$).

Differentiating vectors

The result of differentiating a vector has been assumed to be that vector formed by differentiating its x and y components. This is also consistent with the general definition of differentiation as a rate of change. In a small period of time δt, the particle P moves from a point A, with position vector $\mathbf{r}(t)$, to B with position vector $\mathbf{r}(t + \delta t)$.

From figure 17.7 you can see that the vector AB is $\mathbf{r}(t + \delta t) - \mathbf{r}(t)$. This is the change in position of P:

$$\delta\mathbf{r} = \mathbf{r}(t + \delta t) - \mathbf{r}(t).$$

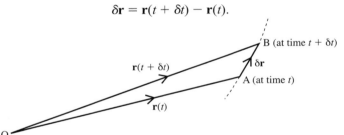

Figure 17.7

The average velocity of P over this small period is the change in position divided by time, $\dfrac{\delta\mathbf{r}}{\delta t}$.

As δt tends to zero, $\dfrac{\delta\mathbf{r}}{\delta t}$ becomes the instantaneous velocity at A. This is written $\dfrac{d\mathbf{r}}{dt}$, where

$$\frac{d\mathbf{r}}{dt} = \lim_{\delta t \to 0} \frac{\mathbf{r}(t + \delta t) - \mathbf{r}(t)}{\delta t}.$$

This definition is independent of any particular co-ordinate system. But $\delta\mathbf{r}$ can be written in terms of its components (figure 17.8):

$$\delta\mathbf{r} = \delta x\mathbf{i} + \delta y\mathbf{j}$$

$$\Rightarrow \quad \frac{\delta\mathbf{r}}{\delta t} = \frac{\delta x}{\delta t}\mathbf{i} + \frac{\delta y}{\delta t}\mathbf{j}.$$

As $\delta t \to 0$

$$\frac{d\mathbf{r}}{dt} = \frac{dx}{dt}\mathbf{i} + \frac{dy}{dt}\mathbf{j}.$$

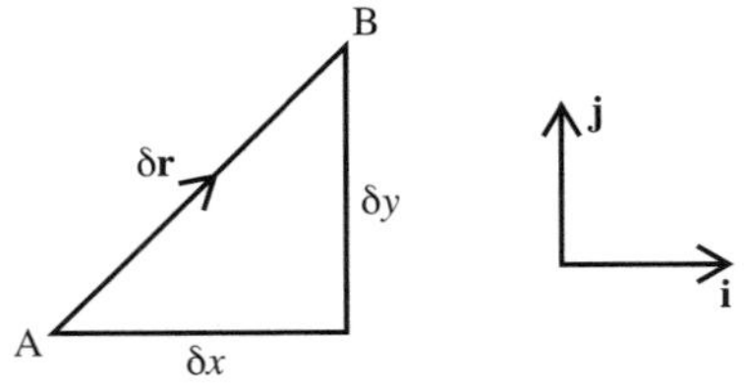

Figure 17.8

This is the result you obtain simply by differentiating the x and y components separately.

Newton's second law in vector form

If $\mathbf{F} = F_x\mathbf{i} + F_y\mathbf{j}$ is a force acting on a body of mass m, then the acceleration along each component axis is derived from the component of force along that axis:

$$F_x = m\ddot{x} \quad \text{and} \quad F_y = m\ddot{y}$$

where (x, y) are the co-ordinates of the body. Thus Newton's second law can be expressed in vector form:

$$\mathbf{F} = m\mathbf{a}.$$

The acceleration $\mathbf{a}$ may be written as $\dfrac{d\mathbf{v}}{dt}$ or $\dfrac{d^2\mathbf{r}}{dt^2}$. The vectors $\mathbf{F}$ and $\mathbf{a}$ may, of course, be three-dimensional, in which case there will be a z component also:

$$\mathbf{F} = F_x\mathbf{i} + F_y\mathbf{j} + F_z\mathbf{k} = m\ddot{x}\,\mathbf{i} + m\ddot{y}\,\mathbf{j} + m\ddot{z}\,\mathbf{k}.$$

EXAMPLE 17.4

A particle P of mass 2 kg moves under the action of a force, which at time t is given by

$$\mathbf{F} = 8\cos 2t\,\mathbf{i} + 8\sin 2t\,\mathbf{j}.$$

At time $t = 0$, P is at rest at the origin.
i)   Find the position vector of P relative to O at time t.
ii) Find at what times the kinetic energy is a maximum.

SOLUTION

i)   By Newton's second law

$$m\frac{d\mathbf{v}}{dt} = \mathbf{F}.$$

So

$$2\frac{d\mathbf{v}}{dt} = 8\cos 2t\,\mathbf{i} + 8\sin 2t\,\mathbf{j}.$$

$$\frac{d\mathbf{v}}{dt} = 4\cos 2t\,\mathbf{i} + 4\sin 2t\,\mathbf{j}.$$

Integrating gives:

$$\mathbf{v} = 2\sin 2t\,\mathbf{i} - 2\cos 2t\,\mathbf{j} + \mathbf{p}.$$

> Each component on the right-hand side has been integrated and the vector $\mathbf{p}$ is the constant of integration. The right-hand side might equally have been written with separate constants for each compoent of $\mathbf{p}$:
> $$\mathbf{v} = (2\sin 2t + p_1)\mathbf{i} - (2\cos 2t + p_2)\mathbf{j}.$$

The body is at rest when $t = 0$, so

$$0 = -2\mathbf{j} + \mathbf{p} \quad \Rightarrow \quad \mathbf{p} = 2\mathbf{j}.$$

So

$$\mathbf{v} = 2\sin 2t\,\mathbf{i} + (2 - 2\cos 2t)\mathbf{j}.$$

Writing $\mathbf{v} = \dfrac{d\mathbf{r}}{dt}$ and integrating again gives

$$\mathbf{r} = -\cos 2t\mathbf{i} + (2t - \sin 2t)\mathbf{j} + \mathbf{q}.$$

The constant of integration $\mathbf{q}$ is again determined by looking at the initial conditions. When $t = 0$, the particle is at the origin, so

$$0 = -\mathbf{i} + \mathbf{q} \qquad \Rightarrow \qquad \mathbf{q} = \mathbf{i} .$$

The position at time t is thus

$$\mathbf{r} = (1 - \cos 2t)\mathbf{i} + (2t - \sin 2t)\mathbf{j}.$$

ii)   The kinetic energy is

$$\tfrac{1}{2}mv^2 = \tfrac{1}{2} \times 2 \times |\mathbf{v}|^2 \longleftarrow \left\{ \begin{array}{l} \text{Sum of the squares of} \\ \text{the components of } v \end{array} \right.$$

$$= m\mathbf{v}(2 \sin 2t)^2 + (2 - 2 \cos 2t)^2$$

$$= 4 \sin^2 2t + 4 - 8 \cos 2t + 4 \cos^2 2t$$

$$= 8(1 - \cos 2t) \qquad (\text{since } \sin^2 2t + \cos^2 2t = 1)$$

$$= 16 \sin^2 t.$$

This is maximum when $\sin t = \pm 1$, that is $t = \pi/2, 3\pi/2, 5\pi/2, \ldots$

Historical note

It was more than 60 years after Newton's discoveries before it was realised that his second law could be conveniently expressed in the form of components of acceleration and force along Cartesian axes. Euler was the first to represent it in this way, in 1736. The vector notation was invented much later, by 19th century mathematical physicists (notably the American Willard Gibbs and the Scot James Clerk Maxwell) who used vectors to give a much simpler formulation of the equations describing electric and magnetic fields. The elegant and powerful notation then spread to mechanics and many other branches of physics.

Why use vectors?

In the foregoing examples, the x and y components of the vectors were treated independently. So you may wonder why write, for example,

$$\mathbf{a} = \cos t\mathbf{i} + \sin t\mathbf{j}$$

rather than

$$a_x = \cos t \qquad a_y = \sin t.$$

The vector form is more compact, but is this the only reason to use vectors? The answer is no. There are vector operations which make vector methods a much simpler way of dealing with certain problems. The *scalar product*, given on the following page, is one such operation. The *vector product*, which you will meet later, is another example, almost essential in three-dimensional work.

17.4 The scalar product

The properties of the *scalar* or *dot* product between two vectors are summarised here, together with some simple applications in mechanics.

The scalar product is written $\mathbf{p} \cdot \mathbf{q}$, where $\mathbf{p}$ and $\mathbf{q}$ are vectors. The result is a number (or scalar) *not* a vector. It is defined as the product of the magnitudes of the vectors and the cosine of the angle between them (figure 17.9):

$$\mathbf{p} \cdot \mathbf{q} = |\mathbf{p}|\,|\mathbf{q}|\cos\theta.$$

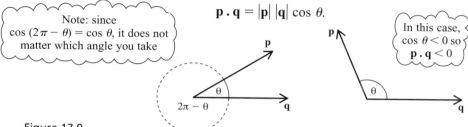

Note: since $\cos(2\pi - \theta) = \cos\theta$, it does not matter which angle you take

In this case, $\cos\theta < 0$ so $\mathbf{p} \cdot \mathbf{q} < 0$

Figure 17.9

From the definition, it is clear that $\mathbf{p} \cdot \mathbf{q} = \mathbf{q} \cdot \mathbf{p}$ (commutative law). It can also be shown that

$$\mathbf{p} \cdot (\mathbf{q} + \mathbf{r}) = \mathbf{p} \cdot \mathbf{q} + \mathbf{p} \cdot \mathbf{r} \quad \text{(distributive law)}$$

so that the scalar product on vectors can be used rather like ordinary multiplication on numbers.

The value of the scalar product is easy to work out if the components of the vectors in the direction of x and y (and z in the three-dimensional case) are known. Given that

$$\mathbf{p} = p_x\mathbf{i} + p_y\mathbf{j}$$
$$\mathbf{q} = q_x\mathbf{i} + q_y\mathbf{j}$$

then

$$\mathbf{p} \cdot \mathbf{q} = p_xq_x + p_yq_y.$$

In the three-dimensional case, when each vector has a z component, the result is

$$\mathbf{p} \cdot \mathbf{q} = p_xq_x + p_yq_y + p_zq_z.$$

The scalar product of two non-zero vectors is zero if and only if the two vectors are perpendicular (because $\cos 90°$ is zero).

The scalar product of a vector with itself is the square of its magnitude:

$$\mathbf{p} \cdot \mathbf{p} = |\mathbf{p}|^2 = p_x^2 + p_y^2 \quad (p_x^2 + p_y^2 + p_z^2 \text{ in three dimensions}).$$

The component of a force in a given direction

One use of the scalar product is the calculation of the component of a force in any given direction. If $\mathbf{F}$ is a force and $\hat{\mathbf{n}}$ is a *unit* vector, then $\mathbf{F} \cdot \hat{\mathbf{n}}$ is the component (or resolved part) of a force $\mathbf{F}$ in the direction of $\hat{\mathbf{n}}$. You can see from figure 17.10 that $\mathbf{F} \cdot \hat{\mathbf{n}}$ has the value $|\mathbf{F}|\cos\theta$ (since $|\hat{\mathbf{n}}| = 1$), that is the resolved part in the direction $\hat{\mathbf{n}}$.

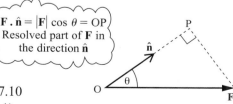

$\mathbf{F} \cdot \hat{\mathbf{n}} = |\mathbf{F}|\cos\theta = OP$
Resolved part of $\mathbf{F}$ in the direction $\hat{\mathbf{n}}$

Figure 17.10

EXAMPLE 17.5

A small ring of mass 100 g slides on a smooth straight wire in the direction $\mathbf{i} + 2\mathbf{j} + 2\mathbf{k}$. Apart from the normal reaction of the wire, forces on the ring are its weight $-\mathbf{k}$ N and a constant force due to the wind of $\frac{1}{2}\mathbf{i} + \frac{1}{2}\mathbf{j}$ N. The ring has an initial speed of 2 m s^{-1}.

i)   Find the unit vector in the direction of the wire.
ii) Find the resolved part of the total forces on the ring in the direction of motion.
iii) Hence find the acceleration of the ring and the distance travelled before coming to rest.

SOLUTION

i)
$$|\mathbf{i} + 2\mathbf{j} + 2\mathbf{k}| = \sqrt{(1^2 + 2^2 + 2^2)} = 3.$$

Hence the unit vector in the direction of motion is

$$\hat{\mathbf{n}} = \tfrac{1}{3}(\mathbf{i} + 2\mathbf{j} + 2\mathbf{k}) = \tfrac{1}{3}\mathbf{i} + \tfrac{2}{3}\mathbf{j} + \tfrac{2}{3}\mathbf{k}.$$

ii) The normal reaction R of the wire is perpendicular to the wire and has no component parallel to it. The resultant of the other forces is

$$\mathbf{F} = \tfrac{1}{2}\mathbf{i} + \tfrac{1}{2}\mathbf{j} - \mathbf{k}.$$

The resolved part in the direction of motion is

$$\mathbf{F} \cdot \hat{\mathbf{n}} = \left(\tfrac{1}{6} + \tfrac{1}{3} - \tfrac{2}{3}\right) = -\tfrac{1}{6}\,\text{N}.$$

iii) Applying $F = ma$ in the direction of motion gives

$$-\tfrac{1}{6} = \tfrac{1}{10}a \quad \text{(mass is } \tfrac{1}{10}\text{ kg)}$$
$$\Rightarrow \quad a = -\tfrac{5}{3}.$$

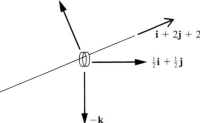

The ring decelerates at $\frac{5}{3}$ m s^{-2}.

To find when the ring comes to rest, apply $v^2 = u^2 + 2as$:

$$0 = 4 - 2 \times \tfrac{5}{3}s$$
$$\Rightarrow \quad s = \tfrac{6}{5}.$$

The ring comes to rest after $\frac{6}{5}$ m.

Work and power as scalar products

When a constant for $\mathbf{F}$ acts on a body which moves through a displacement $\mathbf{s}$ (figure 17.11), the work done is the component of $\mathbf{F}$ in the direction of motion, multiplied by the distance travelled:

$$\text{work done} = |\mathbf{F}| \cos \theta |\mathbf{s}| = \mathbf{F} \cdot \mathbf{s}.$$

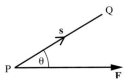

Figure 17.11

In general, forces are variable and objects move through curves, rather than straight lines, in two and three dimensions; the calculation of the work done is then more difficult. However, it still involves the scalar product.

When a force $\mathbf{F}$ moves a body through a displacement $\delta\mathbf{s}$ in time δt, the work done is

$$\delta W = \mathbf{F} \cdot \delta\mathbf{s}.$$

The rate of working, i.e. power, is thus

$$\frac{\delta W}{\delta t} = \mathbf{F} \cdot \left(\frac{\delta\mathbf{s}}{\delta t}\right).$$

In the limit as $\delta t \to 0$, this gives

$$\text{power} = \frac{\mathrm{d}W}{\mathrm{d}t} = \mathbf{F} \cdot \mathbf{v}.$$

The power expended by a force on a moving body is its speed multiplied by the component of the force in the direction of motion. This is the equivalent of the Fv formula you use when the force is in the same direction as the motion.

With reference to figure 17.12, the power at P is

$$|\mathbf{F}| \cos \theta |\mathbf{v}| = \mathbf{F} \cdot \mathbf{v}.$$

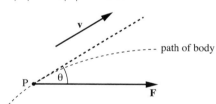

path of body

Figure 17.12

EXAMPLE **17.6**

A particle of mass 2 kg, starting from the rest at the origin, moves under the action of a force, which at time t is given by

$$\mathbf{F} = 8 \cos 2t\mathbf{i} + 8 \sin 2t\mathbf{j}.$$

In the example on page 383, it was shown that at time t, the velocity is

$$\mathbf{v} = 2 \sin 2t\mathbf{i} + (2 - 2 \cos 2t)\mathbf{j}$$

the acceleration is

$$\mathbf{a} = 4 \cos 2t\mathbf{i} + 4 \sin 2t\mathbf{j}$$

and that the kinetic energy is

$$E = 8(1 - \cos 2t).$$

i)   Find the first time $t\,(t > 0)$ that the velocity is perpendicular to the acceleration.
ii) Find the rate of working at time t.
iii) Verify that the power is equal to the rate of change of kinetic energy.

SOLUTION

i) When the acceleration is perpendicular to the velocity, $\mathbf{a} \cdot \mathbf{v} = 0$.

$$\mathbf{a} \cdot \mathbf{v} = (4 \cos 2t)(2 \sin 2t) + (4 \sin 2t)(2 - 2 \cos 2t)$$
$$= 8 \sin 2t \cos 2t + 8 \sin 2t - 8 \sin 2t \cos 2t$$
$$= 8 \sin 2t$$

The first time for which $\mathbf{a} \cdot \mathbf{v} = 0$ is thus when $\sin 2t = 0$, that is $t = \pi/2$.

ii)

$$\text{Power} = \mathbf{F} \cdot \mathbf{v} = [8 \cos 2t \mathbf{i} + 8 \sin 2t \mathbf{j}] \cdot [2 \sin 2t \mathbf{i} + (2 - 2 \cos 2t)\mathbf{j}]$$
$$= 16 \sin 2t \cos 2t + 8 \sin 2t(2 - 2 \cos 2t)$$
$$= 16 \sin 2t$$

iii) Given the kinetic energy $E = 8(1 - \cos 2t)$, the rate of change of kinetic energy is

$$\frac{dE}{dt} = 16 \sin 2t$$

which is the power.

> This is a general result. Work done by all forces = increase in kinetic energy
> Rate of working = power = rate of increase in kinetic energy

17.5 The vector product

The use of the scalar product $\mathbf{a} \cdot \mathbf{b}$ of two vectors in finding scalars such as the component of a vector in a given direction, the work done by a force, have already been considered in this chapter. There is another important operation, the vector product (or cross product), where the result of combining two vectors is also a vector.

Representing an area by a vector

Consider a rectangular box, with sides x, y and z as shown in figure 17.13, containing a gas under a pressure p. Pressure means force per unit area, so the total force of the gas on the top face of the box is

$$\text{force} = \text{pressure} \times \text{area} = pxy.$$

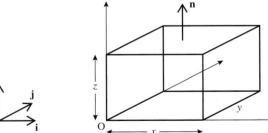

Figure 17.13

The force is at right angles to the face, so the total force on the face is, as a vector,

$$\mathbf{F} = pxy\mathbf{k} \qquad \text{where } \mathbf{k} \text{ is the unit vector in direction } Oz$$
$$= p\mathbf{n} \qquad \text{where } \mathbf{n} = xy\mathbf{k}.$$

The *magnitude* of the vector $\mathbf{n}$ is the area of the top face, and its *direction* is normal to the face, and directed outwards from the box. Thus $\mathbf{n}$ may be thought of as a 'vector area'.

Activity

Write down the 'vector area' of each of the other faces in terms of x, y, z and the standard unit vectors.

Apply this notion of a vector area to a parallelepiped: this is a box whose six faces are al parallelograms, as shown in figure 17.14. Every edge can be represented by one of the three vectors $\mathbf{a}(\overrightarrow{OA})$, $\mathbf{b}(\overrightarrow{OB})$ and $\mathbf{c}(\overrightarrow{OC})$.

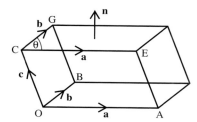

Figure 17.14

The force on, say, the top face can again be written $\mathbf{F} = p\mathbf{n}$, where $\mathbf{n}$ is a vector whose magnitude is the area of this face and whose direction is perpendicular to it, pointing outwards. So

$$\left|\mathbf{n}\right| = \left|\mathbf{a}\right|\left|\mathbf{b}\right| \sin \theta$$

where θ is the angle ECG. The vector $\mathbf{n}$ depends simply on $\mathbf{a}$ and $\mathbf{b}$, and is known as the *vector product* of $\mathbf{a}$ and $\mathbf{b}$, written

$$\mathbf{n} = \mathbf{a} \times \mathbf{b}.$$

Introducing the vector product as a 'vector area' may help to illuminate the normal definition, which is given below.

Definition and properties of the vector product

Given two vectors $\mathbf{a}$ and $\mathbf{b}$, inclined at an angle θ (where $\theta < 180°$), the vector product $\mathbf{a} \times \mathbf{b}$ (you say 'a cross b') is defined as a vector which
● is perpendicular to both $\mathbf{a}$ and $\mathbf{b}$;
● has a magnitude $\left|\mathbf{a}\right|\left|\mathbf{b}\right| \sin \theta$;
● has a direction such that $\mathbf{a}$, $\mathbf{b}$ and $\mathbf{a} \times \mathbf{b}$ in that order form a right-handed set of vectors.

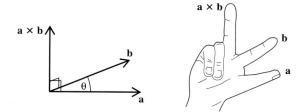

Figure 17.15

This is illustrated in figure 17.15. Without the last condition, there would be two possible candidates, pointing up or down from the plane containing **a** and **b**. The right-hand rule means that if you point your right-hand thumb in the direction of **a** and your index finger in the direction of **b**, then your second finger gives the direction of **a** × **b** (assuming it is perpendicular to the other two).

Thus **b** × **a** has the same magnitude but the opposite direction (figure 17.16):

$$\mathbf{a} \times \mathbf{b} = -(\mathbf{b} \times \mathbf{a}).$$

The vector product is *not commutative*, unlike the scalar product (**a . b** = **b . a**).

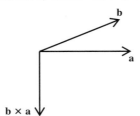

Figure 17.16

Parallel and perpendicular vectors

When two vectors are parallel, θ is zero, so the vector product is zero. In particular,

$$\mathbf{a} \times \mathbf{a} = \mathbf{0}.$$

Note that the bold **0** is written for the zero vector.

When two vectors are perpendicular, $\sin \theta$ is 1, so in this case

$$|\mathbf{a} \times \mathbf{b}| = |\mathbf{a}||\mathbf{b}|.$$

Then **a**, **b** and **a** × **b** form a set of mutually perpendicular vectors.

Note the following relationships between the standard unit vectors **i**, **j** and **k**:

$\mathbf{i} \times \mathbf{j} = \mathbf{k}$	$\mathbf{j} \times \mathbf{k} = \mathbf{i}$	$\mathbf{k} \times \mathbf{i} = \mathbf{j}$
$\mathbf{j} \times \mathbf{i} = -\mathbf{k}$	$\mathbf{k} \times \mathbf{j} = -\mathbf{i}$	$\mathbf{i} \times \mathbf{k} = -\mathbf{j}$
$\mathbf{i} \times \mathbf{i} = 0$	$\mathbf{j} \times \mathbf{j} = 0$	$\mathbf{k} \times \mathbf{k} = \mathbf{0}.$

there is no negative sign when the vectors follow in the cyclic order

Area of a parallelogram and triangle

You saw from the preceding section that $|\mathbf{a} \times \mathbf{b}|$ is the area of the parallelogram OAPB whose sides OA and OB can be represented by vectors **a** and **b**. Similarly, the area of the triangle OAB is $\frac{1}{2}|\mathbf{a} \times \mathbf{b}|$ (figure 17.17).

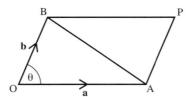

Figure 17.17

Volume of parallelepiped

The volume of a parallelepiped is given by (area of base) × (perpendicular height).

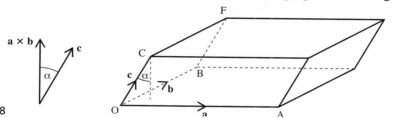

Figure 17.18

Consider $(\mathbf{a} \times \mathbf{b}) . \mathbf{c}$ (figure 17.18). By the definition of the scalar product, this has a value $|\mathbf{a} \times \mathbf{b}||\mathbf{c}| \cos \alpha$. But $|\mathbf{a} \times \mathbf{b}|$ is the area of the base parallelogram and $|\mathbf{c}| \cos \alpha$ is the height. So volume $= (\mathbf{a} \times \mathbf{b}) . \mathbf{c}$.

Distributive law for the vector product

Like the scalar product, the vector product is *distributive* over addition:

$$\mathbf{a} \times (\mathbf{b}_1 + \mathbf{b}_2 + \dots + \mathbf{b}_n) = (\mathbf{a} \times \mathbf{b}_1) + (\mathbf{a} \times \mathbf{b}_2) + \dots + (\mathbf{a} \times \mathbf{b}_n).$$

This useful property is not easy to see from the geometrical definition.

Triple scalar product

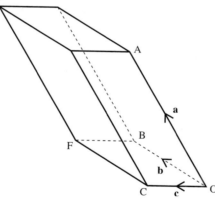

Figure 17.19

The same argument as above, but using OCFB (figure 17.19) as the base of the parallelepiped, would lead to the formula for the volume as $(\mathbf{b} \times \mathbf{c}) . \mathbf{a}$.

Similarly starting from a side face, you would get $(\mathbf{c} \times \mathbf{a}) . \mathbf{b}$. Thus

$$\text{volume} = (\mathbf{a} \times \mathbf{b}) . \mathbf{c} = (\mathbf{b} \times \mathbf{c}) . \mathbf{a} = (\mathbf{c} \times \mathbf{a}) . \mathbf{b}.$$

The expression $(\mathbf{a} \times \mathbf{b}) . \mathbf{c}$ is called a *triple scalar product*. It is positive when $\mathbf{a}$, $\mathbf{b}$, $\mathbf{c}$ form a right-handed set: for example, $(\mathbf{i} \times \mathbf{j}) . \mathbf{k} = \mathbf{k} . \mathbf{k} = +1$. Otherwise it will give the negative of the volume enclosed by the parallelepiped formed by the vectors. This illustrates the fact that the triple scalar product is the same if the vectors are permuted, but they must retain the same cyclic order. For example:

$$(\mathbf{b} \times \mathbf{a}) . \mathbf{c} = -(\mathbf{a} \times \mathbf{b}) . \mathbf{c}.$$

When two of the vectors are the same (or parallel), the triple scalar product is zero:

$$\mathbf{a} . (\mathbf{a} \times \mathbf{b}) = (\mathbf{a} \times \mathbf{a}) . \mathbf{b} = 0 \quad \text{since } \mathbf{a} \times \mathbf{a} \text{ is zero.}$$

Note

Other notations for the triple scalar product are: $\mathbf{a} \times \mathbf{b} \cdot \mathbf{c}$ (without any brackets) or $[\mathbf{a}, \mathbf{b}, \mathbf{c}]$.

Components of the vector product

Given two vectors in component form

$$\mathbf{a} = a_1\mathbf{i} + a_2\mathbf{j} + a_3\mathbf{k}$$
$$\mathbf{b} = b_1\mathbf{i} + b_2\mathbf{j} + b_3\mathbf{k}$$

then the vector product is

$$\mathbf{a} \times \mathbf{b} = (a_1\mathbf{i} + a_2\mathbf{j} + a_3\mathbf{k}) \times (b_1\mathbf{i} + b_2\mathbf{j} + b_3\mathbf{k}).$$

Since the vector product is distributive, this can be worked out term by term in the normal algebraic way. Three of the terms disappear since $\mathbf{i} \times \mathbf{i} = \mathbf{j} \times \mathbf{j} = \mathbf{k} \times \mathbf{k} = 0$. This leaves

$$a_1b_2(\mathbf{i} \times \mathbf{j}) + a_1b_3(\mathbf{i} \times \mathbf{k}) + a_2b_1(\mathbf{j} \times \mathbf{i}) + a_2b_3(\mathbf{j} \times \mathbf{k}) + a_3b_1(\mathbf{k} \times \mathbf{i}) + a_3b_2(\mathbf{k} \times \mathbf{j}).$$

Using the relationships $\mathbf{i} \times \mathbf{j} = \mathbf{k}$ etc. given above, this reduces to

$$\mathbf{a} \times \mathbf{b} = (a_2b_3 - a_3b_2)\mathbf{i} + (a_3b_1 - a_1b_3)\mathbf{j} + (a_1b_2 - a_2b_1)\mathbf{k}.$$

This important formula is used to work out the components of a vector product when you know the components of its constituents $\mathbf{a}$ and $\mathbf{b}$. The $\mathbf{i}$ component of the result involves only the $\mathbf{j}$ and $\mathbf{k}$ components of $\mathbf{a}$ and $\mathbf{b}$, similarly with the other two components of the vector product. Notice how the corresponding subscripts in the brackets follow in cyclic order. For example, the first terms in the brackets start with a_2, a_3, a_1.

Determinant notation for vector products

You will often see the calculation of a vector product written in the following way.

Given $\qquad \mathbf{a} = a_1\mathbf{i} + a_2\mathbf{j} + a_3\mathbf{k} \qquad$ and $\qquad \mathbf{b} = b_1\mathbf{i} + b_2\mathbf{j} + b_3\mathbf{k}$:

$$\mathbf{a} \times \mathbf{b} = \begin{vmatrix} \mathbf{i} & a_1 & b_1 \\ \mathbf{j} & a_2 & b_2 \\ \mathbf{k} & a_3 & b_3 \end{vmatrix} = \mathbf{i}(a_2b_3 - a_3b_2) + \mathbf{j}(a_3b_1 - a_1b_3) + \mathbf{k}(a_1b_2 - a_2b_1).$$

The components are written vertically alongside the **ijk** symbols, all enclosed in vertical lines. If you are familiar with determinants, you will understand how appropriate this notation is; the calculation of the components of the vector product is exactly the calculation you make when working out a determinant.

To find the $\mathbf{i}$ component, cover up the $\mathbf{i}$ row, and look at the remaining components:

$$\begin{vmatrix} a_2 & b_2 \\ a_3 & b_3 \end{vmatrix}.$$

The $\mathbf{i}$ component is the 2×2 determinant $a_2b_3 - a_3b_2$.

For the $\mathbf{j}$ component, cover up the $\mathbf{j}$ row, leaving

$$\begin{vmatrix} a_1 & b_1 \\ a_3 & b_3 \end{vmatrix}.$$

This time you want the negative of the 2×2 determinant $-(a_1b_3 - a_3b_1)$.

Finally, for the $\mathbf{k}$ component cover up the $\mathbf{k}$ row and take the positive determinant $a_1b_2 - a_2b_1$.

EXAMPLE 17.7

i) Work out $\mathbf{a} \times \mathbf{b}$, where $\mathbf{a} = 4\mathbf{i} + \mathbf{j} + 2\mathbf{k}$ and $\mathbf{b} = 3\mathbf{i} - 3\mathbf{k}$.

ii) Confirm that the result is perpendicular to $\mathbf{a}$ and $\mathbf{b}$.

> write down the minus and work out the determinant $\begin{vmatrix} 4 & 3 \\ 2 & -3 \end{vmatrix}$

SOLUTION

i) using the determinant notation:

$$\mathbf{a} \times \mathbf{b} \begin{vmatrix} \mathbf{i} & 4 & 3 \\ \mathbf{j} & 1 & 0 \\ \mathbf{k} & 2 & -3 \end{vmatrix} = [1 \times (-3) - 2 \times 0]\mathbf{i} - [4 \times (-3) - 2 \times 3]\mathbf{j} + (4 \times 0 - 1 \times 3)\mathbf{k}$$

$$\Rightarrow \quad \mathbf{a} \times \mathbf{b} = -3\mathbf{i} + 18\mathbf{j} - 3\mathbf{k}.$$

ii) To confirm this is perpendicular to $\mathbf{a}$ and $\mathbf{b}$, take the scalar product with each in turn.

$$\mathbf{a} \cdot (\mathbf{a} \times \mathbf{b}) = \begin{pmatrix} 4 \\ 1 \\ 2 \end{pmatrix} \cdot \begin{pmatrix} -3 \\ 18 \\ -3 \end{pmatrix} = (4)(-3) + (1)(18) + (2)(-3) = 0$$

$$\mathbf{b} \cdot (\mathbf{a} \times \mathbf{b}) = \begin{pmatrix} 3 \\ 0 \\ -3 \end{pmatrix} \cdot \begin{pmatrix} -3 \\ 18 \\ -3 \end{pmatrix} = (3)(-3) + (0)(18) + (-3)(-3) = 0$$

This is as expected and confirms the answer to part i).

Once you get used to working out vector products, you will do much of the working in your head. *But always work out the scalar product with at least one of the original vectors to check the answer is zero.* It is very easy to make mistakes when evaluating numerical vector products.

EXAMPLE 17.8

The diagram shows a unit cube ABCDEFGH. The line CB is extended to X so that $BX = 1$. Show that $\overrightarrow{AX} = \overrightarrow{AC} \times \overrightarrow{AG}$

i) from the geometrical definition of the vector product;

ii) by working out the components in a Cartesian frame of reference.

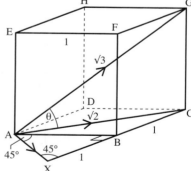

SOLUTION

i) The direction of $\overrightarrow{AC} \times \overrightarrow{AG}$ is at right angles to both AC and AG, i.e. to the plane ACG. The right-hand rule says that $\overrightarrow{AC} \times \overrightarrow{AG}$ is out of the page in the diagram. Thus it is in the direction $\overrightarrow{AX}$.

By definition $|\overrightarrow{AC} \times \overrightarrow{AG}| = |\overrightarrow{AC}||\overrightarrow{AG}| \sin \theta$

$$= \sqrt{2} \times \sqrt{3} \times (1/\sqrt{3})$$
$$= \sqrt{2} = |\overrightarrow{AX}|.$$

So $\overrightarrow{AX} = \overrightarrow{AC} \times \overrightarrow{AG}$.

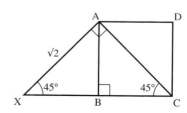

ii) Taking **i** along AB, **j** along AD, **k** along AE, then

$$\overrightarrow{AC} = \mathbf{i} + \mathbf{j} \qquad \overrightarrow{AG} = \mathbf{i} + \mathbf{j} + \mathbf{k}$$

$$\overrightarrow{AC} \times \overrightarrow{AG} = \begin{vmatrix} \mathbf{i} & 1 & 1 \\ \mathbf{j} & 1 & 1 \\ \mathbf{k} & 0 & 1 \end{vmatrix} = (1 - 0)\mathbf{i} + (0 - 1)\mathbf{j} + (1 - 1)\mathbf{k}$$

$$= \mathbf{i} - \mathbf{j}$$

This is the position vector of X with respect to A.

EXAMPLE 17.9

Find a unit vector at right angles to the plane containing $\mathbf{a} = 2\mathbf{i} - 6\mathbf{j} - 3\mathbf{k}$ and $\mathbf{b} = 4\mathbf{i} + 3\mathbf{j} - \mathbf{k}$.

SOLUTION

The vector product $\mathbf{a} \times \mathbf{b}$ is perpendicular to both $\mathbf{a}$ and $\mathbf{b}$ and therefore to the plane containing them.

$$\mathbf{a} \times \mathbf{b} = \begin{vmatrix} \mathbf{i} & 2 & 4 \\ \mathbf{j} & -6 & 3 \\ \mathbf{k} & -3 & -1 \end{vmatrix} = \mathbf{i}(6 + 9) + \mathbf{j}(-12 + 2) + \mathbf{k}(6 + 24)$$

$$= 15\mathbf{i} - 10\mathbf{j} + 30\mathbf{k}$$

(Check: $\mathbf{a} \cdot (\mathbf{a} \times \mathbf{b}) = (2\mathbf{i} - 6\mathbf{j} - 3\mathbf{k}) \cdot (15\mathbf{i} - 10\mathbf{j} + 30\mathbf{k}) = 30 + 60 - 90 = 0$, as expected.)

$$|\mathbf{a} \times \mathbf{b}| = \sqrt{(15^2 + 10^2 + 30^2)} = 35$$

So the unit vector in the direction of $\mathbf{a} \times \mathbf{b}$ is

$$\frac{15\mathbf{i} - 10\mathbf{j} + 30\mathbf{k}}{35} = \frac{3}{7}\mathbf{i} - \frac{2}{7}\mathbf{j} + \frac{6}{7}\mathbf{k}.$$

Notice that $-\frac{3}{7}\mathbf{i} + \frac{2}{7}\mathbf{j} - \frac{6}{7}\mathbf{k}$ would also be at right angles to the plane.

Differentiation of scalar and vector products

When modelling in mechanics, the value of a vector, for instance a force or a velocity, often depends on a variable, such as time t. You saw earlier in this chapter that a vector can be differentiated by differentiating its Cartesian components:

$$\frac{d\mathbf{F}}{dt} = \frac{dF_x}{dt}\mathbf{i} + \frac{dF_y}{dt}\mathbf{j} + \frac{dF_z}{dt}\mathbf{k}$$

where **i**, **j** and **k** are the usual unit vectors in the direction of the axes.

When two vectors, **F** and **G** depend on t, their scalar and vector products can be differentiated using the familiar product rule of calculus applied to the scalar and vector product operations:

$$\frac{d}{dt}(\mathbf{F} \cdot \mathbf{G}) = \frac{d\mathbf{F}}{dt} \cdot \mathbf{G} + \mathbf{F} \cdot \frac{d\mathbf{G}}{dt}$$

$$\frac{d}{dt}(\mathbf{F} \times \mathbf{G}) = \frac{d\mathbf{F}}{dt} \times \mathbf{G} + \mathbf{F} \times \frac{d\mathbf{G}}{dt}.$$

These can be shown by resolving the vectors into components in the **i**, **j** and **k** directions and differentiating each component. It is recommended that you work through one of these for yourself.

EXAMPLE 17.10   The position vector **r** of a particle relative to an origin satisfies the vector differential equation $\dot{\mathbf{r}} = \boldsymbol{\omega} \times \mathbf{r}$, where $\boldsymbol{\omega}$ is a constant vector.

i)   By taking the scalar product of this equation with **r**, show that the path of the particle lies on a sphere.

ii)   By taking the scalar product of the equation with the unit vector $\hat{\boldsymbol{\omega}}$ show that the path must also lie in a plane.

iii)   Hence deduce that the path is a circle and show that the particle moves with constant speed.

SOLUTION

i) $$\dot{\mathbf{r}} = \boldsymbol{\omega} \times \mathbf{r}$$

$$\Rightarrow \quad \mathbf{r} \cdot \dot{\mathbf{r}} = \mathbf{r} \cdot (\boldsymbol{\omega} \times \mathbf{r}) = 0 \quad \text{(two equal vectors in a triple scalar product)}$$

The left-hand side is $\dfrac{1}{2} \dfrac{\mathrm{d}}{\mathrm{d}t} (\mathbf{r} \cdot \mathbf{r})$ by the product rule for differentiating dot products.

So $\mathbf{r} \cdot \mathbf{r} = $ constant, i.e. $|\mathbf{r}|$ is constant, so the particle lies on a sphere.

ii) $$\hat{\boldsymbol{\omega}} \cdot \dot{\mathbf{r}} = \hat{\boldsymbol{\omega}} \cdot (\boldsymbol{\omega} \times \mathbf{r}) = 0$$

$\hat{\boldsymbol{\omega}}$ and $\boldsymbol{\omega}$ are parallel, so the triple scalar product is zero. The left-hand side is

$\dfrac{\mathrm{d}}{\mathrm{d}t} (\hat{\boldsymbol{\omega}} \cdot \mathbf{r})$, again by the product rule ($\hat{\boldsymbol{\omega}}$ is constant). So $\hat{\boldsymbol{\omega}} \cdot \mathbf{r} = $ constant.

This is the standard vector equation of a plane which is normal to $\hat{\boldsymbol{\omega}}$.

iii)   The particle lies on both a sphere and a plane, and must therefore lie on the intersection of these, i.e. a circle.

Differentiating $\dot{\mathbf{r}} = \boldsymbol{\omega} \times \mathbf{r}$ gives

$$\ddot{\mathbf{r}} = \boldsymbol{\omega} \times \dot{\mathbf{r}} \quad \text{(by the product rule for vector products: } \boldsymbol{\omega} \text{ is constant)}$$

$$\Rightarrow \quad \dot{\mathbf{r}} \cdot \ddot{\mathbf{r}} = \dot{\mathbf{r}} \cdot (\boldsymbol{\omega} \times \dot{\mathbf{r}}) = 0.$$

The left-hand side is $\dfrac{1}{2} \dfrac{\mathrm{d}}{\mathrm{d}t} (\dot{\mathbf{r}} \cdot \dot{\mathbf{r}})$. Hence $\dot{\mathbf{r}} \cdot \dot{\mathbf{r}} = $ constant $= |\dot{\mathbf{r}}|^2$ so the speed $|\dot{\mathbf{r}}|$ is constant.

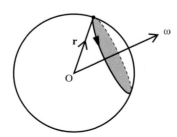

This example shows the power of vector manipulations, although you may find it difficult to see the physical situation modelled by this problem. The vector $\boldsymbol{\omega}$ is the *angular velocity vector* of the particle, through the origin.

Chapter 17 Exercises

17.1 Three forces $\mathbf{F}_1 = 2\mathbf{i} + \mathbf{j}$, $\mathbf{F}_2 = -\mathbf{i} - 5\mathbf{j}$ and $\mathbf{F}_3 = 2\mathbf{i}$ act on a particle at a point with position vector $\mathbf{i} + 2\mathbf{j}$.
  i) Find the resultant force.
  ii) What is the vector equation of the line of action of the resultant force?
  iii) What is the moment of the resultant force about the origin?

17.2 At time t, the position vector of a body of mass m is $t^2\mathbf{i} + \sin t\mathbf{j}$.
  i) Calculate the force $\mathbf{F}$ acting on the body at time t.
  ii) Use $\mathbf{F} \cdot \mathbf{v}$ to work out the rate at which $\mathbf{F}$ is doing work at time t.
  iii) Integrate to find the total work done between times 0 and t and verify this is the same as the increase in kinetic energy of the body.

17.3 At time t, a charged particle moving in an electric field has a position vector

$$\mathbf{r} = l(\omega t - \sin \omega t)\mathbf{i} + l(\omega t + \cos \omega t)\mathbf{j}$$

where ω and l are constants. Show that the acceleration has constant magnitude $l\omega^2$.

17.4 A charged particle of mass m is constrained to move inside a smooth straight tube whose direction is parallel to $3\mathbf{i} - 4\mathbf{j}$, where $\mathbf{i}$ is taken as horizontal and $\mathbf{j}$ is vertically upwards. As well as gravity, the particle is acted on at time t by an electrical force $-\frac{1}{3}mgt\mathbf{i}$. The particle is at rest at time $t = 0$.
  i) Write down the resolved part of the total force in the direction of the tube.
  ii) By applying Newton's second law in the direction of the tube, show that the particle is next at rest after 8 seconds.

17.5 A particle P moves so that at time t its position vector $\mathbf{r}$ relative to a fixed origin O is

$$\mathbf{r} = (a - a \cot^2 \omega t)\mathbf{i} + 2a \cot \omega t\mathbf{j}, \qquad 0 < \omega t \leqslant \frac{\pi}{2}$$

where a and ω are positive constants.
  i) Show that $|\mathbf{r}| = a \operatorname{cosec}^2 \omega t$.
  ii) Show that $\mathbf{v}$, the velocity of P, is given by

$$\mathbf{v} = 2|\mathbf{r}|\omega[\cot \omega t\mathbf{i} - \mathbf{j}]$$

and find, in terms of a and ω, the minimum speed of P.
  iii) Find t when P is moving perpendicular to OP and show that P never moves parallel to OP.

17.6 Find a vector of magnitude 3 perpendicular to both $2\mathbf{i} + \mathbf{j} - 3\mathbf{k}$ and $\mathbf{i} - 2\mathbf{j} + \mathbf{k}$.

17.7 The three sides of a triangle are represented by vectors $\mathbf{p}, \mathbf{q}, \mathbf{r}$ (see diagram). Show that

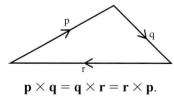

$$\mathbf{p} \times \mathbf{q} = \mathbf{q} \times \mathbf{r} = \mathbf{r} \times \mathbf{p}.$$

What well-known geometrical formula is represented by this result?

17.8   Given points A(1, 2, 1), B(1, 0, 3), C(−1, 2, −1),

i)   write down the vectors $\overrightarrow{AB}, \overrightarrow{BC}, \overrightarrow{CA}$;

ii) use the formula $\frac{1}{2}|\overrightarrow{AB} \times \overrightarrow{AC}|$ to work out the area of the triangle ABC;

iii) confirm the result is the same as $\frac{1}{2}|\overrightarrow{BC} \times \overrightarrow{BA}|$.

Personal Tutor

 17.1   A body of unit mass is initially at the origin moving with a velocity of $-2\mathbf{j}$. It is acted on by a force $\mathbf{F} = 2 \cos t\mathbf{i} + 2 \sin t\mathbf{j}$

i)   Calculate the acceleration and velocity of the body at time t.

ii) Show that the velocity is always perpendicular to the force and that the speed is constant.

iii) Calculate the position at time t.

iii) Show that the body moves in a circle and find its centre and radius.

KEY POINTS

17.1   Given two vectors $\mathbf{a}$ and $\mathbf{b}$, inclined at an angle θ, the vector product $\mathbf{a} \times \mathbf{b}$ is defined as a *vector* which

● is perpendicular to both $\mathbf{a}$ and $\mathbf{b}$;

● has a magnitude $|\mathbf{a}||\mathbf{b}| \sin \theta$;

● has a direction such that $\mathbf{a}$, $\mathbf{b}$ and $\mathbf{a} \times \mathbf{b}$ form a right-handed set of vectors.

17.2   For any vectors $\mathbf{a}$, $\mathbf{b}$, $\mathbf{c}$

$\mathbf{a} \times \mathbf{b} = -(\mathbf{b} \times \mathbf{a})$

$(\mathbf{a} \times \mathbf{b}).\mathbf{c} = (\mathbf{b} \times \mathbf{c}).\mathbf{a} = (\mathbf{c} \times \mathbf{a}).\mathbf{b}$

$\mathbf{a} \times (\mathbf{b} + \mathbf{c}) = (\mathbf{a} \times \mathbf{b}) + (\mathbf{a} \times \mathbf{c})$.

17.3   Given two vectors

$$\mathbf{a} = a_1\mathbf{i} + a_2\mathbf{j} + a_3\mathbf{k} \qquad \text{and} \qquad \mathbf{b} = b_1\mathbf{i} + b_2\mathbf{j} + b_3\mathbf{k}$$

then $\mathbf{a} \times \mathbf{b} = \begin{vmatrix} \mathbf{i} & a_1 & b_1 \\ \mathbf{j} & a_2 & b_2 \\ \mathbf{k} & a_3 & b_3 \end{vmatrix} = (a_2b_3 - a_3b_2)\mathbf{i} + (a_3b_1 - a_1b_3)\mathbf{j} + (a_1b_2 - a_2b_1)\mathbf{k}$.

17.4   Where two vectors, $\mathbf{F}$ and $\mathbf{G}$ depend on t, their vector and scalar products can be differentiated using the product rule:

$$\frac{d}{dt}(\mathbf{F}.\mathbf{G}) = \frac{d\mathbf{F}}{dt}.\mathbf{G} + \mathbf{F}.\frac{d\mathbf{G}}{dt};$$

$$\frac{d}{dt}(\mathbf{F} \times \mathbf{G}) = \frac{d\mathbf{F}}{dt} \times \mathbf{G} + \mathbf{F} \times \frac{d\mathbf{G}}{dt}.$$

18 Variable forces

Is it possible to fire a projectile up to the moon?

The Earth to the Moon *by Jules Verne (1865)*

In his book, Jules Verne says that this is possible … 'provided it possesses an initial velocity of 12000 yards per second. In proportion as we recede from the Earth the action of gravitation diminishes in the inverse ratio of the square of the distance; that is to say at three times a given distance the action is nine times less. Consequently the weight of a shot will decrease and will become reduced to zero at the instant that the attraction of the moon exactly counterpoises that of the Earth; at $\frac{47}{52}$ of its journey. There the projectile will have no weight whatever; and if it passes that point it will fall into the moon by the sole effect of lunar attraction.'

QUESTION 18.1   If an *unpowered* projectile could be launched from the Earth with a high enough speed in the right direction, it would reach the moon.

What forces act on the projectile during its journey?

How near to the moon will it get if its initial speed is not *quite* enough?

In Jules Verne's story, three men and two dogs were sent to the moon inside a projectile fired from an enormous gun. Although this is completely impracticable, the basic mathematical ideas in the passage above are correct. As a projectile moves further from the Earth and nearer to the moon, the gravitational attraction of the Earth decreases and that of the moon increases. In many of the dynamics problems you have met so far it has been assumed that forces are *constant*, whereas on Jules Verne's space missile the total force *varies* continuously as the motion proceeds.

You have already met problems involving variable force. When an object is suspended on a spring and bounces up and down, the varying tension in the spring leads to *simple harmonic motion*. You will also be aware that air resistance depends on velocity.

Gravitation, spring tension and air resistance all give rise to variable force problems, the subject of this chapter.

18.1 Newton's second law as a differential equation

Calculus techniques are used extensively in mechanics and you will already have used differentiation and integration in earlier work. In this chapter you will see how essential calculus methods are in the solution of a variety of problems.

To solve variable force problems, you can use Newton's second law to give an equation for the *instantaneous* value of the acceleration. When the mass of a body is constant, this can be written in the form of a *differential equation*:

$$F = m\frac{dv}{dt}.$$

It can also be written as

$$F = mv\frac{dv}{ds}.$$

This follows from the chain rule for differentiation.

$$\frac{dv}{dt} = \frac{dv}{ds} \times \frac{ds}{dt}$$

$$= v\frac{dv}{ds}$$

Note

Here and throughout this chapter the mass, *m*, is assumed to be constant. Jules Verne's spacecraft was a projectile fired from a gun. It was not a *rocket* whose mass varies, due to ejection of fuel. Rocket motion is considered in the next chapter.

Deriving the constant acceleration formulae

To see the difference in use between the $\frac{dv}{dt}$ and $v\frac{dv}{ds}$ forms of acceleration, it is worth looking at the case where the force, and therefore the acceleration, $\frac{F}{m}$ is *constant* (say *a*).

Starting from the $\frac{dv}{dt}$ form,

$$\frac{dv}{dt} = a$$

Integrating gives

$$v = u + at$$

where *u* is the constant of integration ($v = u$ when $t = 0$).

Since $v = \frac{ds}{dt}$, integrating again gives

$$s = ut + \tfrac{1}{2}at^2 + s_0$$

assuming the displacement is s_0 when $t = 0$.

These are the familiar formulae for motion under constant acceleration.

Starting from the $v\frac{dv}{ds}$ form,

$$v\frac{dv}{ds} = a.$$

Separating the variables and integrating gives

$$\int v \, dv = \int a \, ds$$

$$\Rightarrow \quad \tfrac{1}{2}v^2 = as + k$$

where k is the constant of integration.

Assuming $v = u$ when $s = 0$, $k = \tfrac{1}{2}u^2$, so the formula becomes

$$v^2 = u^2 + 2as.$$

This is another of the standard constant acceleration formulae. Notice that *time is not involved* when you start from the $v\dfrac{dv}{ds}$ form of acceleration.

Deriving equations for simple harmonic motion

Another case where the $v\dfrac{dv}{ds}$ form of the acceleration is useful is in finding solutions to the simple harmonic motion equation. This example also illustrates how trigonometrical substitutions are used to perform some integrations.

When the force on a body is proportional to its displacement from a fixed point, O, and is directed towards O, Newton's second law gives an equation of this form.

$$\frac{d^2x}{dt^2} = -\omega^2 x$$

Writing the acceleration as $v\dfrac{dv}{dx}$ gives

$$v\frac{dv}{dx} = -\omega^2 x.$$

Separating variables gives

$$\int v \, dv = -\int \omega^2 x \, dx$$

$$\Rightarrow \quad \tfrac{1}{2}v^2 = -\tfrac{1}{2}\omega^2 x^2 + c$$

Using $x = a$ when $v = 0$ gives $c = \tfrac{1}{2}\omega^2 a^2$ and rearranging gives the familiar

$$v^2 = \omega^2(a^2 - x^2).$$

To find x in terms of t write

$$v = \sqrt{a\omega^2 - x^2}$$

or

$$\frac{dx}{dt} = \sqrt{a\omega^2 - x^2}.$$

Then

$$\int \frac{dx}{\sqrt{a^2 - x^2}} = \int \omega \, dt = \omega t + \varepsilon \qquad \qquad ①$$

where ε is the constant of integration.

The first integral can be found by substituting $x = a \sin u$ so that

$$\frac{dx}{du} = a \cos u \quad \text{and} \quad dx = a \cos u \, du.$$

> Notice that x is written as a function of u rather than u as a function of x. This means that you have to differentiate with respect to u.

Substituting for x and dx in the integral in terms of u gives

$$\int \frac{dx}{\sqrt{a^2 - x^2}} = \int \frac{a \cos u \, du}{\sqrt{a^2 - a^2 \sin^2 u}}.$$

$\sqrt{a^2 - a^2 \sin^2 u} = a \cos u$ so the integral becomes $\int du = u = \arcsin \dfrac{x}{a}$.

Hence $\arcsin \dfrac{x}{a} = \omega t + \varepsilon$ (from ①)

giving $x = a \sin (\omega t + \varepsilon)$.

This equation can be written in the different forms

$$x = a \cos (\omega t + \varepsilon')$$

and
$$x = A \sin \omega t + B \cos \omega t$$

if required (see earlier chapters).

Activity

18.1   Use the substitution $x = a \cos u$ to find an expression for $\displaystyle\int \frac{-dx}{\sqrt{a^2 - x^2}}$.

Solving *F* = *ma* for variable force

When the force is continuously *variable*, you write Newton's second law in the form of a differential equation and then solve it using one of the forms of acceleration $v\dfrac{dv}{dx}$ or $\dfrac{dv}{dt}$. The choice depends on the particular problem. Some guidelines are given below and you should check these with the examples which follow.

Normally, the resulting differential equation can be solved by separating the variables.

The force is a function of time

When the force is a function, F(t), of time you use $a = \dfrac{dv}{dt}$.

$$F(t) = m \frac{dv}{dt}$$

Separating the variables and integrating gives

$$m\int dv = \int F(t)\ dt.$$

Assuming you can solve the integral on the right-hand side, you then have v in terms of t. Writing v as $\dfrac{ds}{dt}$, the displacement as a function of time can be found by integrating again.

The force is a function of displacement

When the force is a function, $F(s)$, of displacement, you normally start from

$$F(s) = mv\frac{dv}{ds};$$

then

$$\int F(s)ds = m\int vdv.$$

The force is a function of velocity

When the force is given as a function, $F(v)$, of velocity, you have a choice. You can use

$$F(v) = m\frac{dv}{dt}$$

or

$$F(v) = mv\frac{dv}{ds}.$$

You can separate the variables in both forms; use the first if you are interested in behaviour over time and the second when you wish to involve displacement.

Using the second derivative

When F is a *linear function* of velocity or displacement, or both, another possible starting point is to use the second derivative, $\dfrac{d^2s}{dt^2}$.

$$F = m\frac{d^2s}{dt^2}$$

This gives a second-order linear differential equation which can be solved using standard techniques for differential equations. The displacement is then given in terms of time without finding the velocity first. You can use the same method for the simple harmonic motion equation.

Variable force examples

The three examples that follow show the approaches used when the force is given respectively as a function of time, displacement and velocity.

When you are solving these problems, it is important to be clear about which direction is positive *before* writing down an equation of motion.

EXAMPLE 18.1   A crate of mass m is freely suspended at rest from a crane. When the operator begins to lift the crate further, the tension in the suspending cable increases uniformly from mg newtons to $1.2\,mg$ newtons over a period of 2 seconds.

i)   What is the tension in the cable t seconds after the lifting has begun ($t \leqslant 2$)?

ii)   What is the velocity after 2 seconds?

iii)   How far has the crate risen after 2 seconds?

Assume the situation may be modelled with air resistance and cable stretching ignored. Take g as $10\,\text{ms}^{-2}$.

SOLUTION

When the crate is at rest it is in equilibrium and so the tension, T, in the cable equals the weight mg of the crate. After time $t = 0$, the tension increases, so there is a net upward force and the crate rises, see figure 18.1.

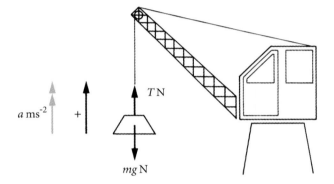

Figure 18.1

i)   The tension increases uniformly by $0.2mg$ newtons in 2 seconds, i.e. it increases by $0.1mg$ newtons per second, see figure 18.2. After t seconds, the tension is

$$T = mg + 0.1mgt.$$

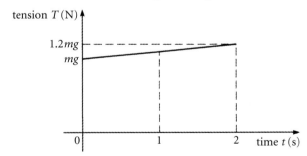

Figure 18.2

ii)   As the force is a function of time use $a = \dfrac{dv}{dt}$. Then at any moment in the 2-second period, $F = ma$ gives

$$(mg + 0.1mgt) - mg = m\frac{dv}{dt}$$

upwards is positive

$$\Rightarrow \qquad \frac{dv}{dt} = 0.1gt.$$

Integrating gives

$$v = 0.05gt^2 + k$$

where k is the constant of integration.

When $t = 0$, the crate has not quite begun to move, so $v = 0$. This gives $k = 0$ and $v = 0.05gt^2$.

When t is 2,

$$v = 0.5 \times 10 \times 4$$
$$= 2.$$

The velocity after 2 seconds is 2 ms^{-1}.

iii) To find the displacement s, write v as $\dfrac{ds}{dt}$ and integrate again.

$$\frac{ds}{dt} = 0.05gt^2$$
$$s = 0.05gt^2 \, dt$$
$$s = 0.05g \times \tfrac{1}{3}t^3 + c$$

When $t = 0$, $s = 0 \Rightarrow c = 0$.

When $t = 2$ and $g = 10$, $s = \tfrac{4}{3}$.

The crate moves $\tfrac{4}{3}$ m in 2 seconds.

QUESTION 18.2   The displacement cannot be obtained by the formula $s = \tfrac{1}{2}(u + v)t$, which would give the answer 2 m. Why not?

EXAMPLE 18.2   A prototype of Jules Verne's projectile, mass m, is launched vertically upwards from the Earth's surface but only just reaches a height of one tenth of the Earth's radius before falling back. When the height, s, above the surface is small compared with the radius of the Earth, R, the magnitude of the earth's gravitational force on the projectile may be modelled as $mg\left(1 - \dfrac{2s}{R}\right)$, where g is gravitational acceleration at the Earth's surface.

Assuming all other forces can be neglected

i)   write down a differential equation of motion involving s and velocity, v

ii)   integrate this equation and hence obtain an expression for the loss of kinetic energy of the projectile between its launch and rising to a height s

iii)   show that the launch velocity is $0.3\sqrt{2gR}$.

SOLUTION

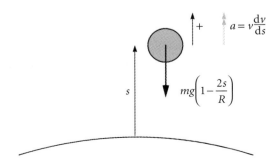

Figure 18.3

i) Taking the upward direction as positive, the force on the projectile is $-mg\left(1 - \dfrac{2s}{R}\right)$.

The force is a function of s, so start from the equation of motion in the form

$$mv\frac{dv}{ds} = -mg\left(1 - \frac{2s}{R}\right).$$

ii) Separating the variables and integrating gives

$$\int mv\,dv = -\int mg\left(1 - \frac{2s}{R}\right)ds$$

$$\Rightarrow \quad \frac{1}{2}mv^2 = -mgs + \frac{mgs^2}{R} + k.$$

> You would normally divide the equation by m, but it is useful to leave it in here in order to get kinetic energy directly from $\int mv\,dv$.

Writing v_0 for the launch velocity, $v = v_0$ when $s = 0$, so $k = \frac{1}{2}mv_0^2$ and rearranging gives

$$\tfrac{1}{2}mv_0^2 - \tfrac{1}{2}mv^2 = mgs - \frac{mgs^2}{R}. \qquad \text{①}$$

The left-hand side is the loss of kinetic energy, so

$$\text{loss of KE} = mgs - \frac{mgs^2}{R}.$$

> You can check that this is dimensionally consistent to give you confidence that your working is correct.

iii) Dividing equation ① by m and multiplying by 2 gives

$$v_0^2 - v^2 = 2gs - \frac{2gs^2}{R}.$$

If the projectile just reaches a height $s = \dfrac{R}{10}$, then the velocity v is zero at that point.

Substituting $s = \dfrac{R}{10}$ and $v = 0$ gives

$$v_0^2 = 2g\left(\frac{R}{10}\right) - \frac{2gR^2}{100R}$$

$$= \frac{18gR}{100}$$

$$\Rightarrow \quad v^0 = \frac{3}{10}\sqrt{2gR}.$$

So the launch velocity is $0.3\sqrt{2gR}$.

EXAMPLE 18.3

A body of mass 2 kg, initially at rest on a smooth horizontal plane, is subjected to a horizontal force of magnitude $\dfrac{1}{2v + 1}$ N, where v is the velocity of the body ($v > 0$).

i) Find the time when the velocity is 1 ms^{-1}.

ii) Find the displacement when the velocity is 1 ms^{-1}.

SOLUTION

i) Using $F = ma = m\dfrac{dv}{dt}$

$$\frac{1}{2v+1} = 2\frac{dv}{dt}.$$

> Write acceleration in $\dfrac{dv}{dt}$ form since time is required.

Separating the variables gives

$$\int dt = \int 2(2v+1)\, dv$$

$$\Rightarrow \quad t = 2v^2 + 2v + k.$$

When $t = 0$, $v = 0$ so $k = 0$ and therefore

$$t = 2v^2 + 2v.$$

When $v = 1$, $t = 4$. That is, when the velocity is $1\ \text{ms}^{-1}$, the time is 4 seconds.

ii) $F = ma = mv\dfrac{dv}{ds}$

$$\frac{1}{2v+1} = 2v\frac{dv}{ds}.$$

> Write acceleration in $v\dfrac{dv}{ds}$ form since displacement is required.

Separating the variables gives

$$\int ds = \int 2v(2v+1)\, dv$$

$$\Rightarrow \quad s = \tfrac{4}{3}v^3 + v^2 + k.$$

When $s = 0$, $v = 0$ so $k = 0$ and therefore

$$s = \tfrac{4}{3}v^3 + v^2.$$

When $v = 1$, $s = \tfrac{7}{3}$. When the velocity is $1\ \text{ms}^{-1}$, the displacement is $2\tfrac{1}{3}\ \text{m}$.

18.2 Gravitational force

Gravity is one of the fundamental forces of the universe, responsible for the motions of the planets, satellites and comets and indeed for the large-scale structure of the universe. All bodies attract each other by a gravitational force. It is very tiny for pairs of everyday objects, but large and important for objects of an astronomical size, such as the sun, the moon and the planets. Two particles attract each other with a force proportional to the product of their masses, $m_1 m_2$, and inversely proportional to the square of the distance, r, between them.

$$\text{force} = \frac{Gm_1 m_2}{r^2}$$

The constant G is known as the gravitational constant. For uniform spherical bodies, the force is along the line joining their centres and the distance is measured between their centres.

At the Earth's *surface*, an object of mass m is attracted to the centre by a force $\dfrac{GMm}{R^2}$, where M is the mass of the Earth and R is its radius.

Applying Newton's second law gives

$$\frac{GMm}{R^2} = ma.$$

So a, the acceleration of a body at the Earth's surface, is $\dfrac{GM}{R^2}$. This is g, the acceleration due to gravity. The Earth is so large that g can be assumed to be constant near to the surface. However, for objects such as meteorites or returning spacecraft, the continuous change in the gravitational force due to the changing distance from the centre of the Earth must be taken into account when the motion is analysed.

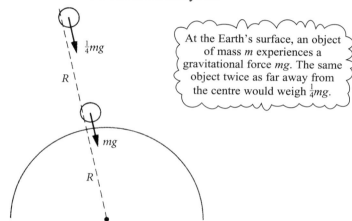

At the Earth's surface, an object of mass m experiences a gravitational force mg. The same object twice as far away from the centre would weigh $\frac{1}{4}mg$.

Figure 18.4

EXAMPLE 18.4

i) Denoting the Earth's radius by R metres, show that the gravitational force on a body of mass m kg above the Earth's surface and a distance s metres from the Earth's centre ($s > R$) is $\dfrac{mgR^2}{s^2}$.

ii) A projectile is fired vertically from the Earth's surface with an initial velocity of u ms^{-1} and reaches a maximum height of h m.
Derive from Newton's second law an expression giving u^2 in terms of R, g and h. (Neglect air resistance.)

iii) For what launch speed would the projectile just reach a height equal to the radius of the Earth, 6400 km? (Use $g = 9.8$ ms^{-2}.)

iv) What is the minimum launch speed if the projectile is never to return?

SOLUTION

i) At the Earth's surface, the force on a body of mass m is $\dfrac{GMm}{R^2}$ newtons. So

$$\frac{GMm}{R^2} = \text{mass} \times \text{acceleration} = mg$$

$$\Rightarrow \quad GM = gR^2.$$

Above the Earth's surface at a distance s metres from the centre, the force on a body of mass m is

$$\frac{GMm}{s^2} = \frac{gR^2m}{s^2} \qquad \text{(substituting } GM = gR^2\text{)}$$

$$= \frac{mgR^2}{s^2}.$$

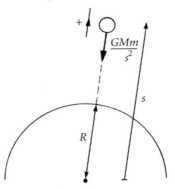

Figure 18.5

ii) Take the positive direction (increasing s) upwards. The force is acting downwards, hence the negative sign when applying the equation of motion.

$$-\frac{mgR^2}{s^2} = mv\frac{\mathrm{d}v}{\mathrm{d}s}$$

The $v\dfrac{\mathrm{d}v}{\mathrm{d}s}$ form for acceleration is used since the problem involves distances and velocities.

Dividing by m and separating the variables gives

$$\int v \,\mathrm{d}v = -\int \frac{gR^2}{s^2}\,\mathrm{d}s$$

$$\Rightarrow \qquad \frac{1}{2}v^2 = \frac{gR^2}{s} + c.$$

s is the distance from the centre of the Earth, so $v = u$ when $s = R$, giving $c = \frac{1}{2}u^2 - gR$,

$$\Rightarrow \qquad \frac{1}{2}v^2 = \frac{gR^2}{s} + \frac{1}{2}u^2 - gR.$$

$v = 0$ when $s = h + R$ gives

$$\frac{1}{2}u^2 = gR - \frac{gR^2}{h + R} \qquad\qquad \textcircled{1}$$

$$u^2 = \frac{2gRh}{h + R}.$$

iii) When $h = R$, $u^2 = gR$

$$\Rightarrow \qquad u = \sqrt{gR} = 7920. \blacktriangleleft \overbrace{\qquad}^{} \boxed{R = 6400 \times 10^3 \text{ in metres.}}$$

The launch speed is approximately $7.9 \,\mathrm{km\,s^{-1}}$.

iv) $\dfrac{gR^2}{h+R} \to 0$ as $h \to \infty$ so equation ① gives $\dfrac{1}{2}u2 \to gR$ for large h.

In order never to return, the minimum launch speed is $\sqrt{2gR} = 11\,200\ \text{ms}^{-1}$.

This is $11.2\,\text{km\,s}^{-1}$. For obvious reasons, this value of u is known as the *escape velocity*.

An alternative solution using energy methods is given in Example 18.6 on pages 415 to 416.

Activity

18.2 *Escape velocity*

Assume that all of a group of planets, including the Earth, can be modelled as spheres of equal densities.

i) Show that the escape velocity from the surface of a planet is proportional to the radius.

$$\left(\text{Remember: } g = \frac{GM}{R^2} \right)$$

ii) What would be the escape velocity from the surface of a planet with a radius $\frac{1}{1000}$th that of the Earth's?

iii) Many asteroids (minor planets between the orbits of Mars and Jupiter) are only a few kilometres in radius.

Could you hit a tennis ball into space from the surface of an asteroid?

Could you jump off into space?

Variation of g on the Earth's surface

The apparent weight of a body varies slightly at different parts of the Earth's surface. One reason is that the Earth is rotating. Unless you are at one of the poles, you are rotating in a circle round the Earth's axis and a small part of the gravitational force supplies the central acceleration for circular motion. You would otherwise be thrown into space. This has nothing to do with gravity but does affect what is measured as g.

In addition, the Earth is not spherical but bulges at the equator. The equatorial radius is about 1 part in 300 greater than the polar radius. Thus someone at the equator is further from the centre than someone at a pole.

These effects combine to make g about 0.5% less at the equator than at the poles.

QUESTION 18.3 How much faster does a pendulum swing at the poles?

Historical note

The radius of the Earth has been known since ancient times; the first reasonable estimate was by Eratosthenes (born in 276 BC). He was curator of the library at Alexandria and measured the elevation of the sun at noon on midsummer day to be about 7°. Due south at Syene (now Aswan), at the same time, on the same day, the sun was known to be overhead. So a 7° arc along the circumference of the Earth corresponded to the known distance between Alexandria and Syene. Thus the circumference of the Earth and hence its radius could be estimated.

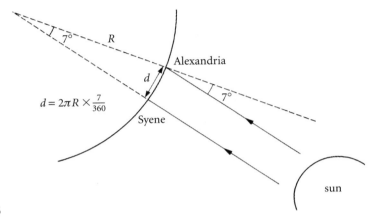

Figure 18.6

The value of g can be measured directly so, if the gravitational constant G can be determined, the mass of the Earth can be deduced using $g = \dfrac{GM}{R^2}$. Henry Cavendish, a brilliant but reclusive British physicist, performed a classic experiment to measure G in 1798. A rod with lead weights at each end was suspended on a fine fibre. When large weights are brought near the suspended ones, the tiny gravitational attraction causes a minute twist of the rod. This can be amplified and measured by the movement of a beam of light reflected from a mirror attached to the fibre.

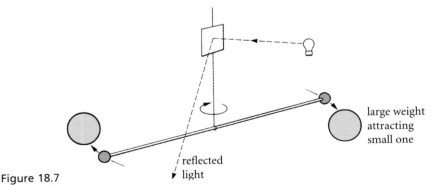

Figure 18.7

18.3 Work, energy and impulse with variable forces

The concepts of work, energy and impulse are very valuable in the context of variable forces. In particular, the principle of conservation of mechanical energy, which has been previously used to solve problems without having to calculate acceleration explicitly, often enables you to solve variable-force problems which would not be easily dealt with by integration.

You have already partly covered this topic. In earlier chapters, we gave the definition of the impulse of a variable force and showed how to calculate the work against the tension when stretching an elastic string. This section reviews the definitions of work, energy and impulse when variable forces are involved and applies them to the resistive and gravitational forces you have already met in this chapter.

Work done by a variable force

When a body moves a short distance δs along a line solely under the action of a parallel force F, you know that the force has done *an element of work* $\delta W = F\,\delta s$. When the force is *varying*, the total work is the sum of all these elements, that is $\int F\,\mathrm{d}s$.

Note

You may not be satisfied with this and similar informal arguments. A more rigorous treatment is as follows.

Assuming the force increases from F to $F + \delta F$ as the body moves; then

$$F\delta s < \delta W < (F + \delta F)\delta s$$

$$\Rightarrow \quad F < \frac{\delta W}{\delta s} < F + \delta F.$$

As $\delta s \to 0$, $\delta F \to 0$ and $\dfrac{\delta W}{\delta s} \to \dfrac{\mathrm{d}W}{\mathrm{d}s}$; hence $\dfrac{\mathrm{d}W}{\mathrm{d}s} = F$.

By integration, the total work done over the period of the motion is $W = \int F\,\mathrm{d}s$.

Previously, $\int F\,\mathrm{d}s$ was used to find the work done in extending an elastic string, of stiffness k, by an amount s. When extended by x, the tension in the string is, by Hooke's law, kx. Hence the total work starting from zero extension is

$$\int_0^s kx\,\mathrm{d}x = \tfrac{1}{2}ks^2$$

which is the elastic energy in a stretched spring.

Work done = increase in kinetic energy

You have seen this important result, which follows from Newton's second law

$$F = ma = mv\frac{dv}{ds}.$$

Separating the variables and taking u and v as the starting and finishing velocities respectively, you get

$$\int F\,ds = \int mv\,dv$$
$$= \tfrac{1}{2}mv^2 - \tfrac{1}{2}mu^2$$

work done by force = increase in kinetic energy.

EXAMPLE 18.5

A particle of mass 1 kg moves along a line with a velocity v ms^{-1} under the influence of a resistive force of magnitude kv^2 N, where k is a constant.

Initially, the velocity of the particle is 10 ms^{-1} and the force continues to act until the particle has slowed down to 5 ms^{-1}.

i) Use kinetic energy considerations to write down the work done by the resistive force.

ii) Solve the equation of motion of the particle and express the displacement x in terms of v.

   Hence show that the particle travels a distance $s = \dfrac{1}{k}\ln 2$ m while the force acts.

iii) From part ii) express v and hence F in terms of x.

   Hence confirm by integration the result obtained in part i).

SOLUTION

i) Work done = final K.E. − initial K.E.

$$= \tfrac{1}{2} \times 5^2 - \tfrac{1}{2} \times 10^2 \text{ J}$$
$$= -37.5 \text{ J}.$$

This is negative because the force is in the opposite direction to the displacement.

ii) Applying $F = ma$ with $m = 1$ and $a = v\dfrac{dv}{dx}$, where x is displacement,

$$-kv^2 = v\frac{dv}{dx}.$$

Figure 18.8

Separating the variables gives

$$-k \int dx = \int \frac{dv}{v}$$

$$\Rightarrow \quad -kx = \ln v + c.$$

When $x = 0$, $v = 10$, hence $c = -\ln 10$. So

$$-kx = \ln v - \ln 10$$

$$= -\ln \frac{10}{v}$$

$$\Rightarrow \quad x = \frac{1}{k} \ln \frac{10}{v} \qquad \qquad \text{①}$$

When $v = 5$ (final velocity)

$$x = s = \frac{1}{k} \ln 2.$$

iii) Making v the subject in ① gives $v = 10\,e^{-kx}$. Therefore

$$F = -kv^2$$

$$= -100k\,e^{-2kx}.$$

Work done is $\int F\,dx$:

$$\int F\,dx = \int_0^s -100k\,e^{-2kx}\,dx$$

$$= 50 \int_0^s -2k\,e^{-2kx}\,dx$$

$$= 50\Big[e^{-2kx}\Big]_0^s.$$

Now $s = \dfrac{1}{k} \ln 2$ so

$$e^{-2ks} = e^{-2\ln 2}$$

$$= e^{\ln \frac{1}{4}}$$

$$\Rightarrow \quad e^{-2ks} = \tfrac{1}{4}.$$

Thus the work done $= 50\left(\tfrac{1}{4} - 1\right)$

$$= -37.5.$$

This is -37.5 J, as in part i).

Force–distance graph

Note that if you plot force against distance (as in figure 18.9), the work done $\int F\,dx$ is the area under the graph.

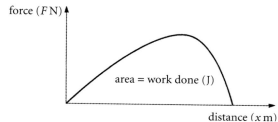

Figure 18.9

Activity

18.3 ***Simple harmonic motion by the energy method***

Earlier, in experiment 14.1 (see figure 14.26), you investigated the spring-mass oscillator: a particle of mass m suspended on a spring oscillating vertically about its equilibrium position. You can derive the same results by the energy method.

A particle of mass m is suspended from a spring of natural length l and stiffness k. Without referring to the equilibrium position, write down an expression for the total energy when the spring is extended by x and moving with velocity v.

The total energy is constant throughout the motion. Differentiate the energy equation with respect to x and show that the mass oscillates with simple harmonic motion. Deduce the equilibrium position *from* the simple harmonic motion equation.

Gravitational potential energy

For a body moving near the Earth's surface, the work done by the gravitational force is often expressed as the change in the *potential energy* of the body. When a body of mass m rises a distance h, the work done *against* the gravitational force, mg, is mgh and you know this is the increase in potential energy of the body.

When the gravitational force *changes*, such as in the case of a missile launched from the Earth's surface into space, the work done is obtained by integration as follows.
The gravitational force on a body of mass m at a distance r_1 from a body of mass M is

$$F = -\frac{GMm}{r_1^2}.$$

The positive direction will be that of increasing r, hence the minus sign. Suppose a body of mass m is pulled away from a distance r_1 to a distance r_2 (as in figure 18.10), what work is done *against* the gravitational force?

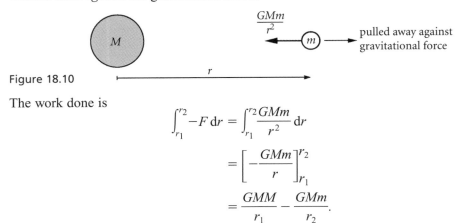

Figure 18.10

The work done is

$$\int_{r_1}^{r_2} -F \, dr = \int_{r_1}^{r_2} \frac{GMm}{r^2} \, dr$$

$$= \left[-\frac{GMm}{r} \right]_{r_1}^{r_2}$$

$$= \frac{GMM}{r_1} - \frac{GMm}{r_2}.$$

This is positive ($r_2 > r_1$), since increasing the separation between the bodies involves positive work against the gravitational force.

Just as in the case of constant gravitation, the *increase in potential energy* is defined as the work done against the force. With this type of motion, the zero level of potential energy is normally taken at infinity so r_1 is infinite. Then the gravitational potential energy of a body of mass m at a distance r from a body of mass M is defined as the work done in bringing them together from infinity, namely

$$-\frac{GMm}{r} \qquad (r = r_2 < r_1).$$

The work done against gravitation in moving a spacecraft away from a planet is thus the increase in potential energy, which according to the formula depends only on the initial and final values of r. And since

$$\text{work done against force} = \text{loss in kinetic energy}$$

it follows once more that

$$\text{gain in potential energy} = \text{loss in kinetic energy.}$$

The following example shows the use of gravitational potential energy on the problem solved in Example 18.4 using the equation of motion.

EXAMPLE 18.6

A ballistic missile, fired vertically from the Earth's surface with an initial velocity of u just reaches a height h.

Using energy methods, derive an expression giving u^2 in terms of R (the radius of the Earth), g and h. Assume the gravitational acceleration at a distance r from the centre of the Earth is $\dfrac{GM}{r^2}$, where $GM = gR^2$.

SOLUTION

At launch, the kinetic energy of the missile is $\frac{1}{2}mu^2$ and its potential energy is $-\dfrac{GMm}{R}$.

At the missile's highest point, its velocity is zero and since it is a distance $(R + h)$ from the centre of the Earth, the potential energy is $-\dfrac{GMm}{R + h}$.

So, using the principle of conservation of energy:

$$\text{loss in kinetic energy} = \text{gain in potential energy}$$

$$\frac{1}{2}mu^2 = GMm\left(-\frac{1}{R + h} + \frac{1}{R}\right)$$

$$= GMm\left(\frac{h}{R(R + h)}\right)$$

$$\Rightarrow \quad u^2 = \frac{2gRh}{(R + h)} \qquad \text{since } GM = gR^2.$$

This is the result obtained previously on page 409.

Note

Conservative forces

It is shown on the previous page that the work done against the gravitational force *depends only on the initial and final position of the body*. This means that when the body returns to its starting point, no *net* work has been done, so the kinetic energy will be the same as before. That is, the force conserves the total mechanical energy. Forces which have this property are known as *conservative*. It can be shown that conservation of mechanical energy is equivalent to the fact that the work done is dependent only on the initial and final positions.

Potential energy is associated only with conservative forces. The gravitational force, the electrical force between charged particles, the tension in a spring (obeying Hooke's law) – these are all examples of conservative forces. Frictional forces and the forces involved in non-elastic collisions are not conservative. They are known as *dissipative* as they result in a reduction of the total mechanical energy.

Power with variable forces

As you will be aware, power is defined as the rate at which work is done. The definition is independent of whether or not the force is varying.

$$P = \frac{dW}{dt} \qquad \Rightarrow \qquad W = \int P \, dt$$

where the integration is performed over the total time period. This implies that the area under the power–time graph (see figure 18.11) is equal to the work done.

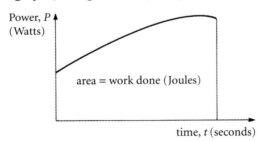

Figure 18.11

Power = force × velocity

You have seen that $F = \dfrac{dW}{ds}$, where W is the work done. Hence by the chain rule

$$\frac{dW}{dt} \times \frac{dt}{ds} = F$$

$$\Rightarrow \qquad \frac{dW}{dt} = F \frac{ds}{dt}.$$

Therefore, $$P = Fv$$

where v is the velocity of the body on which F acts.

This is a familiar result. Note that *constant power does not imply constant force*. When a car engine exerts a thrust of F N and the car is travelling with a speed v, the rate of working, i.e. power, is Fv Js^{-1}. If the power is constant, and v varies then F must vary.

EXAMPLE 18.7

When working at a constant power of 2.5 kW against a resistance proportional to the square of its speed, the maximum speed a vehicle can attain on a level road is 50 ms^{-1}. If the vehicle accelerates from rest under the same conditions, how far does it travel before it attains half the maximum speed? The mass of the vehicle is 1500 kg.

SOLUTION

The power is 2500 W = force × velocity, so the engine's driving force has a magnitude of $\dfrac{2500}{\text{velocity}}$.

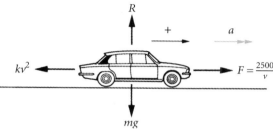

Figure 18.12

When the speed is v ms^{-1} and the resistance is kv^2 N, Newton's second law gives

$$\frac{2500}{v} - kv^2 = 1500a. \qquad \textcircled{1}$$

The maximum speed of 50 ms^{-1} occurs when the acceleration is zero. So

$$\frac{2500}{50} = k(50)^2$$

$$\Rightarrow \quad k = \frac{1}{50}.$$

Substituting this value of k in $\textcircled{1}$ and using $a = v\dfrac{dv}{dx}$ since x is required gives

$$\frac{2500}{v} - \frac{v^2}{50} = 1500v\frac{dv}{dx}$$

$$\Rightarrow \quad 125\,000 - v^3 = 75\,000v^2\frac{dv}{dx}.$$

This can be solved by separating the variables and integrating.

$$\int dx = \frac{75\,000v^2}{125\,000 - v^3}\,dv$$

> This integral has been evaluated by inspection but could be found by substituting $z = 125000 - v^3$.

$$\Rightarrow \quad x = \frac{75\,000}{3}\ln(125\,000 - v^3) + c$$

$$x = 0 \text{ when } v = 0 \quad \Rightarrow \quad c = 25\,000 \ln 125\,000$$

so

$$x = 25\,000 \ln\left(\frac{125\,000}{125\,000 - v^3}\right)$$

when

$$v = 25, \quad x = 3338.$$

The distance is therefore approximately 3.34 km.

Impulse of a variable force

For a constant force, the *impulse* is force $\times$ time for which it acts.

When the force varies, the impulse over a small time interval δt is $F \, \delta t$, and so the total impulse over a period is defined as $\int F \, \mathrm{d}t$ (see figure 18.13).

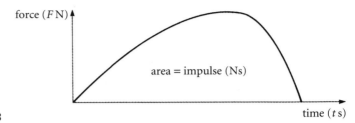

force (F N)

area = impulse (Ns)

time (t s)

Figure 18.13

Over a period T in which the velocity changes from U to V,

$$\int_0^T F \, \mathrm{d}t = \int_U^V m \frac{\mathrm{d}v}{\mathrm{d}t} \, \mathrm{d}t \qquad \left(\text{since } F = m \frac{\mathrm{d}v}{\mathrm{d}t} \right)$$

$$\int_0^T F \, \mathrm{d}t = \int_U^V m \, \mathrm{d}v$$

$$= mV - mU$$

when m is constant.

This is the result which you saw in earlier, that the *impulse of a force is equal to the change in momentum*. This applies even when the force is varying.

Any collision involves variable forces. When two snooker balls, A and B, collide, they are in contact for a very short time. During that time the force between them is not constant. It is zero just as they touch, builds up to a maximum while they deform slightly and then goes down to zero again as the balls rebound. But, by Newton's third law, the force of A on B is equal and opposite to that of B on A *at every moment*. So, if the total impulse on A is $\int F \, \mathrm{d}t$, that on B is

$$\int -F \, \mathrm{d}t = -\int F \, \mathrm{d}t$$

The sum of the impulses on A and on B is zero and so the total momentum change is zero. The principle of conservation of momentum applies even though forces are variable.

18.4 Resisted motion

A strong enough wind can blow you over; the force exerted by the air obviously depends on wind speed. You feel the same effect when you cycle or ski quickly – except that it is now called *air resistance* rather than wind. Any object moving through a fluid, such as a gas or a liquid, encounters a *resistive force* opposing its motion.

Experiments have shown that the air resistance on a moving object, such as a falling pebble, is approximately proportional to the square of its speed. The resistance is caused mainly because of the force required to push the particles of air into motion, in other words to change the momentum of the air particles. It is similar when the object is at rest and the wind blows against it – the force is produced by the loss of momentum of the air as it hits the object.

However, when the object is tiny (a particle of dust, for example), is slow-moving, or when it is moving through a liquid rather than a gas, then most of the resistive force is caused by *viscosity* – friction between layers of the liquid and between the liquid and the object. The viscous force (on a falling leaf or a lump of sugar dropped into a cup of coffee, for example) is directly proportional to speed. In general, motion through a fluid will be opposed by resistive forces which depend on a mixture of both *velocity* and *velocity squared*, but it is often the case that one dominates and the other can be ignored.

Terminal velocity

Imagine that a pebble of mass m, falling through the air, is subjected to a resistance proportional to the square of its speed, say kv^2, where k is a constant. This is a reasonable modelling assumption for such an object. The net downward force is $mg - kv^2$ (see figure 18.14). This is clearly a variable force. Applying Newton's second law at any instant gives

$$mg - kv^2 = m\frac{\mathrm{d}v}{\mathrm{d}t}.$$

$$a = \frac{\mathrm{d}v}{\mathrm{d}t} \quad + \quad \begin{array}{c} \uparrow kv^2 \\ \bigcirc \\ \downarrow mg \end{array}$$

Figure 18.14

You can get some idea of the behaviour of the pebble simply by looking at the differential equation. Initially v is zero, so the acceleration is g. Thus v will begin to increase. As v increases the acceleration $\dfrac{dv}{dt}$ gets smaller. As v^2 approaches $\dfrac{mg}{k}$, the left-hand side of the equation approaches zero, and therefore so does the acceleration. If v^2 ever becomes $\dfrac{mg}{k}$, the forces balance and there is no acceleration. The pebble would then continue at the speed $v = \sqrt{\dfrac{mg}{k}}$. This is known as its *terminal velocity*: it is the limiting speed that the pebble can reach. When an object is dropped from rest, its speed gets nearer and nearer to the terminal velocity but in theory does not quite achieve it, as the later worked examples show.

For situations that can be modelled by a resistive force proportional to v, for example a tiny raindrop falling in the air, Newton's second law gives

$$mg - cv = m\frac{dv}{dt}$$

where the resistive force is cv. You can see that in this case the terminal velocity has a value $\dfrac{mg}{c}$.

Modelling air resistance

Air resistance is quite significant in everyday situations. For example, a 2 cm diameter pebble dropped off a high cliff (don't do it!) has a terminal velocity of about $35\ \text{ms}^{-1}$, and the velocity is close to this after about 7 seconds.

QUESTION 18.4   The graph shows how the speed of a granite pebble varies with time when it is dropped in air. The pebble has a diameter of 2 cm and a mass of 11.3 g.

What is the gradient of the graph at the origin?

How would you use the graph to estimate how far the stone has fallen before reaching 90% of its terminal velocity?

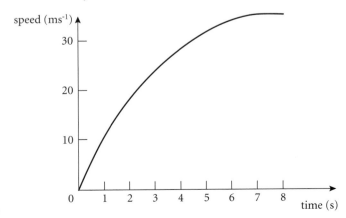

Figure 18.15

Solving the resistive motion equations

To solve differential equations of resistive motion, write the acceleration in the form $\dfrac{dv}{dt}$ or $v\dfrac{dv}{ds}$, depending on whether you require *time* in the answer.

You can separate the variables and then integrate. You will find that many of the integrations result in logarithms or inverse trigonometrical functions.

EXAMPLE 18.8

A raindrop of mass m starts from rest and falls vertically. When falling with velocity v, it experiences a resistive force of magnitude kv.
i) Express the terminal velocity V_T of the raindrop in terms of m, k and g.
ii) Show that, if the raindrop starts from rest, its velocity v after time t is given by
$$v = V_T\!\left(1 - e^{-gt/V_T}\right).$$
iii) Assuming the raindrop has a terminal velocity of $5\,\mathrm{ms}^{-1}$, how long does it fall before it has 99% of the terminal velocity?
(Take g as $10\,\mathrm{ms}^{-2}$.)
iv) Derive an expression for the distance fallen in terms of time.

SOLUTION

i) The forces on the raindrop are mg downwards and kv upwards. Applying Newton's second law at any instant:
$$mg - kv = ma.$$

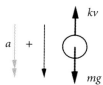

Figure 18.16

If the raindrop were moving at the terminal velocity, the acceleration would be zero, so
$$V_T = \frac{mg}{k}.$$

ii) It is helpful at this stage to divide the first equation by k and write $\dfrac{mg}{k} = V_T$ and $\dfrac{m}{k} = \dfrac{V_T}{g}$. Since time is required use $a = \dfrac{dv}{dt}$. The equation then becomes
$$V_T - v = \frac{V_T}{g} \times \frac{dv}{dt}.$$

Now separate the variables and integrate.
$$\frac{dv}{(V_T - v)} = \frac{g}{V_T}\,dt$$
$$\Rightarrow \qquad -\ln\left|V_T - v\right| = \frac{gt}{V_T} + c$$

$V_T > v$ throughout the motion so $\left|V_T - v\right| = (V_T - v)$.

$v = 0$ when $t = 0$ so $c = -\ln V_T$

$$\Rightarrow \qquad -\ln(V_T - v) = \frac{gt}{V_T} - \ln V_T$$

$$\Rightarrow \qquad \ln\left[\frac{V_T - v}{V_T}\right] = -\frac{gt}{V_T}$$

$$\frac{V_T - v}{V_T} = e^{-gt/V_T}$$

> This makes it clear that, as you would expect, the velocity, v, is equal to the terminal velocity minus a small and decreasing amount.

which gives

$$v = V_T\left(1 - e^{-gt/V_T}\right)$$

or $\qquad\qquad v = V_T - V_T e^{-gt/V_T}.$ ①

iii) When v is 99% of the terminal velocity, $\dfrac{v}{V_T} = 0.99.$

$$0.99 = 1 - e^{-gt/V_T} \qquad\qquad ②$$

$$= 1 - e^{-10t/5} \qquad (g = 10, V_T = 5)$$

$$= 1 - e^{-2t}$$

$$\Rightarrow \qquad e^{-2t} = 0.01$$

$$e^{2t} = 100$$

$$\Rightarrow \qquad t = \tfrac{1}{2} \ln 100 = 2.3$$

> Note that the time taken to reach a given *percentage* of V_T does depend on the value of V_T. $V_T = 5$ was substituted in ②.

The raindrop falls for 2.3 seconds before achieving 99% of V_T.

iv) Having derived ①, write v as $\dfrac{ds}{dt}$ and integrate to get

$$\int ds = \int \left(V_T - V_T e^{-gt/V_T}\right) dt$$

$$s = V_T t + \left(\frac{V_T^2}{g}\right) e^{-gt/V_T} + c$$

where c is the constant of integration.

Since $s = 0$ when $t = 0$, $c = -\dfrac{V_T^2}{g}$, so

$$s = V_T t - \frac{V_T^2}{g}\left(1 - e^{-gt/V_T}\right).$$

> The first term $V_T t$ is the distance that the raindrop would fall in a time t if travelling at speed V_T the whole time. s is always less than this.

Using a second-order equation

Example 18.8 involved integration twice, once to find v and again to find s. If you are not interested in v, and want s directly, you can express acceleration as $\dfrac{d^2s}{dt^2}$ and solve the second-order differential equation using standard techniques.

Deciding on the positive direction

It is easy to get confused with the signs when writing down Newton's second law. Decide which is the direction in which you are measuring *positive displacement* and take this as your positive direction for *all variables*. A positive value of v means the body is moving in this direction. In the equation of motion, resistive forces will be in a direction opposite to that in which the body is moving. Thus, when the resistive force is proportional to velocity, write it as $-kv$ where k is a positive constant. If in fact the velocity is negative, the value of $-kv$ will be positive, still in the opposite direction to the motion. Your equation is correct whichever direction the body is actually moving in.

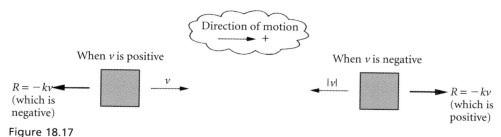

Figure 18.17

However, this does not work when the resistive force is proportional to v^2. You need different equations depending on which way the body is travelling, and the two cases have to be considered separately as in the next example.

EXAMPLE 18.9

A heavy ball is retarded by air resistance which is modelled by kv^2, where $v\,\text{ms}^{-1}$ is the speed of the ball and k is a positive constant. When falling under gravity in air, its terminal velocity is $V\,\text{ms}^{-1}$.

i) Write an expression for the total force on the ball when it is descending and hence derive an expression for V^2 in terms of k, g and the mass of the ball.

ii) The ball is projected vertically upwards with an initial speed $u\,\text{ms}^{-1}$.
   Derive a differential equation, involving s and V, but not k, covering its ascent.

iii) Show that, when $u = 50$ and $V = 100$, the ball reaches a height of almost 114 m.

SOLUTION

i) Assume that the mass of the ball is m and the resistance is kv^2. When descending, the net downward force is $mg - kv^2$.

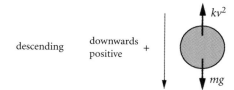

Figure 18.18

The terminal velocity, V, is obtained when the force is zero, so $mg = kV^2$.

$$\Rightarrow \quad V^2 = \frac{mg}{k}$$

ii) When ascending, both the weight and the air resistance are in the same direction. Take the positive direction as *upwards*, so the forces are negative and the initial velocity is positive.

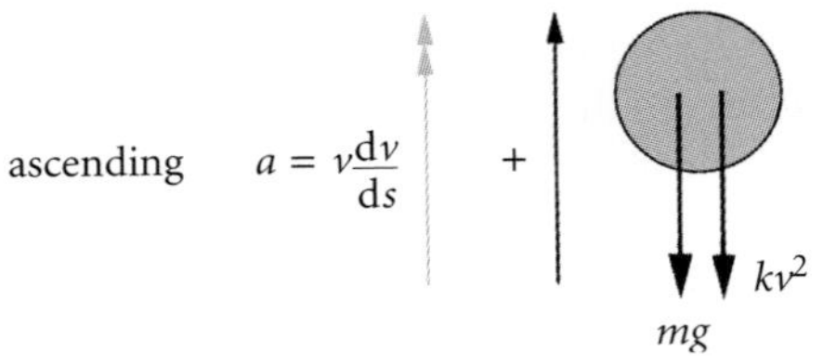

ascending $a = v\dfrac{dv}{ds}$ +

kv^2

mg

Figure 18.19

The equation of motion is then

$$-mg - kv^2 = mv\frac{dv}{ds}$$

> For an equation in s write $a = v\dfrac{dv}{ds}$.

$$-\frac{mg}{k} - v^2 = \frac{m}{k}\,v\frac{dv}{dx}.$$

Substituting $\dfrac{mg}{k} = V^2$ gives

$$-V^2 - v^2 = \left(\frac{V^2}{g}\right)v\frac{dv}{ds}$$

$$v\frac{dv}{ds} = \frac{-g(V^2 + v^2)}{V^2}.$$

iii) Separating the variables and integrating gives

$$\frac{v\,dv}{V^2 + v^2} = -\frac{g}{V^2}\,ds$$

$$\frac{1}{2}\ln(V^2 + v^2) = \frac{gs}{V^2} + c.$$

If $v = u$ when $s = 0$, $c = \frac{1}{2}\ln(V^2 + u^2)$

$$\Rightarrow \qquad \frac{gs}{V^2} = \frac{1}{2}\ln(V^2 + u^2) - \frac{1}{2}\ln(V^2 + v^2)$$

$$s = \frac{V^2}{2g}\ln\left(\frac{V^2 + u^2}{V^2 + v^2}\right).$$

> This is the general solution for s in terms of v.

When $v = 0$, $s = h$ so

$$h = \frac{V^2}{2g}\ln\left(\frac{V^2 + u^2}{V^2}\right).$$

> It is useful to check the dimensions. The expression $\ln\left(\dfrac{V^2 + u^2}{V^2}\right)$ is dimensionless so, on the right-hand side, $\dfrac{V^2}{2g}$ gives
> $$\frac{(LT^{-1})^2}{LT^{-2}} =$$
> $L^2T^{-2}L^{-1}T^2 = L,$ as required.

Substituting the values $V = 100$, $u = 50$ given in the question:

$$h = \frac{100^2}{2g}\ln 1.25 = 113.8.$$

The ball reaches a height of 113.8 m.

Question 18.5 Can a body descending under air resistance start off with a speed *greater* than the terminal velocity? How? What happens subsequently?

When a ball is thrown up in the air, does it take longer to go up or come down? Bear in mind that energy is dissipated by the air resistance and consider the velocity at corresponding points on the way up and the way down.

Historical note

The first thorough treatment of the motion of bodies in a resisting medium was given by Newton, who devoted the second of the three books of his master work *Philosophiae Naturalis Principia Mathematica* (first edition 1687) to this topic. He considered motion where the fluid resistance is proportional to the velocity, the square of the velocity and a combination of the two. He also investigated the physical causes of fluid resistance.

Terminal velocity under resistive force was an important part of R.A. Millikan's classic experiment, published in 1910, to measure the charge on an electron (the smallest quantum of charge). He observed tiny charged droplets of oil rising or falling in an electric field and was able to measure the terminal velocity very accurately. The field causes an electric force on the charged particle in addition to the gravitational force and viscous force of the air. By observing, over many hours, the terminal velocity of the same droplet using different values of the electric field, Millikan was able to eliminate various other unknown quantities (e.g. the droplet's mass) and calculate the charge on the droplet. Such measurements on hundreds of droplets showed that the charge was always an integer multiple of the smallest charge: the charge on an electron.

Vector forms

All the work in this chapter has been in one dimension but many of the expressions you have used have equivalent forms which can be used for working in two or three dimensions. These are given below for completeness.
- In vector form, Newton's second law is $\mathbf{F} = m\mathbf{a}$. Two-dimensional problems can be solved by regarding this as two equations for the $\mathbf{i}$ and $\mathbf{j}$ directions.
- Work, impulse and power are defined for variable forces in vector form as follows.

Work done	$\int \mathbf{F} \cdot \mathrm{d}s$
Impulse	$\int \mathbf{F} \, \mathrm{d}t$
Power	$\mathbf{F} \cdot \mathbf{v}$

Chapter 18 Exercises

18.1   A horse pulls a 500 kg cart from rest until the speed, v, is about 5 ms^{-1}. Over this range of speeds, the magnitude of the force exerted by the horse can be modelled by $500(v + 2)^{-1}$ N. Neglecting resistance,

   i)   write down an expression for $v\dfrac{\mathrm{d}v}{\mathrm{d}s}$ in terms of v

ii)   show by integration that when the velocity is $3\,\text{ms}^{-1}$, the cart has travelled $18\,\text{m}$

iii) write down an expression for $\dfrac{dv}{dt}$ and integrate to show that the velocity is $3\,\text{ms}^{-1}$ after 10.5 seconds

iv)   show that $v = -2 + \sqrt{4 + 2t}$

v)   integrate again to derive an expression for s in terms of t, and verify that after 10.5 seconds, the cart has travelled $18\,\text{m}$.

18.2   The gravitational acceleration on a body at a distance r from the centre of a uniform sphere of mass M has magnitude $\dfrac{GM}{r^2}$, where G is the gravitational constant.

i)   What are the dimensions of G?

ii)   Assuming the Earth can be modelled as a perfect sphere of mass M and radius R, write down an expression for a_h, the acceleration due to gravity at a distance h above the Earth's surface.

   What symbol is normally used for a_0?

iii) Show that

$$a_h = g\left(1 + \frac{h}{R}\right)^{-2}.$$

iv)   Hence show that when h is small compared to the radius of the Earth, so that $\left(\dfrac{h}{R}\right)^2$ is negligible,

$$a_h = g\left(1 - \frac{2h}{R}\right).$$

v)   Given that the radius of the Earth is about 6400 km, by what percentage does gravitational acceleration differ from g at a height of 50 km?

18.3   The gravitational acceleration on a body at a distance x from the centre of the Earth has magnitude $\dfrac{k}{x^2}$, where k is a constant. An artificial satellite is in a circular orbit of radius r about the centre of the Earth.

i)   Show that $k = gR^2$, where R is the radius of the Earth and g has its usual significance.

ii)   Write down the gravitational acceleration of the satellite in terms of R, r and g.

   Hence show that the time T taken for the satellite to orbit the Earth is given by

$$T = \frac{2\pi}{R}\sqrt{\frac{r^3}{g}}$$

iii) Show that a satellite in circular orbit with a period of 24 hours (a geosynchronous satellite) will be about 36 000 km above the Earth's surface. (Take g as $9.8\,\text{ms}^{-2}$ and the radius of the Earth as 6400 km.)

18.4   A body of unit mass *inside* a uniform sphere at a distance r from its centre O experiences a gravitational attraction towards the centre of $\dfrac{GM_r}{r^2}$, where G is the gravitational constant and M_r is the mass of material inside the sphere of radius r.

(In other words it is as if the body were on the surface of a sphere of radius r, all the matter further from the centre than r will have no net gravitational effect on the body.)

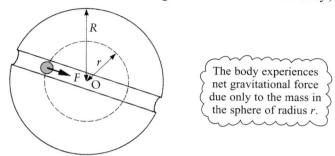

The body experiences net gravitational force due only to the mass in the sphere of radius r.

Suppose a straight tube could be drilled right through the Earth, modelled as a uniform sphere of radius R and total mass M. A ball of unit mass is dropped into the tube at the surface of the Earth.

i) Work out the mass of a sphere of the Earth of radius r, in terms of R, M and r.

ii) Write down an expression for the gravitational force on the ball when at a distance r from the centre.

iii) Hence show that the ball will oscillate with simple harmonic motion with an amplitude equal to the radius of the Earth. Determine the period.

18.5 A car of mass 1000 kg is travelling on level ground at 5 ms^{-1} at time $t = 0$ when it begins to accelerate with its engine working at a constant 10 kW. Assume that its motion can be modelled neglecting friction and other forces.

i) Show that the thrust of the engine is inversely proportional to the speed of the car.
Write down the equation of motion involving velocity, v, and time, t.

ii) Integrate and hence express v in terms of t.
Confirm that after 10 seconds the velocity is 15 ms^{-1}.

iii) By writing v in the form $\dfrac{ds}{dt}$, integrate again and show that the distance travelled after time t is given by

$$s = \frac{1}{30}\left(20t + 25\right)^{\frac{3}{2}} - \frac{25}{6}.$$

iv) Eliminate t between the equations derived in ii) and iii) and hence express s in terms of v.

v) Show that the result in part iv) can be derived directly from another form of the equation of motion.

18.6 A particle of mass m, moving in the sun's gravitational field, at a distance x from the centre of the sun, experiences a force Gmx^{-2} (where G is a constant) directly towards the sun.

i) Show that, if at some time $x = h$ and the particle is travelling directly away from the sun with speed V, then x cannot become arbitrarily large unless $V^2 \geqslant 2Gh^{-1}$.

ii) A particle is initially motionless a great distance from the sun (of radius R). If, at some later time, it is at a distance h from the centre of the sun, how long after that will it take to fall into the sun?

18.7 A motor car of mass m accelerates along a level straight road. Over a period, the engine works at a rate which can be modelled by $pv + q$, where v is the velocity of the car and p and q are positive constants. Its speed at the beginning of the period is $\dfrac{q}{p}$ and at the end of the period is $\dfrac{3q}{p}$. Resistance can be neglected.

i) What is the thrust on the engine when the car is moving with velocity v?

ii) Write down the equation of motion. Show that the variables can be separated to give

$$\int \frac{p}{m}\,\mathrm{d}t = \int \frac{pv}{pv + q}\,\mathrm{d}v.$$

iii) Show that the length of the period is

$$(2 - \ln 2)\frac{mq}{p^2}.$$

18.8 A car of mass m kg, moving at $25\,\mathrm{ms}^{-1}$ on a level road, runs out of petrol and coasts to a lay-by 0.6 km further on. It is acted on by a resistive force of magnitude $\dfrac{m(v^2 + 1)}{200}$ N, where $v\,\mathrm{ms}^{-1}$ is the car's speed. Find the speed of the car when it reaches the lay-by.

18.9 A sphere of radius R m dropped in water is resisted by a viscous force $6\pi R\eta v$ N, where $v\,\mathrm{ms}^{-1}$ is the velocity of the sphere and η is the coefficient of viscosity. Also according to Archimedes' principle, a body immersed in a fluid receives an upthrust equal to the weight of fluid displaced by the volume of the body. Experiments are performed by releasing glass marbles in the water. ρ_g denotes the density of glass and ρ_w the density of water.

i) Find the weight of water which a marble displaces and hence write the equation of motion for the marble as it descends.

ii) Show that the terminal velocity V_T has the value

$$\left(\frac{2R^2 g}{9\eta}\right)(\rho_g - \rho_w)g$$

and that the equation of motion may be written

$$V_T - v = \left(\frac{2R^2 \rho g}{9\eta}\right)\frac{\mathrm{d}v}{\mathrm{d}t}.$$

iii) Taking ρ_g as $3000\,\mathrm{kg\,m}^{-3}$, ρ_w as $1000\,\mathrm{kg\,m}^{-3}$, the value of η as 1 in SI units, and g as $10\,\mathrm{ms}^{-2}$, find the terminal velocity of a 1 cm diameter marble. Deduce the terminal velocity for a 2 cm diameter marble.

iv) How long does it take for the 1 cm marble to reach 99% of its terminal velocity?

18.10 In tests of a new engine for a sports car, the car is driven along a level road. The mass of the car is m and v is the speed of the car at time t. The driving force is equal to mav and the resistance to mbv^2, where a and b are constants. After a time t the car has travelled a distance x.

i) Write down the differential equation for v that describes the motion of the car as a function of t.

Hence write down a differential equation for $\dfrac{dv}{dx}$.

ii) Show that $v = \dfrac{a}{b} + Be^{-bx}$, where B is a constant.

Give interpretations of the quantities $\dfrac{a}{b}$ and $\dfrac{a}{b} + B$ in terms of the speed of the car.

At the start of a particular trial, the speed of the car increases from V to $2V$ as x increases from 0 to L, where $V < \dfrac{a}{2b}$.

iii) Find B in terms of V, a and b. Show that $L = \dfrac{1}{b}\ln\!\left(\dfrac{\dfrac{a}{b} - V}{\dfrac{a}{b} - 2V}\right)$.

Show further that the car has its maximum acceleration when $v = \dfrac{a}{2b}$ and that this occurs when it has moved a distance $\dfrac{1}{b}\ln\!\left(2 - \dfrac{2bV}{a}\right)$.

18.11 A golf ball of mass 0.04 kg lies at rest on a golf course. It is struck by a golf club which exerts upon it a force of magnitude $F = 5000\,(e^{-1000t} - e^{-2000t})$ newtons for a period $0 \leqslant t \leqslant 0.01$ second after which the ball and the club separate. Assume that the force acts along a straight line.

Find, correct to 3 significant figures,
i) the magnitude of the maximum force exerted on the ball
ii) the speed of the ball at the instant the ball and the club separate
iii) the distance the ball travels while in contact with the club
iv) the work done on the ball by the club.

18.12 The speed limiter on a test vehicle operates by reducing the driving force F as the speed increases. The force is given by
$$F = mk(A^2 - v^2),$$
where v is the speed, m the mass and k, A are constants.

When moving on level ground, the resistance to motion is Bmv^2, where B is a constant. The greatest speed that the vehicle can reach is V_0.
i) Write down the equation of motion for the vehicle.
ii) Show that $V_0^2 = \dfrac{kA^2}{k + B}$, and deduce that the equation of motion can be written as $\dfrac{dv}{dt} = c(V_0^2 - v^2)$, where $c = k + B$ and t is time.

The vehicle starts from rest at $t = 0$ and after a time t has moved a distance of x.
iii) Show that the speed v at time t is given by
$$v = V_0\left(\frac{e^{cV_0t} - e^{-cV_0t}}{e^{cV_0t} + e^{-cV_0t}}\right).$$

iv) Hence, or otherwise, deduce an expression for x as a function of t.

Personal Tutor

 18.1   The gravitational force on a particle of mass m at a distance s from the centre of the Earth is given by $\dfrac{mgR^2}{s^2}$ where R is the radius of the Earth and $s \geqslant R$.

The particle is projected vertically from the Earth's surface with an initial speed U, and the maximum height reached above the surface of the Earth is H. When it is at a height x above the surface of the Earth, the speed of the particle is v.

Resistance to motion and the rotation of the Earth can be neglected.

i)   Write down an equation for the upward motion of the particle. Hence show that
$$v^2 = U^2 - \frac{2gRx}{R + x}.$$

ii)   Deduce an expression for H.

iii)   Find an expression for the potential energy P at a height x, taking P to be equal to $-mgR$ when $x = 0$.

iv)   Explain carefully, with reference to both H and P, what happens
   a)   when $U^2 < 2gR$.   b)   when $U^2 \geqslant 2gR$.

KEY POINTS

18.1   When a particle is moving along a line under a variable force F, Newton's second law gives a differential equation. It is generally solved by writing acceleration as

$\dfrac{dv}{dt}$ when F is given as a function of time, t

$v\dfrac{dv}{ds}$ when F is given as a function of displacement, s

$\dfrac{dv}{dt}$ or $v\dfrac{dv}{ds}$ when F is given as a function of velocity, v.

18.2   Where forces are conservative, variable force problems may be handled using the principle of conservation of energy.

18.3   A *resistive* force is opposite to the direction of motion, and its magnitude often depends on the speed: typically it is proportional to v or v^2.

18.4   The *terminal velocity* of a falling body is reached when the resistive force is balanced by the force of gravity.

18.5   The gravitational force between two particles is proportional to the product of their masses and inversely proportional to the square of the distance between them: $F = \dfrac{Gm_1 m_2}{r^2}$.

18.6   Work, impulse and power are defined for variable forces as follows.

Work done	$\int F \, ds$
Impulse	$\int F \, dt$
Power	Fv

19 Variable mass

Plus ça change, plus c'est la même chose

Alphonse Karr

QUESTION 19.1

1 What causes the rocket to move?
2 A wagon is rolling freely along a level track at constant speed when it passes under a hopper which dumps coal into it. A short way further on, the bottom of the wagon opens up to allow the coal to drop out.

Assuming friction can be neglected, will the speed of the wagon at the end be greater than, less than or the same as at the beginning?

A rocket loses mass as it expels fuel. For example, each of the two launch rockets of Ariane 5 uses over 1.8 tonnes of fuel in just over 2 minutes. A falling raindrop may shrink because of evaporation or, when falling in mist, grow because of condensation. The motion of the rocket and that of the raindrop are affected not only by external forces but by the fact that their mass is changing. This chapter deals with such situations. The topic is usually called *variable mass* because you form an equation of motion referring to a system (e.g. the rocket) whose mass is changing. Of course, mass is not (normally) created or destroyed. The mass of the rocket indeed decreases but the lost mass is fuel expelled into space. As you will see, the velocity of this fuel is a crucial factor in the motion of the rocket.

19.1 Obtaining the equations of motion

In this chapter you will meet three types of problem in which the mass of an object changes continuously as it moves in a straight line.

● The object picks up stationary matter as it goes along, for example, a raindrop falling through a cloud.
● The object drops matter as it goes along, for example, a truck dropping tarmac on to a road.
● The object expels matter in order to accelerate, for example, a rocket.

The equation of motion depends on the situation and it is best to approach it from first principles using the momentum–impulse equation. In the next two sections you will see how this works in the first two situations. Rocket motion is covered later in the chapter.

19.2 Matter is picked up continuously from rest

The diagram shows an object of variable mass m which picks up a small mass, δm, over a short time, δt, during which its velocity changes from v to $v + \delta v$. There is an external force F acting in the direction of motion.

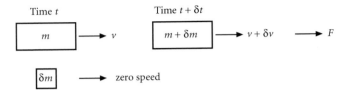

Figure 19.1

The gain in momentum is equal to the impulse of the force so:

$$(m + \delta m)(v + \delta v) - mv = F\delta t$$

$$\Rightarrow \quad m\delta v + \delta mv + \delta m\delta v = F\delta t$$

$$\Rightarrow \quad m\frac{\delta v}{\delta t} + \frac{\delta m}{\delta t}v + \frac{\delta m}{\delta t}\delta v = F,$$

> $\delta v \to 0$ as $\delta t \to 0$, so the term $\frac{\delta m}{\delta t}\delta v$ disappears.
> In the same way, any small variation in F over the time δt would also disappear, so the force, F, *does not have to be constant*. It could be mg for example.

In the limit as $\delta t \to 0$, this gives

$$m\frac{dv}{dt} + v\frac{dm}{dt} = F$$

$$\Rightarrow \quad \frac{d(mv)}{dt} = F.$$

This is the more general form of Newton's second law.

Force = rate of change of momentum

When m is constant it becomes the familiar

$$F = m\frac{dv}{dt} = ma.$$

When $F = 0$, the rate of change of momentum is zero, so momentum is conserved.

EXAMPLE 19.1 A canal barge is drifting without power under a hopper which is filling it with coal at a rate of $r\,\text{kg s}^{-1}$. The barge was initially moving with a velocity $u\,\text{ms}^{-1}$ and its mass without the coal is $M\,\text{kg}$. The resistance of the water may be neglected.

i) Denoting by m and v respectively the mass and velocity of the barge plus coal at time t, show that $m\dfrac{dv}{dt} = -v\dfrac{dm}{dt}$.

ii) Show that the velocity of the barge t seconds after the coal begins to drop into the barge is given by $v = \dfrac{Mu}{(M + rt)}$.

iii) Show that the distance travelled in this time is $S = \left(\dfrac{mu}{r}\right)\ln\left(\dfrac{1 + r}{Mt}\right)$.

SOLUTION

i) Consider an instant at time t when the barge plus coal has mass m moving with velocity v. At a time δt later, a mass δm (whose initial velocity is zero) has been added, and the whole is now moving with a velocity $v + \delta v$. The change in momentum is

$$(m + \delta m)(v + \delta v) - mv = mv + v\delta m + m\delta v + \delta m \delta v - mv$$

$$= v\delta m + m\delta v + \delta m \delta v.$$

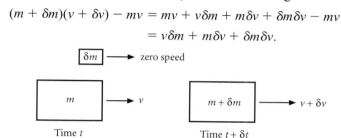

Figure 19.2   Time t Time $t + \delta t$

There is no net horizontal force, so there is no change in the total linear momentum of the coal plus barge: $v\delta m + m\delta v + \delta m \delta v = 0$.

Dividing through by δt and letting $\delta t \to 0$ gives

$$v\frac{dm}{dt} + m\frac{dv}{dt} = 0$$

$$\Rightarrow \quad m\frac{dv}{dt} = -v\frac{dm}{dt}.$$

$$\left(\delta m \frac{\delta v}{\delta t} \to 0 \text{ as } \delta t \to 0 \right)$$

As each δm of coal is added, its momentum increases from zero to $v\delta m$ so $v\dfrac{dm}{dt}$ is the rate of increase in momentum of the added mass. Hence the equation can be interpreted as

mass × acceleration = $-$(rate of increase of momentum from added mass).

Compare with $ma = -F$ and you can see that the effect of the coal being added is the same as a *resisting force of magnitude equal to the rate of increase in momentum* of the new coal.

ii) This equation can be solved by separating the variables, but it is easier to go back to

$$m\frac{dv}{dt} + v\frac{dm}{dt} = 0$$

$$\Rightarrow \quad \frac{d}{dt}(mv) = 0.$$

Integrating gives $mv = $ constant.

After time t, a mass rt of coal has been added. The mass is then $m = M + rt$. Hence

$$(M + rt)v = \text{constant} = Mu \quad \text{(since when } t = 0, v = u)$$

$$\Rightarrow \quad v = \frac{Mu}{(M + rt)}.$$

It is a good idea to check results by setting variables to extreme values. $r = 0$ implies no added mass, so the barge should continue to move at constant speed, hence $v = u$.

iii) To work out distance travelled, v must be written as $\dfrac{ds}{dt}$.

$$\frac{ds}{dt} = \frac{Mu}{(M + rt)}$$

$$\Rightarrow \quad s = \left(\frac{Mu}{r}\right) \ln(M + rt) + k$$

where k is the constant of integration. When $t = 0$, $s = 0$, giving

$$k = -\left(\frac{Mu}{r}\right) \ln M$$

$$s = \left(\frac{Mu}{r}\right) \left[\ln(M + rt) - \ln M\right]$$

$$= \left(\frac{Mu}{r}\right) \ln\left(\frac{1 + rt}{M}\right).$$

19.3 Matter is dropped continuously

You can work from first principles using a method similar to that on page 432 to deal with this situation. In this case the mass of the object decreases, so δm is negative. However, it makes the problem easier to visualise if you use $|\delta m|$ as the mass dropped in time δt. Figure 19.3 summarises the situation.

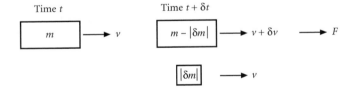

Time t Time $t + \delta t$

m v $m - |\delta m|$ $v + \delta v$ F

$|\delta m|$ v

Figure 19.3

The gain in momentum is equal to the impulse of the force so:

$$(m - |\delta m|)(v = \delta v) + |\delta m|v = mv = F\delta t$$

$$\Rightarrow \quad m\delta v - |\delta m|\delta v = F\delta t$$

$$\Rightarrow \quad m\frac{\delta v}{\delta t} + \frac{\delta m}{\delta t}\delta v = F.$$

> As noted on page 432, any small variation in F over the time δt disappears, so the force, F, does not have to be constant.

In the limit as $\delta t \to 0$, this gives

$$m\frac{\delta v}{\delta t} = F.$$

> $-|\delta m| = +\delta m$ when $\delta m < 0$.

This looks more like the form $F = ma$, of Newton's second law.

When $F = 0$, the velocity is constant, but the momentum of the object decreases because its mass decreases. This is not a contradiction however, because the *total* momentum of the whole system is conserved. The mass which is dropped has momentum until it is acted on by some other external force.

The following example uses the same method with $|\delta m|$ written in terms of the time, δt.

EXAMPLE 19.2

A hopper truck containing gravel is moving along a horizontal railway line. The gravel is dropping out of the bottom at a constant rate of k kg s^{-1}.

i) By considering the total linear momentum before and after a small interval of time from t to $t + \delta t$, show that the truck does not accelerate if there are no horizontal forces on it.

ii) The truck is actually being pulled by a variable force F.

Derive the relationship between F and the acceleration of the truck when the mass of the truck plus gravel is m.

SOLUTION

i) Denote by v the velocity of the truck at a time t when the mass of the truck and remaining gravel is m. The total linear momentum is thus mv.

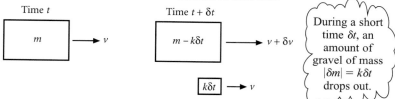

Figure 19.4

During a short time δt, an amount of gravel of mass $|\delta m| = k\delta t$ drops out.

When the gravel begins dropping, its horizontal velocity is the same as that of the truck, i.e. v. The small change in this velocity over the time δt can be ignored when considering the momentum of this small mass. At the end of the period δt, the velocity of the truck will have increased to $v + \delta v$. The total momentum is now

$$(m - k\delta t)(v + \delta v) + vk\delta t = mv + m\delta v - k\delta t\,\delta v.$$

Note that the gravel dropping out retains its momentum until it hits the ground, of course, when other forces, not relevant to the problem, come into play.

The *increase* in linear momentum of the whole system is thus $m\delta v - k\delta t\,\delta v$. When there is no horizontal force, there is no change in momentum, so

$$m\delta v - k\delta t\,\delta v = 0.$$

Dividing through by the time interval δt gives

$$\frac{m\delta v}{\delta t} = k\delta v.$$

In the limit as $\delta t \to 0$ and $\delta v \to 0$, the equation becomes $m\dfrac{dv}{dt} = 0$. The acceleration is zero.

ii) When the truck is pulled by a force F, the impulse of the force over this period, $F\delta t$, is equal to the increase of momentum over the period.

$$F\delta t = m\delta v - k\delta t\,\delta v$$

$$\Rightarrow \quad F = m\frac{\delta v}{\delta t} - k\delta v$$

In the limit as $\delta t \to 0$, and $\delta v \to 0$, the equation becomes $F = m\dfrac{dv}{dt}$.

You can see from the last two examples that, in variable mass problems, you have to be careful when using Newton's second law in the form $F = \dfrac{\mathrm{d}}{\mathrm{d}t}(mv)$. The momentum, mv, refers to the *total momentum of the system*. In Example 19.1, the coal has no linear momentum (in the direction of the barge motion) before it drops on to the barge, so that the momentum of the barge plus contents is equal to the momentum of the whole system. So, to solve part **i)** of this problem, you could simply have written.

$$0 = \frac{\mathrm{d}}{\mathrm{d}t}(mv) = m\frac{\mathrm{d}v}{\mathrm{d}t} + v\frac{\mathrm{d}m}{\mathrm{d}t}.$$

But in Example 19.2, the gravel dropping off the truck retained its velocity, so applying $F = \dfrac{\mathrm{d}}{\mathrm{d}t}(mv)$ only to the truck would have been wrong. The lost mass still had its momentum. Of course, the gravel eventually hits the ground and loses its velocity, but then other forces are involved.

The following is another example where it is safe to apply the full form of Newton's second law. The mass being added to a moving object has no initial velocity and therefore $F = \dfrac{\mathrm{d}}{\mathrm{d}t}(mv)$ can be applied simply to the moving object.

EXAMPLE 19.3

A spherical droplet, initially of radius a, falls from rest under the influence of a gravitational field g, through a stationary light mist. The mass of the droplet increases due to the condensation of mist on its surface. Air resistance may be ignored.

i)   Denoting the mass and velocity of the droplet at any time t by m and v respectively, write down Newton's second law for the motion.

Deduce that the equation is consistent with an initial acceleration of g.

ii)   Given that the mass increases at a rate proportional to the instantaneous surface area of the droplet, show that its radius r increases linearly with time t, i.e. can be expressed in the form $r = ct + a$, where c is a constant.

iii)   Find v in terms of r, a, c and g.

iv)   Deduce that when $r \gg a$, the acceleration is approximately $\dfrac{g}{4}$.

SOLUTION

i)   The extra mass is at rest before condensing; therefore force = rate of change of momentum.

$$mg = \frac{\mathrm{d}}{\mathrm{d}t}(mv) = m\frac{\mathrm{d}v}{\mathrm{d}t} + v\frac{\mathrm{d}m}{\mathrm{d}t}$$

When $v = 0$, $mg = m\dfrac{\mathrm{d}v}{\mathrm{d}t} \implies \dfrac{\mathrm{d}v}{\mathrm{d}t} = g.$

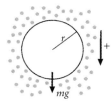

Figure 19.5

The initial acceleration is g.

ii) Mass of droplet = volume × density

$$m = \frac{4\pi r^3}{3}\rho$$

$$\frac{dm}{dt} = (4\pi r^2\rho)\frac{dr}{dt}$$

The rate of increase of m is proportional to the surface area. Therefore

$$\frac{dm}{dt} = k \times 4\pi r^2$$

for some constant k. Hence

$$4\pi r^2 k = (4\pi r^2\rho)\frac{dr}{dt}$$

$$\frac{dr}{dt} = \frac{k}{\rho} = c$$

where c is a constant.

Integrating gives

$$r = ct + a$$

where the constant of integration is a, since $r = a$ when $t = 0$.

Therefore r increases linearly with t.

iii) $\dfrac{d}{dt}(mv) = mg = \dfrac{4}{3}\pi\rho g r^3 = \dfrac{4}{3}\pi\rho g\,(ct + a)^3$

Integrating gives

$$mv = \frac{4\pi\rho g}{3}(ct + a)^4 \times \frac{1}{4c} + d \qquad (d \text{ is the constant of integration})$$

$$= \left(\frac{\pi\rho g}{3c}\right)r^4 + d \qquad\qquad (r = ct + a).$$

When $t = 0$, $r = a$ and $v = 0$ giving $d = -\left(\dfrac{\pi\rho g}{3c}\right)a^4$.

Hence $\qquad mv = \dfrac{\pi\rho g}{3c}(r^4 - a^4)$.

But $m = \dfrac{4}{3}\pi r^3\rho$; thus

$$\frac{4\pi r^3\rho v}{3} = \frac{\pi\rho g}{3c}(r^4 - a^4)$$

$$\Rightarrow \qquad v = \frac{g}{4c}\left(r - \frac{a^4}{r^3}\right).$$

iv) Acceleration $\dfrac{dv}{dt} = \dfrac{dv}{dr} \times \dfrac{dr}{dt}$

$$= \frac{g}{4c}\left(1 + \frac{3a^4}{r^4}\right)\frac{dr}{dt}$$

$$= \frac{g}{4} + \frac{3g}{4}\left(\frac{a}{r}\right)^4 \qquad \text{since } \frac{dr}{dt} = c$$

when $r = a$, $\dfrac{dv}{dt} = g$ as expected (initial acceleration).

When $r \gg a$, the second term is negligible; thus the acceleration is approximately $\dfrac{g}{4}$.

19.4 Rockets

One of the most interesting applications of variable mass methods is the motion of rockets. The rocket stores a large amount of fuel which is burnt and ejected at very high speed and it gains forward momentum to compensate for the backward momentum of the expelled fuel. Rockets are normally designed so that the fuel leaves the rocket at a constant mass rate and at a constant velocity *relative to the rocket*. This is more complicated than Example 19.2, where the lost mass simply had the *same* velocity as the truck.

The motion of a rocket

You can analyse rocket motion by working from first principles as before. In this case assume that fuel is ejected at a constant speed of u relative to the rocket so its speed in the direction of motion is $v - u$. Again the mass, m, of the rocket is decreasing and this means that δm is negative, so use $|\delta m|$ for the mass ejected over time δt. Figure 19.6 summarises the situation.

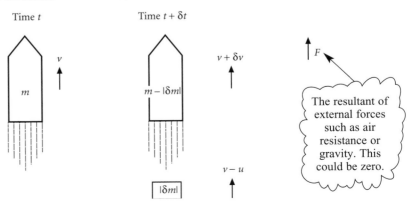

Figure 19.6

The gain in momentum is equal to the impulse of the force so

$$(m - |\delta m|)(v + \delta v) + |\delta m|\,(v - u) - mv = F\delta t$$

$$\Rightarrow \quad m\delta v - |\delta m|\,\delta v - |\delta m|\,u = F\delta t$$

$$\Rightarrow \quad m\frac{\delta v}{\delta t} - \frac{|\delta m|}{\delta t}\,\delta v - \frac{|\delta m|}{\delta t}\,u = F.$$

> As before, any small variation in F over the time δt disappears, so the force, F, does not have to be constant.

In the limit as $\delta t \to 0$, this gives

$$m\frac{dv}{dt} - \left|\frac{dm}{dt}\right|u = F$$

$$m\frac{dv}{dt} = F + \left|\frac{dm}{dt}\right|u$$

There is a forward thrust, $\left|\dfrac{dm}{dt}\right|u$, caused by the ejection of the fuel.

Notice that the force, F, is assumed to be in the direction of motion of the rocket, so it is *negative* when the rocket is leaving the Earth's surface.

In the following example $\left|\dfrac{dm}{dt}\right|$ is replaced by the rate, k, at which the fuel is burnt. This example demonstrates the essential mathematical principles of rocket motion.

EXAMPLE 19.4

At time t, a rocket in one-dimensional motion has mass m, is travelling with speed v and is burning fuel at a constant mass rate k. The burnt fuel is ejected with a constant exhaust speed u relative to the rocket. Such a rocket has initial mass m_0, of which a fraction α is fuel, and is fired vertically upwards from rest in a constant gravitational field g.

i) By considering the momentum over a small period, derive the equation of motion

$$(m_0 - kt)\frac{dv}{dt} = ku - (m_0 - kt)g.$$

ii) Write down a condition for the rocket to start rising when its motor is fired.

iii) For how long will the motor burn?

iv) Assuming that the rocket does start to rise immediately, find its final speed when all the fuel has burnt.

v) Use the expression derived in part **iv)** to suggest ways in which the final speed might be increased, assuming that the initial mass m_0 is fixed.

SOLUTION

i) At time t, the mass of the rocket is $m = m_0 - kt$ and the momentum is mv.

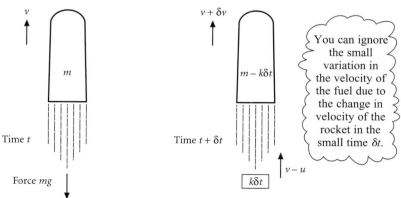

Figure 19.7

At a small time δt later, an amount of mass $k\delta t$ has been expelled at a speed of u relative to the rocket. The velocity of the fuel is thus $(v - u)$ upwards.

At time $t + \delta t$, the velocity of the rocket has increased to $v + \delta v$. The momentum at time $t + \delta t$ is

$$(m - k\delta t)(v + \delta v) + k\delta t\,(v - u) = mv + m\delta v - k\delta tu - k\delta t\delta v.$$

The *increase* in momentum over this period is thus approximately $m\delta v - k\delta tu - k\delta t\delta v$.

This is equal to the impulse of the external force which is $-mg\delta t$.

$$-mg\delta t = m\delta v - k\delta tu - k\delta t\delta v$$

$$\Rightarrow \qquad -mg = m\frac{\delta v}{\delta t} - ku - ku\delta v$$

As $\delta t \to 0$, $\dfrac{\delta v}{\delta t} \to \dfrac{dv}{dt}$. The term $-ku\delta v$ tends to zero, so

$$-mg = m\frac{dv}{dt} - ku$$

> Compare with $ma = F$. The left-hand side is mass $\times$ acceleration of the rocket. The right-hand side, in addition to the external force of gravity, includes a term ku which can be interpreted as the forward force (thrust) of the engine; it is the *rate of change of momentum of the expelled mass*.

$$\Rightarrow \qquad m\frac{dv}{dt} = ku - mg.$$

Substituting $m = m_0 - kt$, gives the required result:

$$(m_0 - kt)\frac{dv}{dt} = ku - (m_0 - kt)g.$$

ii) $(m_0 - kt)\dfrac{dv}{dt} = ku - (m_0 - kt)g$

When the rocket is fired, $t = 0$. For the rocket to start rising, the acceleration must be positive, i.e. $\dfrac{dv}{dt} > 0$. Hence

$$ku - m_0g > 0$$

$$\Rightarrow \qquad ku > m_0g.$$

The thrust of the engine, ku, must be greater than the gravitational force m_0g.

iii) The motor burns as long as there is fuel. Since the total fuel is αm_0, then when all the fuel is burnt

$$\alpha m_0 = kt \quad \Rightarrow \quad t = \frac{\alpha m_0}{k}$$

iv) From part i)

$$\frac{dv}{dt} = \frac{ku}{m_0 - kt} - g.$$

Integrating gives

$$v = -u \ln(m_0 - kt) - gt + c \qquad \text{①}$$

where c is the constant of integration. When $t = 0$, $v = 0$, so $c = u \ln m_0$.

Substituting $t = \dfrac{\alpha m_0}{k}$ gives the final velocity

$$v_f = u \ln\left(\frac{m_0}{m_0 - \alpha m_0}\right) - \frac{g\alpha m_0}{k}$$

$$= u \ln\left(\frac{1}{1 - \alpha}\right) - \frac{g\alpha m_0}{k} \qquad \text{②}$$

v) An increase in v_f can be achieved by increasing the first term or decreasing the second. The possibilities are
- increase u: use a fuel with a high exhaust rate
- make α near 1: increase the proportion of the total mass that is fuel
- make k larger: increase the burn rate.

Rocket motion with no external force

If the rocket is fired in space where there is no appreciable external force, the rocket equation is simply

$$m\frac{dv}{dt} = ku$$

where, using the notation of Example 19.4, u is the relative exhaust speed, m is the rocket mass at a given time and $k = -\dfrac{dm}{dt}$ is the rate at which fuel is burned.

The effective thrust of the engine is:

(rate of expulsion of mass, relative to the rocket) × (relative speed of the fuel).

Equation ② from Example 19.4 is

$$v_f = u \ln\left(\frac{1}{1 - \alpha}\right)$$

$$\text{Replace } g \text{ by zero}$$

or

$$v_f = u \ln\left(\frac{m_0}{m_0 - \alpha m_0}\right)$$

$$= u \ln\left(\frac{\text{initial mass}}{\text{final mass}}\right).$$

The final velocity depends only on u, the relative velocity of the fuel, and on the proportion of the total mass that is fuel. Note that k does not appear in this expression. When there is no external force, the rate at which mass is burned does not affect the *final* velocity, although it obviously determines how long the rocket takes to reach this velocity.

If the rocket begins with some initial velocity v_0, equation ① gives $c = v_0 + u \ln m_0$. Then the result is

$$v_f - v_0 = u \ln\left(\frac{\text{initial mass}}{\text{final mass}}\right).$$

$(v_f - v_0)$ is the *velocity gained* over the burn period.

EXAMPLE 19.5   A spaceship is pointing towards a planet. The initial total mass of the ship is M_0 kg and the fuel is ejected at a rate of k kg s^{-1} with a speed u ms^{-1} relative to the spaceship. The ship starts at rest relative to the planet, and its speed, when all the fuel is exhausted, is $2u$ ms^{-1}. However, it is still so far from the planet that the gravitational force can be neglected.

i)   From the equation of motion derive an expression for v in terms of the remaining mass m at any time.

Hence show that the total mass at the end is $\dfrac{M_0}{e^2}$ kg.

ii)   Find the velocity of the *ejected fuel* relative to the planet just before the end of the period. Comment on this result.

iii) Show that the distance travelled during the period of the rocket firing is

$$\frac{M_0 u}{ke^2}(e^2 - 3).$$

SOLUTION

i)   After time t, the mass of the ship is $m = M_0 - kt$. Since there are no external forces, the equation of motion is

$$m\frac{dv}{dt} = ku$$

$$\frac{dv}{dt} = \frac{ku}{(m_0 - kt)}.$$

Integrating gives

$$v = -u \ln(M_0 - kt) + c.$$

When $t = 0$, $v = 0$, so $c = u \ln (M_0)$. This gives

$$v = -u \ln\left(\frac{M_0}{M_0 - kt}\right) = u \ln\left(\frac{M_0}{m}\right). \qquad ①$$

At the end of the burn, $v = 2u$, giving

$$\ln\left(\frac{M_0}{m}\right) = 2$$

$$\frac{M_0}{m} = e^2$$

$$\Rightarrow \quad m = \frac{M_0}{e^2}.$$

ii) At the end of the burn period, the velocity of the fuel relative to the planet is
$2u - u = u\,\text{ms}^{-1}$. The expelled fuel is actually travelling *towards* the planet, although
not as fast as the rocket! There is no paradox here; all velocities are relative.

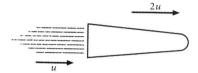

Figure 19.8

iii) From ①,

$$\frac{ds}{dt} = u \ln\left(\frac{M_0}{m_0 - kt}\right)$$

where s is the displacement from the start. Integrating gives

$$s = u \int \ln\left(\frac{M_0}{m_0 - kt}\right) dt$$

$$= u \int \ln M_0 \, dt - u \int \ln(M_0 - kt) \, dt. \qquad \text{②}$$

The second integral is easier if you make the substitution $m = M_0 - kt$.

$$\int \ln(M_0 - kt) \, dt = -\frac{1}{k} \int \ln m \, dm$$

You can use integration
by parts with $u = \ln x$,
$\dfrac{dv}{dx} = 1$ to show that
$\int \ln x \, dx = x \ln x - x$.

$$= -\frac{1}{k} (m \ln m - m) + \text{constant}.$$

Substituting this in ② gives

$$s = ut \ln M_0 + \frac{mu}{k} (\ln m - 1) + c.$$

When $s = 0$, $t = 0$, $m = M_0$. So

$$c = -\frac{M_0 u}{k} (\ln M_0 - 1).$$

Hence

$$s = ut \ln M_0 + \frac{mu}{k} (\ln m - 1) - \frac{M_0 u}{k} (\ln M_0 - 1). \qquad \text{③}$$

At burn out, $m = \dfrac{M_0}{e^2}$ (proved in part i))

$$\Rightarrow \quad M_0 - kT = \frac{M_0}{e^2}$$

where T is time of burn out

$$\Rightarrow \quad T = \frac{M_0}{ke^2} (e^2 - 1).$$

Substituting $m = \dfrac{M_0}{e^2}$ and $t = T = \dfrac{M_0}{ke^2}(e^2 - 1)$ in ③ gives

$$s = \frac{uM_0}{ke^2}(e^2 - 1)\ln M_0 + \frac{M_0 u}{ke^2}(\ln(\frac{M_0}{e^2}) - 1) - \frac{M_0 u}{k}(\ln M_0 - 1)$$

$$= \frac{M_0 u}{ke^2}[e^2 \ln M_0 - \ln M_0) + (\ln M_0 - 2 - 1) - (e^2 \ln M_0 - e^2)]$$

$$= \frac{M_0 u}{ke^2}(e^2 - 3).$$

$\ln e^2 = 2$

Historical note

The first rockets were probably bamboo tubes packed with 'black powder', a precursor of gunpowder. These were used by the Chinese in the 10th century. The technology spread both for warfare and entertainment; from the 16th century onwards, firework displays with rockets were used widely in Europe. Rockets were successfully used against the British by Indian forces during the battles of Seringapatam in 1792 and 1799. As a result, the British army adopted rockets for military purposes and used them for most of the next century. A typical military rocket of that time was described as an iron cylinder, 200 mm long, 40 mm wide with a 3 m guiding stick. By the end of the 19th century, better artillery superseded the rocket and the military lost interest. A few enthusiasts maintained development in rocket technology during the following decades until World War 2, when Germany developed the V2. This could reach a target some 300 km away in 5 minutes. Despite many technical problems, it was used to devastating effect, with about 1500 landing in southern England. With the end of the war, and the beginning of the Cold War, most work on the development of rockets moved to the USA and USSR.

In 1957, the space age began, with the USSR putting the first artificial satellite, Sputnik, into orbit. This led to manned space launches, the most spectacular of which was the moon landing in 1969. Huge rockets were built: the first stage of the enormous Saturn V rocket which put the first men on the moon had a thrust of 3.4×10^7 N.

The huge growth in the need for satellites for communication purposes has meant that rocket technology has continued to develop, and the space rocket is now a standard part of contemporary technology.

Jet aircraft

Figure 19.9

A jet aircraft is propelled in a similar way to a rocket, in the sense that propulsion is achieved by expelling mass. However, the engine uses the surrounding medium in that it takes in air at the front, compresses it and ejects it at high speed (see figure 19.9). The only change in mass is the expenditure of some fuel also expelled as exhaust gases, but this is small compared with the mass of air. Assuming that air is entering and leaving at a mass rate k, the aircraft is moving at a velocity v relative to the air, and expels it at a rate u relative to the aircraft, the effective thrust is the rate of change of momentum of this air, which is $k(u - v)$.

Unlike a rocket, it is impossible for the aircraft, in still air, to achieve a speed greater than the relative speed of ejection of the fuel.

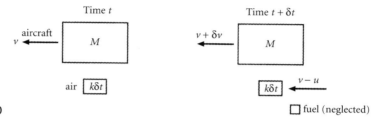

Figure 19.10

A mass of still air $k\delta t$ is picked up and expelled in a time period δt. Equating linear momentum before and after this period (see figure 19.10):

$$Mv + k\delta t \times 0 = M(v + \delta v) - k\delta t(u - v)$$

$$M\delta v = k\delta t(u - v)$$

$$M\frac{\delta v}{\delta t} = k(u - v)$$

In the limit, the effective thrust $M\dfrac{\delta v}{\delta t}$ is $k(u - v)$.

Chapter 19 Exercises

19.1   A charged droplet of mass M, initially at rest, is under the influence of a constant horizontal electrical force of magnitude F. The droplet evaporates so that at time t its mass is Me^{-kt}, where k is a positive constant.

i)   Show that the equation of motion in the direction of the electrical force is

$$F = Me^{-kt}\frac{dv}{dt}.$$

where v is the speed of the droplet in this direction.

ii)   Solve this equation to express v in terms of t.

19.2   As it moves forward, a dumper truck is releasing topsoil at a rate of r kg s^{-1}. The net forward force on the truck works at a constant rate of P kW. Initially the truck is at rest and the total load (truck plus soil) is M_0 kg. The soil drops out of the truck with no horizontal velocity relative to the truck.

i)   Show that the equation of motion may be written

$$1000P = (M_0 - r_t)v\frac{dv}{dt}.$$

ii) Solve the equation to express v^2 in terms of t.

Hence show that when the initial load is 25 tonnes, $r = 50$ and $P = 20$, the truck moves at just over $7\,\mathrm{ms}^{-1}$ after half a minute.

19.3 A trailer releasing fertiliser is being pulled across a field by a constant force F. In order to give an even spread, the releasing mechanism ensures that the rate of release, $r\,\mathrm{kg\,s}^{-1}$ is proportional to the speed of the trailer: $r = kv$, where k is constant. Fertiliser release starts when the mass of the trailer plus load is M and its speed is u.

i) Show that the total mass of the trailer after it has moved a distance s is $M - ks$.

ii) Write the equation of motion in terms of v and s.

iii) Show that

$$v^2 = u^2 - \frac{2F}{K}\ln\left(1 - \frac{ks}{M}\right).$$

iv) Confirm that as $k \to 0$, the formula becomes

$$v^2 = u^2 + \frac{2F}{M}s.$$

(Use the series expansion of $\ln(1 + x)$.)

19.4 A train of mass M moving under a force with constant power P picks up stationary water at a rate k. The initial velocity is v_0. Show that when the total mass (train plus water) is m, its velocity v satisfies the equation

$$k\left(m^2 v^2 - M^2 v_0^2\right) = P\left(m^2 - M^2\right)$$

19.5 An electron of mass m_0 is initially at rest when an electric field is applied to it. This field subjects the electron to a force of magnitude E in a constant direction. When the electron is moving very fast, its mass as measured by a stationary observer increases to

$$m = \frac{m_0}{1 - \left(\dfrac{v}{c}\right)^2}$$

where the constant c is the velocity of light. (This is a consequence of Einstein's special theory of relativity.)

i) Assuming that Newton's second law applies, that force is rate of change of momentum, show that

$$v^2 = \frac{c^2 k^2 t^2}{1 + k^2 t^2} \quad \text{where} \quad k = \frac{E}{cm_0}.$$

Deduce that the velocity of the electron never reaches the velocity of light.

ii) Find the distance travelled by the electron after time T.

iii) Use the binomial expansion of $(1 + x)^{\frac{1}{2}}$ to show that when $kt \ll 1$, the distance travelled is approximately what it would have been if the mass had remained constant at m_0.

19.6 A particle of initial mass M falls from rest under gravity through a stationary cloud. The particle picks up mass from the cloud at a rate equal to mkv, where m and v are the mass and speed of the particle at time t and k is a constant. Resistance to motion can be neglected.

    i)   Write down differential equations which describe

        a)   the increase in mass of the particle

        b)   the motion of the particle.

        Hence show that the speed satisfies the differential equation

$$v\frac{dv}{dx} + kv^2 = g, \quad \text{where } x \text{ is the distance fallen.}$$

    ii) By solving the equation in part **i)** find v in terms of g, k and x.

        Deduce that the speed tends to the limiting value $\dfrac{g}{k}$.

    iii) Show that $\dfrac{dm}{dx} = km$. Hence show that the mass of the particle is $2M$ when

        its speed is a fraction $\dfrac{\sqrt{3}}{2}$ of its limiting value.

19.7   A balloon is blown up and let go so that it shoots away in a horizontal direction. Assuming that air is expelled at a constant speed, c, relative to the balloon and that the mass of air in the balloon decreases at a constant rate, k, use a momentum–impulse equation over a short time δt to obtain the equation of motion of the balloon.

       There is an increase in kinetic energy when the balloon is released. Where does this energy come from?

19.8   A rocket of initial total mass M burns fuel at a constant rate rM and expels the fuel with a constant relative velocity u. When the rocket is launched vertically under gravity, the thrust is just sufficient for the rocket to rise.

    i)   Show that $ru = g$.

    ii) Show that when a mass M' remains, the rocket's velocity is

$$u \ln\left(\frac{M}{M'}\right) - u\left(1 - \frac{M'}{M}\right).$$

19.9   A rocket-propelled vehicle starts from rest with mass M and ejects fuel at a constant mass rate r per unit time with a constant relative speed u. When travelling horizontally with speed v the total resistance to motion is kv.

    i)   Show that its acceleration at time t is $\dfrac{ru - kv}{m}$ where $m = M - rt$.

    ii) Show that its speed at time t is $\dfrac{ru}{k}\left[1 - \left(\dfrac{m}{M}\right)^{k/r}\right]$.

19.10 An amateur designer creates a small rocket which burns fuel so that the rocket loses mass at the constant rate k. The initial total mass of the rocket and fuel is M_0 and the initial mass of fuel is $\frac{1}{2}M_0$.

    i)   Write down an expression for the mass m of the rocket and remaining fuel at any later time t, while the fuel is still burning.

The burnt fuel is expelled from the rocket with a constant speed c (backwards) relative to the rocket. The rocket is fired vertically upwards and air resistance may be neglected. At time $t = 0$, the rocket lifts from the ground. After a time t, the speed of the rocket is denoted by v.

ii) Show that the differential equation describing the motion of the rocket is

$$m\frac{dv}{dt} = kc - mg, \quad \text{provided that } M_0 g < kc.$$

Why is this inequality necessary?

iii) Show that the speed of the rocket while the fuel is still burning is given by

$$v = c \ln\left(\frac{M_0}{M_0 - kt}\right) - gt.$$

iv) Find the greatest speed of the rocket while it is still moving upwards.

Personal Tutor

19.1 At time t a rocket in one-dimensional motion has mass M, is travelling with speed v and is burning fuel at a constant mass rate k. The burnt fuel is ejected with a constant exhaust speed c relative to the rocket. The rocket has an initial mass of M_0, of which a fraction α is fuel, and is fired from rest in a gravity-free environment.

i) Write down the mass of the rocket after time t and find the time for which the fuel burns.

ii) Derive the equation of motion

$$M\frac{dv}{dt} = kc.$$

The mass ratio R of a rocket is defined to be the mass of the rocket when fuelled divided by the mass when empty.

iii) Show that the final speed, V, of the rocket is given by

$$V = -c \ln(1 - \alpha) = c \ln R.$$

A two-stage rocket consists of a first stage of mass M_1 which carries a second stage of mass M_2. Each stage has a fraction α of fuel and $M_1 + M_2 = M_0$. The first stage is fired from rest and carries the second stage until all of the first-stage fuel is burnt. At this point the second stage separates and its motor is fired until all of its fuel is burnt.

iv) Find the mass ratios R_1 and R_2 of the first stage (carrying the second) and the second stage of the rocket.

v) Determine the speed of the two-stage system at separation in terms of c and R_1. Find the final speed of the second stage in terms of c, R_1 and R_2.

In the following, F is the external force on a body whose mass is changing, m is the mass of the body at time t, and v is its velocity.

The first key point is the crucial one, from which the others follow.

19.1 The standard approach to variable mass problems is to apply the principle

$$\text{change in momentum} = \text{impulse of force}$$

over a small time interval t to $t + \delta t$. The change in the force over this interval may be neglected.

19.2 Newton's second law can be used directly in the form

$$\text{external force} = \text{rate of change of } \textit{total} \text{ momentum.}$$

However, you must ensure that this takes into account the *total* momentum change in the system.

19.3 For a body losing mass, where the lost mass maintains its momentum as it leaves, (e.g. coal dropping from a hopper truck), the equation becomes

$$F = m \frac{dv}{dt}.$$

19.4 For a body gaining mass, where the extra mass previously has no velocity (e.g. moisture from still air condensing on to a raindrop), the equation is

$$F = \frac{d}{dt}(mv) = m\frac{dv}{dt} + v\frac{dm}{dt}.$$

19.5 For a rocket expelling mass at a constant rate k with a relative velocity u opposite to the direction of rocket motion

$$m\frac{dv}{dt} = ku + F$$

where ku may be regarded as the thrust on the rocket due to the burning of fuel.

19.6 A rocket under no external force will have a velocity gain given by

$$v_f - v_0 = u \ln\left(\frac{\text{initial mass}}{\text{final mass}}\right).$$

Dynamics of rigid bodies rotating about a fixed axis

The human mind has first to construct forms, independently, before we can find them in things.

Einstein, 1879–1955

QUESTION 20.1   So far you have modelled moving objects as particles. In many circumstances this is reasonable, but how would you model the motion of the sails of a windmill or the other objects illustrated in the pictures above?

Do the two children on the roundabout have the same kinetic energy?

What is the kinetic energy of a rotating wheel?

20.1 A rigid body rotating about a fixed axis

You might not be able to answer all these questions fully now, but the issues involved should become clearer as you work through this chapter.

It is reasonable to treat a large object as a particle when every part of it is moving in the same direction with the same speed, but clearly this is not always the case. The particles in a rotating wheel have different velocities and accelerations and are subject to different forces.

The laws of particle dynamics which you have used so far need to be developed so that they can be applied to the rotation of large objects.

Definitions

You are already familiar with many aspects of rotation such as the angular speed and acceleration of a particle and you have also taken moments to determine the turning effect of a force, but it is as well to be clear about what is meant by some of the terms involved before continuing with the discussion.

Rigid bodies

Wheels can be modelled as rigid bodies. A *rigid body* is such that each point within it is always the same distance from any other point. You are not a rigid body but a hard chair is one (molecular vibrations being ignored).

The axis of rotation

When you lean back on your chair, it might rotate about a point, say A, at the end of one leg. You will have more control, however, if it rotates about the axis formed by the line joining the ends, A and B, of two legs.

The idea of an axis of rotation is important when considering the rotation of rigid bodies. When the only fixed point is A, the axis of rotation might be continually changing; any particle in the chair moves on the surface of a sphere with its centre at A. When the chair rotates about the *fixed axis* AB, however, each particle in it moves in a circle in a plane perpendicular to the axis (see figure 20.1).

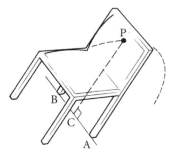

Figure 20.1

Angular speed

In figure 20.2, P is moving in a circle, centre O, with an angular speed $\dfrac{\mathrm{d}\theta}{\mathrm{d}t}$ or $\dot{\theta}$ (often denoted by ω).

It was shown that the speed of P is given by

$$v = r\omega.$$

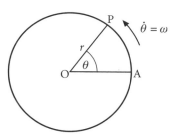

Figure 20.2

In a rigid body rotating about a fixed axis, *all* the particles have the same angular speed, ω, about the axis and the speed of each particle is given by its angular velocity times its perpendicular distance from that axis. Thus if the chair in figure 20.3 is rotating about the axis through the base of the back legs, the speed of the point P_1 is

$$v = AP_1 \times \dot{\theta}.$$

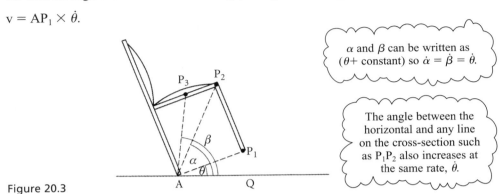

> α and β can be written as $(\theta +$ constant) so $\dot{\alpha} = \dot{\beta} = \dot{\theta}$.

> The angle between the horizontal and any line on the cross-section such as P_1P_2 also increases at the same rate, $\dot{\theta}$.

Figure 20.3

20.2 The kinetic energy of a rigid body rotating about a fixed axis

Because kinetic energy is a scalar quantity, the kinetic energy of a rigid body, such as a wheel, can be found by calculating the energy of each of the separate particles and then adding them. For example, you could find the kinetic energy of the children on the roundabout by treating each one as a separate particle moving in a circle. They could be modelled as a simple rigid body like that in the next example.

EXAMPLE 20.1   A rigid body consists of two particles P_1 and P_2 of masses m_1 and m_2 attached to a light rod AB as shown in figure 20.4. $AP_1 = r_1$ and $AP_2 = r_2$ and P_2 is at B.

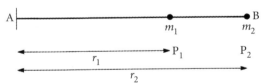

Figure 20.4

Find the kinetic energy of the body when it rotates with angular speed ω about an axis perpendicular to the rod

i)   through A   ii)   through B.

SOLUTION

i)   When the rod rotates with angular speed ω about an axis through A perpendicular to the rod, each particle moves in a circle, centre A.

P_1 has speed $v_1 = r_1\omega$ and P_2 has speed $v_2 = r_2\omega$. The total kinetic energy is, therefore,

$$\tfrac{1}{2}m_1v_1^2 = \tfrac{1}{2}m_2v_2^2 = \tfrac{1}{2}m_1r_1^2\omega^2 + \tfrac{1}{2}m_2r_2^2\omega^2$$
$$= \tfrac{1}{2}(m_1r_1^2 + m_2r_2^2)\omega^2.$$

ii) When the axis is through the end B, the particle P_2 at B does not move so has no kinetic energy. The particle P_1 now moves in a circle of radius $(r_2 - r_1)$, so the total kinetic energy is now $\frac{1}{2}m_1(r_2 - r_1)^2 \, \omega^2$.

You can see that the kinetic energy depends not only on the mass of the body and the angular speed, but also on the distance of the particles of the body from the axis of rotation.

Now consider a more complex rigid body rotating with an angular speed $\dot{\theta}$. Think of it as being made up of a lot of small particles of mass, $m_1, m_2, \dots$.

A typical particle of mass m_p, moving in its circle of radius r_p round the axis of rotation with angular speed $\dot{\theta}$, has a speed $r_p \, \dot{\theta}$ (see figure 20.5). Its kinetic energy is $\frac{1}{2}m_p r_p^2 \dot{\theta}^2$.

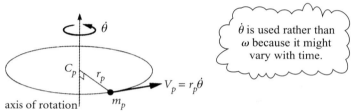

Figure 20.5

Summing over all particles gives the total kinetic energy as

$$\sum_{\text{all } p} \tfrac{1}{2} m_p r_p^2 \dot{\theta}^2 = \tfrac{1}{2} \left(\sum_{\text{all } p} m_p r_p^2 \right) \dot{\theta}^2 \quad \text{(since } \dot{\theta} \text{ is the same for all particles).}$$

Moment of inertia

The quantity $\displaystyle\sum_{\text{all } p} m_p r_p^2$ is called the *moment of inertia* of the body about the axis and is conventionally denoted by the letter I. Moment of inertia has dimensions ML^2 and so its SI unit is $1\,\text{kg m}^2$.

The moment of inertia (I or $\sum m_p r_p^2$) of a body about an axis is a scalar quantity which depends on the manner in which the particles of the body are distributed about that particular axis. Its value varies according to the position and orientation of the axis.

Once you know the moment of inertia of a body about a given axis, you can find its kinetic energy when rotating about that axis. It is $\frac{1}{2}I\dot{\theta}^2$. Notice that there is an analogy between this expression for kinetic energy and the kinetic energy, $\frac{1}{2}mv^2$, of a particle. The mass, m, can be replaced by I and the speed, v, by the angular speed $\dot{\theta}$ (or ω) to give $\frac{1}{2}I\dot{\theta}^2$ (or $\frac{1}{2}I\omega^2$). You will see later that there are similar analogies between other quantities you have used for the motion of particles and those which apply to the rotation of a rigid body.

'Inertia' is a word which is used to describe a resistance to change in motion; it is sometimes used in place of mass. The larger the mass, or inertia, of a particle, the greater the amount of energy required to change its motion. In the same way, the energy required to change *rotational* motion is greater for bodies with large moments of inertia.

EXAMPLE 20.2 A wooden top has a moment of inertia of $2.4 \times 10^{-5}\,\text{kg m}^2$ about its axis. It starts spinning when a string wound round the spindle of the top is pulled with a constant force of 0.5 N. Assuming there is no loss of energy due to friction, find the angular speed attained by the top when the length of the string is 0.3 m.

SOLUTION

The work done in pulling the string is $0.5 \times 0.3 = 0.15\,\text{J}$ and, as no energy is lost in the process, this is equal to the gain in kinetic energy of the top. So the angular speed attained is $\omega\,\text{rad s}^{-1}$, where

$$\tfrac{1}{2} \times 2.4 \times 10^{-5}\omega^2 = 0.15$$
$$\Rightarrow \qquad \omega^2 = 12\,500.$$

The angular speed is $112\,\text{rad s}^{-1}$ correct to 3 significant figures.

The kinetic energy of a rotating wheel

So what is the kinetic energy of a rotating wheel?

As usual when modelling mechanical systems, it is useful to begin with a simple case. The simplest model of a wheel is one in which the mass of the spokes or their equivalent is negligible and all the mass can be considered to be concentrated at the rim in a hoop or ring of radius r. Then every particle is the same distance, r, from the axle. The moment of inertia of the wheel about the axle is then

$$I = \sum m_p r_p^2$$
$$= \left(\sum m_p\right) r^2$$
$$= Mr^2$$

where M is the total mass.

When this wheel is rotating with angular speed $\dot{\theta}$ about an axis through its centre perpendicular to its plane, its kinetic energy is

$$\tfrac{1}{2}I\dot{\theta}^2 = \tfrac{1}{2}Mr^2\dot{\theta}^2.$$

The kinetic energy of a more complex wheel can be found when you know its moment of inertia, I, about the axis of rotation.

Radius of gyration

The moment of inertia, I, of a body about a given axis of rotation depends on the mass of the body, M, and its distribution around that axis – as you have seen, parts of the body further away from the axis contribute more. The *radius of gyration* of the body about the given axis is defined as

$$k = \frac{I}{M}$$

so

$$I = Mk^2.$$

The significance of k is that if all the mass of the body were concentrated as a particle at a distance k from the axis (or in a ring of radius k) it would have the *same* moment of inertia as the original body.

QUESTION 20.2   The T-shaped lamina can be rotated about three axes (perpendicular to its plane) in the positions shown in figure 20.6.

Which positions give the least and the greatest radius of gyration? Explain why.

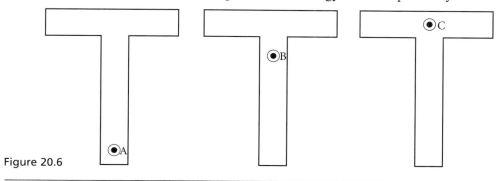

Figure 20.6

20.3 Using integration in moment of inertia calculations

The moment of inertia of a rigid body about an axis is given by

$$I = \sum m_p r_p^2.$$

The sum is calculated over all particles of the body and m_p denotes the mass of a typical particle which is a fixed perpendicular distance, r_p, from the axis. The axis can be anywhere, even outside the body, so long as r_p is constant for each particle, which is therefore restricted to motion in a circle of radius r_p relative to the axis.

You might have guessed that calculus methods are required to work out most moments of inertia. These are very similar to those you have used before where sums are involved, namely: subdivide the body into elementary parts for which you know the moment of inertia and then sum the parts.

EXAMPLE 20.3   Calculate the moment of inertia of a uniform circular disc of radius r, thickness t and mass M about an axis through its centre perpendicular to its plane.

Decide on appropriate elements.

SOLUTION

The disc is divided into elementary rings. A typical ring has radius x, width δx and thickness t. Its volume is approximately $(2\pi x\,\delta x) \times t = 2\pi t x\,\delta x$.

It is useful to use the density, ρ, of the disc so that the mass of each part can be obtained. It can be written in terms of M at the end of the calculation.

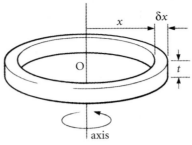 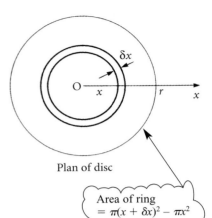

Figure 20.7   Plan of disc

The mass of the ring is then approximately

> Find the mass of a typical element in terms of the density.

$$\delta m = (2\pi t x \delta x) \times \rho$$

$$= 2\pi t \rho x \delta x.$$

> Area of ring
> $= \pi(x + \delta x)^2 - \pi x^2$
> $\approx 2\pi x \delta x$
> for negligible δx^2.

For small δx every particle of such a ring is approximately the same distance, x, from the axis. The moment of inertia of the ring about the axis is therefore approximately

$$\delta m x^2 = (2\pi t \rho x \delta x)x^2$$

$$= 2\pi t \rho x^3 \delta x$$

> Determine the moment of inertia of the element about the axis.

and the moment of inertia of the whole disc about the axis is approximately

> Sum for all elements.

$$I = \sum(2\pi t \rho x^3 \delta x)$$

$$= 2\pi t \rho \sum(x^3 \delta x) \quad \text{(since } \rho \text{ and } t \text{ are the same for all rings).}$$

In the limit as $\delta x \to 0$, this gives

> Write as an integral.

$$I = 2\pi t \rho \int_0^r x^3 \, dx$$

> Evaluate the integral. $\Rightarrow$

$$I = 2\pi t \rho \left[\frac{x^4}{4}\right]_0^r$$

> Note the limits: x takes values between 0 and r.

$$= \tfrac{1}{2}\pi t \rho r^4.$$

It is now necessary to replace ρ in terms of the mass of the disc, which has volume $\pi r^2 t$, so

$$M = \pi r^2 t \rho.$$

> Substitute for the density.

Hence the moment of inertia of a disc about a perpendicular axis through its centre is

$$I = \tfrac{1}{2}(\pi t \rho r^2)r^2$$

$$\Rightarrow \quad I = \tfrac{1}{2}Mr^2.$$

> or you could write
> $\dfrac{I}{M} = \dfrac{\pi t \rho r^4}{2\pi r^2 t \rho} = \dfrac{1}{2}r^2.$

The following points are worth noting at this stage.

● The moment of inertia of the disc ($\tfrac{1}{2}Mr^2$) is less than the moment of inertia (Mr^2) of a ring of the same mass and radius because most of the matter in the disc is nearer the axis.

- The thickness of the disc does not appear in the equation. The disc could be a solid cylinder of any length and the same formula would hold so long as the axis of rotation is the axis of the cylinder. Of course, the moment of inertia of a cylinder is greater than that of a thin disc of the same radius and density, but only because the mass is greater.
- Using the same argument, the moment of inertia of a hollow cylinder of radius r about its axis has the same form as that of the ring, that is Mr^2. Each particle of the cylinder is the same distance, r, from the axis.
- The radius of gyration of the disc about this axis is $\dfrac{r}{\sqrt{2}}$. A ring of the same mass and moment of inertia would have a radius of $\dfrac{r}{\sqrt{2}}$.

| **EXAMPLE 20.4** | Find the moment of inertia of a thin rod of length $2a$ and mass M about an axis through its centre and perpendicular to the rod. |

SOLUTION

Imagine that the rod is subdivided into small elements of width δx. Assume that the area, A, of a cross-section of the rod is so small that every point on it can be regarded as being the same distance from the axis. Then the only variable is the distance x of the elementary portion of the rod from the axis.

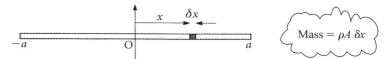

Figure 20.8

The mass of the element is $\rho A\,\delta x$, where ρ is the density, and its moment of inertia about the axis is approximately $(\rho A\,\delta x)x^2 = \rho A x^2\,\delta x$.

The moment of inertia of the rod about the axis is therefore approximately $\sum \rho A x^2\,\delta x$. In the limit as $\delta x \to 0$ this gives

$$I = \int_{-a}^{a} \rho A x^2\,dx$$

$$= \rho A \int_{-a}^{a} x^2\,dx$$

$$\Rightarrow \quad I = \rho A \left[\frac{x^3}{3}\right]_{-a}^{a}$$

$$= \frac{\rho A}{3}\left[a^3 - (-a)^3\right]$$

$$= \tfrac{2}{3}\rho A a^3.$$

The mass of the rod is $M = 2aA\rho = 2\rho A a$. Hence the moment of inertia of a rod about a perpendicular axis through its centre is

$$I = \tfrac{1}{3}(2\rho A a)a^2$$

$$= \tfrac{1}{3}Ma^2.$$

or write $\dfrac{I}{M} = \dfrac{2\rho A a^3}{3 \times 2\rho A a} = \dfrac{1}{3}a^2$

Once the moment of inertia of a body such as a rod or a disc is known, it can be used to find moments of inertia of other bodies.

For example, the thin rectangular lamina shown in figure 20.9 can be thought of as the sum of a large number of elementary rods.

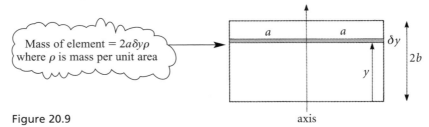

Mass of element $= 2a\delta y\rho$
where ρ is mass per unit area

Figure 20.9

The mass of the elementary rod shown is $(2a\delta y)\rho$, where ρ is now the mass per unit area of the lamina. The moment of inertia of the elementary rod about the axis is then

$$\tfrac{1}{3}(2a\rho\,\delta y)a^2 = \tfrac{2}{3}a^3\,\rho\delta y.$$

So the moment of inertia of the rectangular lamina about an axis of symmetry in its plane is

$$I = \int_0^{2b} \tfrac{2}{3}a^3\rho\,\mathrm{d}y$$

$$= \tfrac{2}{3}a^3\rho\int_0^{2b} \mathrm{d}y$$

$$= \tfrac{2}{3}a^3\rho 2b$$

$$= \tfrac{4}{3}ab\rho a^2.$$

But the mass of the lamina is $M = 4ab\rho$. Hence

$$I = \tfrac{1}{3}Ma^2.$$

Notice that this is independent of b and is in the same form as the moment of inertia of a thin rod about the axis. It is another case where the body is extended in the direction of the axis, leading to the same expression for the moment of inertia although, of course, the mass is greater.

Moments of inertia of selected uniform bodies

	Body of mass M	Axis	Moment of inertia
	Hoop or hollow cylinder of radius r	Through centre perpendicular to circular cross-section	Mr^2
	Disc or solid cylinder, of radius r	Through centre perpendicular to circular cross-section	$\tfrac{1}{2}Mr^2$

	Body of mass M	Axis	Moment of inertia
	Thin rod of length $2l$	Through centre perpendicular to rod	$\frac{1}{3}Ml^2$
	Rectangular lamina	Edge perpendicular to sides of length $2l$	$\frac{4}{3}Ml^2$
	Solid sphere of radius r	Diameter	$\frac{2}{5}Mr^2$

Dividing hollow bodies into elements

When a three-dimensional body is divided into elements, you should take care when deciding the 'width' of a typical element. figure 20.10 shows a typical element, which you might always have considered to be approximately cylindrical. This is appropriate for a *solid body*, but not for a *surface*, especially near a point where it crosses the x axis. Also, because the mass depends on the surface area rather than the volume, a more appropriate approximation for a *thin shell* is to treat the element as part of a hollow cone.

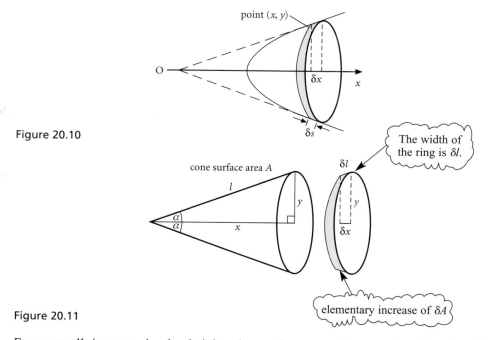

Figure 20.10

Figure 20.11

For a small increase in the height, the surface area increases by $\delta A = 2\pi y \delta l$ or $2\pi r \delta l = 2\pi l \sin \alpha \, \delta l$ in this case. It translates into $2\pi y \delta s$ for any surface of revolution such as the one in figure 20.11.

The moment of inertia of a uniform hollow cone about its axis

You can use the result above to find the moment of inertia of the hollow cone about its axis. It is

$$\int_0^l \sigma 2\pi l \,(\sin \alpha \, dl)r^2 = \int_0^l \sigma 2\pi l \sin \alpha \, dl \,(l \sin \alpha)^2 \quad (\sigma \text{ is the mass per unit area})$$

$$= 2\sigma\pi \sin^3 \alpha \int_0^l l^3 \, dl$$

$$= \tfrac{1}{2}\sigma\pi \sin^3 \alpha l^4$$

$$= \tfrac{1}{2}(\sigma \pi l^2 \sin \alpha)l^2 \sin^2 \alpha$$

$$= \tfrac{1}{2}Mr^2 \quad (M = \sigma A \text{ and } r = l \sin \alpha).$$

Combining bodies

When a rigid body has several parts, its moment of inertia about an axis can be found by adding the moments of inertia of the separate parts about the same axis. This is a direct consequence of the definition of the moment of inertia as a sum taken over all particles of the body; it doesn't matter if the sum is taken separately for different groups of particles. The next example illustrates this principle.

EXAMPLE 20.5   A wheel of mass M has been strengthened using a metal ring. It consists of a uniform disc of radius r and mass σ per unit area surrounded by a uniform solid ring (the rim) of radius r, negligible width, and mass $5r\sigma$ per unit length.
i)   Find the mass of the two parts of the wheel in terms of M.
ii) Write down the moment of inertia of each part about the axle of the wheel.
iii) Find the kinetic energy when the wheel is rotating with an angular speed ω.
iv) Write this kinetic energy as a percentage of the kinetic energy of a hoop with the same mass and radius rotating at the same angular speed.

SOLUTION

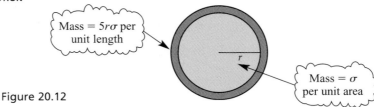

Mass = $5r\sigma$ per unit length

Mass = σ per unit area

Figure 20.12

i)   The area of the disc is πr^2, so its mass is $M_1 = \sigma\pi r^2$.
     The length of the ring is $2\pi r$, so its mass is

$$M_2 = 2\pi r(5r\sigma)$$

$$= 10 \, \sigma\pi r^2.$$

Hence

$$M = 11 \, \sigma\pi r^2$$

$$\Rightarrow \quad M_1 = \frac{M}{11} \quad \text{and} \quad M_2 = \frac{10M}{11}.$$

ii) The moment of inertia of the inside disc about the axle is

$$I_2 = \tfrac{1}{2}M_1 r^2$$

$$= \frac{Mr^2}{22}.$$

The moment of inertia of the rim about the axle is

$$I_2 = M_2 r^2$$

$$= \frac{10Mr^2}{11}.$$

iii) The kinetic energy of the wheel is $\tfrac{1}{2}I\omega^2$, where $I = I_1 + I_2$.

$$I_1 + I_2 = \frac{Mr^2}{22} + \frac{10Mr^2}{11}$$

$$= \frac{21}{22}Mr^2$$

iv) The moment of inertia of the hoop about its axis is Mr^2, so its kinetic energy is $\tfrac{1}{2}Mr^2\omega^2$.

The kinetic energy of the wheel is $\tfrac{21}{22} \times 100\%$ of the kinetic energy of the hoop, namely 95.5%.

Dividing regular bodies

You have already met the idea of elongating bodies in the direction of the axis. The elongated body is formed by combining identical elementary parts which all have the same moment of inertia about the axis. Conversely, when a body can be divided into two or more parts which obviously have the same moment of inertia about a given axis, it is possible to calculate the moments of inertia of the parts quite easily. For example, a sphere can be divided into two hemispheres using any plane through its centre. To each particle P_1 in one half there corresponds a similarly placed particle P_2 in the other half with the same moment of inertia about the axis. These are shown in figure 20.13.

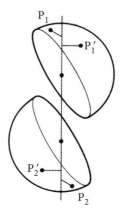

Typical particles P_1 and P_2 have the same moment of inertia about the axis. So have P_1' and P_2'.

Figure 20.13

The moment of inertia of each hemisphere about the axis is therefore

$$\tfrac{1}{2}\left(\tfrac{2}{5}Mr^2\right) = \tfrac{2}{5}\left(\frac{M}{2}\right)r^2$$
$$= \tfrac{2}{5}M_1r^2$$

where M_1 is the mass of each hemisphere. Again, the form of the expression does not change although the mass does.

Conservation of energy

Knowing the moment of inertia of a rotating body enables you to solve some rotation problems purely by energy methods.

EXAMPLE 20.6

A uniform thin rod AB of mass m and length $2a$ has a particle of mass M attached at B and is hinged at its centre O so that it rotates freely in a vertical plane. The rod is held horizontally and released.

i)   Show the moment of inertia of the system about the hinge is $\tfrac{4}{3}Ma^2$.
ii) Find the angular speed when the rod is vertical.

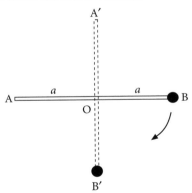

Figure 20.14

SOLUTION

i)   The moment of inertia of the rod alone is $\tfrac{1}{3}Ma^2$ (see page 458).
     The moment of inertia of the particle about the axis is Ma^2.
     So the total moment of inertia of the system about the axis is

$$I = Ma^2 + \tfrac{1}{3}Ma^2$$
$$= \tfrac{4}{3}Ma^2.$$

ii) When in the vertical position, the gain in kinetic energy is:

$$\tfrac{1}{2}I\omega^2 = \tfrac{2}{3}Ma^2\omega^2.$$

The loss in potential energy arises from the descent of the mass at B and is given by Mga. Hence, by conservation of energy:

$$\tfrac{2}{3}Ma^2\omega^2 = Mga$$
$$\Rightarrow \quad \omega = \sqrt{\frac{3g}{2a}}.$$

Note that there is no change in the potential energy of the rod during the rotation because the axis is through its centre of mass. In the subsequent sections you will meet the case where this is not so.

20.4 Further calculations of moments of inertia

Two important theorems are useful in the calculation of moments of inertia.

The perpendicular axes theorem

You know the moment of inertia of a rectangular lamina about an axis of symmetry in its plane. The moments of inertia of the lamina in figure 20.15 are

$$I_y = \tfrac{1}{3}Ma^2 \quad \text{about the } y \text{ axis (AB has length } 2a)$$

and similarly

$$I_x = \tfrac{1}{3}Mb^2 \quad \text{about the } x \text{ axis (BC has length } 2b).$$

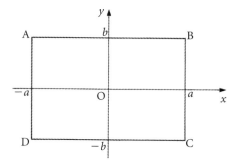

Figure 20.15

But what is its moment of inertia about the z axis through O perpendicular to its plane? This can be found by applying a very useful result known as the perpendicular axes theorem for a lamina, which can be stated as follows.

● The moment of inertia of a lamina about an axis which is perpendicular to the plane of the lamina and passes through a point O in its plane, is equal to the sum of the moments of inertia about two perpendicular axes in the plane of the lamina which also pass through O.

Using the notation above, $I_z = I_x + I_y$.

You will see from the following proof that it is essential for the body to be a lamina.

A particle P of mass m situated at the point (x, y) of any lamina in the xy plane is a distance r from O (and hence the z axis), where $r^2 = x^2 + y^2$ (see figure 20.16).

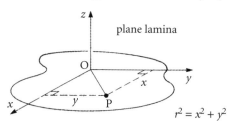

Figure 20.16

The moment of inertia of P about the z axis through O is therefore $m(x^2 + y^2)$. Hence the moment of inertia of the whole lamina about the z axis is

$$I_z = \sum m(x^2 + y^2)$$
$$= \sum mx^2 + \sum my^2.$$

But $\sum mx^2$ is the moment of inertia, I_y, about the y axis and $\sum my^2$ is the moment of inertia, I_x, about the x axis. It follows that

$$I_z = I_x + I_y.$$

This is true for any lamina in the xy plane.

Note that the origin and the x and y axes must always be in the plane of the lamina.

For the rectangle this gives

$$I_z = \tfrac{1}{3}Ma^2 + \tfrac{1}{3}Mb^2$$
$$= \tfrac{1}{3}M(a^2 + b^2) \quad \text{or} \quad \tfrac{1}{3}Md^2$$

where d is the distance from the centre to a vertex of the rectangle.

EXAMPLE 20.7 Use the perpendicular axes theorem to find the moment of inertia of a thin circular disc of radius r about a diameter.

SOLUTION

By symmetry the moment of inertia about every diameter is the same, say I_d. This means that, when the disc is in the xy plane with its centre at the origin,

$$I_x = I_y + I_d.$$

But I_z is the moment of inertia about the axis through O perpendicular to the disc and this is $\tfrac{1}{2}Mr^2$. Hence

$$\tfrac{1}{2}Mr^2 = I_x + I_y$$
$$= 2I_d$$

$I_x = I_y$ and $I_z = \tfrac{1}{2}Mr^2$

$$\Rightarrow \quad I_d = \tfrac{1}{4}Mr^2.$$

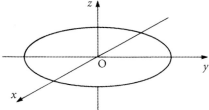

Figure 20.17

The parallel axes theorem

In many circumstances, for example when a disc rotates about a diameter, or when two rigid bodies are combined to form another, the axis of rotation may not be in the most convenient position for the calculation of a moment of inertia. The parallel axes theorem can be used in these circumstances because it gives a relationship between the moments of inertia of the same rigid body about different parallel axes.

The theorem can be stated as follows.

● The moment of inertia of a rigid body of mass M about a fixed axis through a point A is equal to its moment of inertia about a parallel axis through its centre of mass G plus Md^2, where d is the perpendicular distance between the axes:

$$I_A = I_G + Md^2.$$

Figure 20.18 shows two parallel axes which are a fixed distance, d, apart. The axis GR passes through G and the other axis, AQ, passes through a point A such that AG is perpendicular to both axes and hence of length d. The moments of inertia of the body about these two axes are denoted by I_G and I_A. Consider a typical particle P of mass m_p and suppose that the plane through P perpendicular to the two axes meets GR at O and AQ at S as shown. Then $OS = GA = d$.

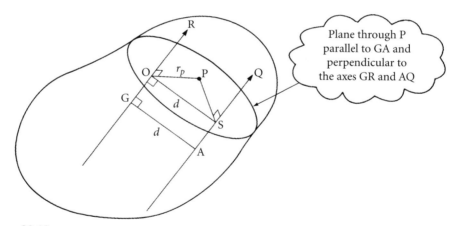

Plane through P parallel to GA and perpendicular to the axes GR and AQ

Figure 20.18

Figure 20.19 shows the plane through P. Co-ordinate axes have been taken through O parallel and perpendicular to OS. The co-ordinates of P relative to these axes are (x_p, y_p) and $r_p^2 = x_p^2 + y_p^2$ as usual.

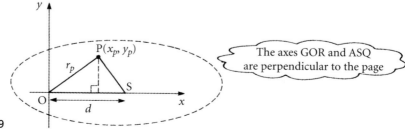

The axes GOR and ASQ are perpendicular to the page

Figure 20.19

$$PS^2 = (d - xp)^2 + y_p^2$$
$$= x_p^2 + y_p^2 + d^2 - 2dx_p$$
$$= r_p^2 + d^2 - 2dx_p$$

The moment of inertia of the particle P about the axis ASQ is

$$m_p PS^2 = m_p(r_p^2 + d^2 - 2dx_p)$$

The moment of inertia of the whole body about the axis AQ is

$$I_A = \sum m_p(r_p^2 + d^2 - 2dx_p) \quad \text{(summed over all particles P)}$$
$$= \sum m_p r_p^2 + \sum m_p d^2 - \sum 2dm_p x_p$$
$$= I_G + Md^2 - 2d\sum m_p x_p.$$

Remember, r_p is the distance of P from the axis GR.

Now x_p is equal to the x co-ordinate of P relative to a new three-dimensional co-ordinate system with origin at G and GA as its x axis (see figure 20.20).

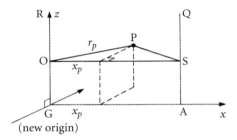

Figure 20.20

But, by the definition of the centre of mass, $\sum m_p x_p = M\bar{x}$ and $\bar{x}$ is zero because G is at the origin. So $2d \sum m_p x_p = 0$. Hence

$$I_A = I_G + Md^2.$$

The parallel axes theorem can be used to extend your repertoire of moments of inertia, but remember that one of the axes must be through the centre of mass. A consequence of this theorem is that the moment of inertia, and hence also the rotational kinetic energy, are least when the axis is through G. However, the minimum kinetic energy possible for a given body and a given value of ω depends on the orientation of the axis.

EXAMPLE 20.8   Use the parallel axes theorem to find the moments of inertia of
i)   a thin uniform rod of mass M and length h about a perpendicular axis through its end
ii)   a thin uniform solid disc of mass M and radius r about
   a)   an axis perpendicular to its plane through a point on its circumference
   b)   a tangent.
iii)   Which of the above is equally applicable to a solid cylinder?

SOLUTION

i)   For the rod $I_G = \frac{1}{3}M\left(\frac{h}{2}\right)^2$ and the axes are a distance $\frac{h}{2}$ apart. Hence

$$I_A = I_G + M\left(\frac{h}{2}\right)^2$$

$$= \tfrac{1}{3}M\left(\frac{h}{2}\right)^2 + M\left(\frac{h}{2}\right)^2$$

$$= \tfrac{1}{3}Mh^2.$$

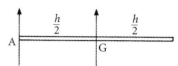

Figure 20.21

ii) a) In this case $I_G = \frac{1}{2}Mr^2$ and the axes
         are a distance r apart.

         Hence $I_A = I_G + Mr^2$
                         $= \tfrac{1}{2}Mr^2 + Mr$
                         $= \tfrac{3}{2}Mr^2.$

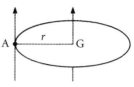

Figure 20.22

     b) Now $I_G = \tfrac{1}{4}Mr^2$
         $\Rightarrow$ $I_A = \tfrac{1}{2}Mr^2 + Mr$
                         $= \tfrac{5}{4}Mr^2.$

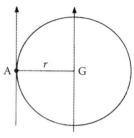

iii) The result is applicable to a cylinder so long as
     this is formed by elongating the body in a direction
     parallel to the axis of rotation. It therefore applies
     in part ii) a) but not in the other cases.

Figure 20.23

The next example illustrates how the parallel axes theorem can be used in conjunction
with calculus methods to find the moment of inertia of a solid, in this case a cylinder,
about an axis which is not an axis of symmetry.

EXAMPLE 20.9   Find the moment of inertia of a uniform solid cylinder of mass M, radius r and height h
about an axis which is perpendicular to the axis of the cylinder and which passes through
the centre of one end.

SOLUTION

Choose axes as shown in figure
20.24 so that the required moment
of inertia is about the x axis.
Let ρ be the density of the cylinder,
so that $M = \pi r^2 h \rho$.
Subdivide the cylinder into
elementary discs as shown.

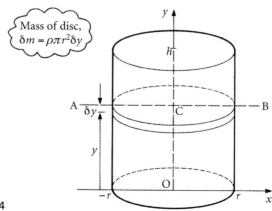

Mass of disc,
$\delta m = \rho \pi r^2 \delta y$

Figure 20.24

A typical disc has thickness δy and centre at $C(0, y)$.

Its mass δm is approximately $\rho \pi r^2 \delta y$.

The moment of inertia of the element about an axis through C parallel to the x axis (i.e. a diameter) is $\frac{1}{4}\delta m r^2$.

C is the centre of mass of the disc, so by the parallel axes theorem, its moment of inertia about the x axis is

$$\tfrac{1}{4}\delta m r^2 + \delta m y^2 = \tfrac{1}{4}(\rho \pi r^2 \delta y)r^2 + (\rho \pi r^2 \delta y)y^2$$
$$= \tfrac{1}{4}\rho \pi r^2(r^2 + 4y^2)\delta y.$$

For the whole cylinder, the moment of inertia about the x axis, I_x, is approximately

$$\sum \tfrac{1}{4}\rho \pi r^2(r^2 + 4y^2)\delta y = \tfrac{1}{4}\rho \pi r^2 \sum (r^2 + 4y^2)\delta y.$$

In the limit as $\delta y \to 0$, this gives

$$I_x = \tfrac{1}{4}\rho \pi r^2 \int_0^h (r^2 + 4y^2)\, \mathrm{d}y$$
$$= \tfrac{1}{4}\rho \pi r^2 \left[r^2 y + 4\frac{y^3}{3} \right]_0^h$$
$$= \tfrac{1}{4}\rho \pi r^4 h + \tfrac{1}{3}\rho \pi r^2 \, h^3$$
$$= \tfrac{1}{4}M r^2 + \tfrac{1}{3}M h^2 \quad (M = \rho \pi r^2 h).$$

It is interesting to note that this has the same form as the sum of the moments of inertia of a disc of radius r about its diameter and a rod of length h about a perpendicular axis through its end: $h = 0$ gives the disc and $r = 0$ gives the rod.

Summary

- $I = \sum m r^2$
- Perpendicular axes theorem: $I_z = I_x + I_y$ for a lamina only.
- Parallel axes theorem: $I_A = I_G + M d^2$

$$= I_G + M(\mathrm{AG})^2$$

when G is the centre of mass and AG is perpendicular to the axes.

20.5 Bodies with variable density

Some bodies are made of several materials of differing densities. The moments of inertia of these can be found by dealing with each part separately. Others have densities which vary continuously and, provided the functions are relatively simple, calculus methods can be used to determine their moments of inertia.

EXAMPLE 20.10

A star is modelled as a sphere of radius a whose density is given in terms of the distance r from the centre by the function $r = r_0\left(1 - \dfrac{r^2}{a^2}\right)$.

i) Given that ρ_0 is constant, find the total mass M of the sphere.
ii) Find its moment of inertia about a diameter.

SOLUTION

Imagine that the sphere is divided into elementary shells and a typical shell has radius r and thickness δr.

i) The mass of the shell is

$$4\pi r^2 \rho \delta r = \frac{4\pi \rho_0}{a^2}(a^2 r^2 - r^4)\,\delta r \qquad \qquad ①$$

$$\Rightarrow \qquad M = \int_0^a \frac{4\pi \rho_0}{a^2}(a^2 r^2 - r^4)\,dr$$

$$= \frac{4\pi \rho_0}{a^2}\left[\frac{a^2 r^3}{3} - \frac{r^5}{5}\right]_0^a$$

$$= \frac{8\pi \rho_0}{15}a^3.$$

ii) The moment of inertia of a hollow shell of mass m about a diameter is $\frac{2}{3}mr^2$, so the moment of inertia of an elementary shell is

$$\frac{2}{3} \times \frac{4\pi \rho_0}{a^2}(a^2 r^2 - r^4)\,\delta r \times r^2 \qquad \text{(from ①)}$$

$$\Rightarrow \qquad \text{moment of inertia of sphere} = \int_0^a \frac{8\pi \rho_0}{3a^2}(a^2 r^4 - r^6)\,dr$$

$$= \frac{8\pi \rho_0}{3a^2}\left[a^2 \frac{r^5}{5} - \frac{r^7}{7}\right]_0^a$$

$$= \frac{16\pi \rho_0}{105}a^5$$

$$= \frac{2}{7}Ma^2 \qquad \text{(from ②).}$$

A baseball bat is thicker at one end than the other, but its thickness is always small compared with its length so it is possible to model it by a rod of varying density, as illustrated in the next example.

EXAMPLE 20.11

A baseball bat has a uniform handle AB of length 0.36 m and mass $M_1 = 0.136$ kg. The remainder of the bat is modelled as a thin rod BC of length 0.7 m and mass per unit length $1.43(x^2 + x + 0.25)$ kg m^{-1}, where x is the distance from B. Assuming the handle can also be modelled as a thin rod, find

i) the mass M kg of the bat
ii) the distance of its centre of mass, G, from B
iii) its moment of inertia about a perpendicular axis through B
iv) its moment of inertia about a perpendicular axis through the end A.

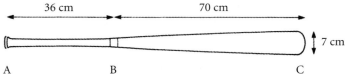

Figure 20.25   A   B   C

SOLUTION

figure 20.26 shows the essential features of the bat.

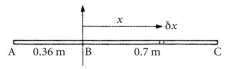

Figure 20.26   mass $= 0.136$ kg density $= 1.43 \ (x^2 + x + 0.25)$ k gm^{-1} $(x \geqslant 0)$

i)   The mass of an element of BC of length δx is $1.43(x^2 + x + 0.25)\delta x$ kg

$$\Rightarrow \quad \text{mass of BC} = \int_0^{0.7} 1.43(x^2 + x + 0.25) \, dx \text{ kg}$$

$$= \left[1.43\left(\frac{x^3}{3} + \frac{x^2}{2} + 0.25x\right) \right]_0^{0.7} \text{kg}$$

$$= 0.764 \text{ kg}$$

$\Rightarrow$ Total mass $M = 0.764 + 0.136 = 0.900$ kg (to 3 significant figures).

ii) To find the distance, $\bar{x}$, of G from B, take moments about an axis through B. Then

$$\left(\sum \delta m\right) \bar{x} = \sum (\delta m x) \quad \text{(summed for all particles of the bat).}$$

For AB $\sum (\delta m x) = -M_1 \times \tfrac{1}{2} \text{AB}$

$$= -0.136 \times 0.18$$

$$= -0.0245$$

and for BC $\sum (\delta m x) = \sum 1.43(x^2 + x + 0.25)x \, \delta x$.

In the limit as $\delta x \to 0$, this becomes

$$\int_0^{0.7} 1.43(x^2 + x + 0.25)x \, dx$$

$$= \int_0^{0.7} 1.43(x^3 + x^2 + 0.25x) \, dx$$

$$= 1.43\left[\frac{x^4}{4} + \frac{x^3}{3} + 0.25\frac{x^2}{2}\right]_0^{0.7}$$

$$= 0.3369.$$

Combining AB and BC gives

$$0.9\bar{x} = -0.0245 + 0.3369$$

$$\Rightarrow \quad 0.9\bar{x} = 0.3124$$

$$\bar{x} = 0.347.$$

The centre of mass is about 35 cm from B.

iii) The moment of inertia, I_1, of the uniform rod AB about a perpendicular axis through its centre is $\frac{1}{3}M_1(0.18)^2$. So, by the parallel axes theorem, its moment of inertia about the end B is

$$I_1 = \tfrac{1}{3}M_1(0.18)^2 + M_1(0.18)^2$$
$$= \tfrac{4}{3}(0.136)(0.18)^2$$
$$= 0.0059 \text{ kg m}^2.$$

The moment of inertia, I_2, of the part BC about the axis through B is

$$I_2 = \lim_{\delta x \to 0} \left(\sum \delta m x^2 \right)$$
$$= \int_0^{0.7} 1.43(x^2 + x + 0.25)x^2 \, dx$$
$$= \int_0^{0.7} 1.43(x^4 + x^3 + 0.25x^2) \, dx$$
$$= 1.43 \left[\frac{x^5}{5} + \frac{x^4}{4} + 0.25\frac{x^3}{3} \right]_0^{0.7}$$
$$= 0.1748.$$

Hence: $I_1 + I_2 = 0.1807.$

The moment of inertia of the bat about a perpendicular axis through B is about 0.18 kg m^2.

iv) By the parallel axes theorem,

$$I_B = I_G + M(BG)^2$$
and $$I_A = I_G + M(AG)^2$$
$$\Rightarrow \quad I_A = I_B - M(BG)^2 + M(AG)^2$$
$$\Rightarrow \quad I_A = I_B + M(AG^2 - BG^2)$$
$$= 0.1807 + 0.9(0.3795)$$
$$= 0.5222$$

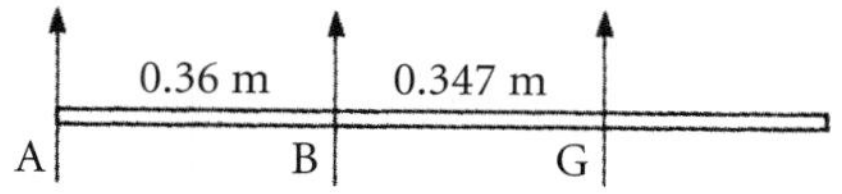

Figure 20.27

The moment of inertia about the end, A, is about 0.52 kg m^2.

⚠ The parallel axes theorem

Part **iv)** of Example 20.11 illustrates the necessity for one of the axes to be through the centre of mass, G, when using the parallel axes theorem. In general, the moments of inertia about parallel axes through two points A and B where AG and BG are perpendicular to the axes are related by the equation

$$I_A = I_B + M(AG^2 - BG^2).$$

(Note also that $AG^2 - BG^2 \neq AB^2$.) The moment of inertia of the bat about the player's elbow or shoulder (assuming that the arm and wrist are kept rigid) can be calculated using this result.

20.6 The equation of motion for the rotation of a rigid body about a fixed axis

QUESTION 20.3   An ice-skater might stretch out her arms or draw them in to change her speed of rotation.

How does this work?

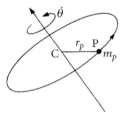

In the previous sections you learnt about the moment of inertia of a rigid body and how to use this to find the kinetic energy of the body as it rotates about a fixed axis. In the following sections you will learn more about fixed-axis rotation, starting from the equation of motion.

Previously, the expression for the kinetic energy of a wheel was obtained by summing the energies of individual particles. The same approach is useful for finding the equation of motion for the rotation of a rigid body about a fixed axis. This is the equivalent of Newton's second law for the linear motion of a particle, so first consider the equation of motion for a typical particle in the body. When the rigid body rotates about a fixed axis, each particle in the body moves in a plane in a circle with its centre on the axis, as

Figure 20.28

shown in figure 20.28. Because the body is rigid and rotates about a fixed axis, all the particles in the body have the same angular velocity $\dot{\theta}$ at all times, so they also have the same angular acceleration $\ddot{\theta}$.

A typical particle P rotates in a circle of radius r_p about a point C on the axis. Its acceleration has radial and transverse components $-r_p\dot{\theta}^2$ and $r_p\ddot{\theta}$. These are shown in figure 20.29, together with the resultant force $\mathbf{F}_p$ which acts on the particle in the plane of the motion and is at an angle α to PC.

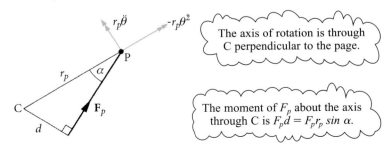

The axis of rotation is through C perpendicular to the page.

The moment of F_p about the axis through C is $F_p d = F_p r_p \sin \alpha$.

Figure 20.29

Newton's second law in the transverse direction (i.e. for circular motion, the tangential direction) gives

$$F_p \sin \alpha = m_p r_p \ddot{\theta}.$$

Multiplying by r_p gives

$$F_p r_p \sin \alpha = m_p r_p^2 \ddot{\theta}.$$

The equation of motion for the rotation of the whole rigid body can be found by summing both sides for all the particles.

$F_p r_p \sin \alpha$ is the moment of the force $\mathbf{F}_p$ about the axis of rotation. This is made up of components which include internal forces as well as the external forces on the body which happen to act on the particular particle.

By Newton's third law, the internal forces are equal and opposite. It is therefore to be expected that, when all the moments of all the forces are summed for all particles, the moments of the internal forces cancel and only the moments of the external forces remain.

Summing both sides of the equation for all particles therefore gives

$$\sum(\text{moments of external forces}) = \sum (m_p r_p^2 \ddot{\theta})$$
$$= \left(\sum m_p r_p^2\right)\ddot{\theta} = I\ddot{\theta}$$

$\ddot{\theta}$ is the same for all particles.

where $I = \sum m_p r_p^2$ as before.

There is no standard notation for a moment and M cannot be used because M is used for mass. In this book C is used to represent the moment of a force about a particular axis (sometimes called a torque) or the moment of a couple. So when the total moment of all forces about the axis is C, the equation of motion for rotation becomes $C = I\ddot{\theta}$.

This is the equivalent of Newton's second law for the rotation of a rigid body about a fixed axis. It is a very concise equation, and the beauty of it is that it can be compared with the equation $F = m\ddot{x}$ for the linear motion of a particle in a similar way to the analogy between kinetic energies met on page 452.

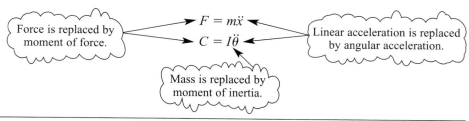

Force is replaced by moment of force.

$F = m\ddot{x}$

$C = I\ddot{\theta}$

Linear acceleration is replaced by angular acceleration.

Mass is replaced by moment of inertia.

When using $F = m\ddot{x}$ for linear motion, it is important to take the positive direction of F in the direction of x increasing. In a similar way, moments can be clockwise or anticlockwise and when you use the equation $C = I\ddot{\theta}$, it is important to remember that the moments of the forces should be positive in the same sense as that of increasing θ.

Remember that, although the moment of inertia of a body about an axis is a scalar quantity, its value depends on the position and direction of the axis and this should always be stated.

Wheels in machines

If you visit a place where old machines are conserved, you will see that they often have a large wheel, called a 'flywheel', as part of the driving mechanism. The next example demonstrates why these large wheels are useful.

EXAMPLE 20.12

Two wheels have the same mass M. One can be modelled by a large hollow cylinder of radius R and the other by a smaller solid cylinder of radius r. When they are rotating, they are each subject to a frictional couple of constant magnitude C. While the wheels are being driven, there is a break in power which lasts for a time t. Assuming all units are compatible, find expressions for

i)   the angular retardation of each wheel while the power is off

ii)   the reduction in the angular velocity of each wheel during this time.

The wheels have the same initial angular velocity, and the radius of the larger is twice that of the smaller.

iii)   Show that the percentage reduction in the angular velocity during the time the power is off is eight times greater for the smaller wheel.

iv)   When the wheels are rotating with the same angular speed, show that the kinetic energy of the larger is eight times that of the smaller.

SOLUTION

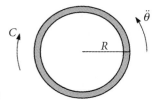

Figure 20.30

i)   When a wheel has an angular acceleration $\ddot{\theta}$ as a result of the action of a couple, $-C$, the equation of motion gives

$$-C = I\ddot{\theta}$$
$$\Rightarrow \quad \ddot{\theta} = \frac{-C}{I}$$

The moment of inertia of the hollow cylinder about its axis is MR^2.

Its acceleration is $\dfrac{-C}{MR^2}$.

The moment of inertia of the solid cylinder about its axis is $\frac{1}{2}Mr^2$.

Its acceleration is $\dfrac{-2C}{Mr^2}$.

ii)   For constant angular acceleration α, the new angular velocity, ω, of a wheel after t seconds is

$$\omega = \omega_0 + \alpha t.$$

So the reduction in the angular speed is

$$\omega_0 - \omega = \frac{Ct}{I}.$$

$$\alpha = \frac{-C}{I}$$

The reductions in the angular speeds of the wheels are

$$\frac{Ct}{Mr^2} \quad \text{and} \quad \frac{2Ct}{Mr^2}.$$

iii) When the initial angular speeds are the same, the percentage reductions in the angular speeds are proportional to the actual reductions.

These are in the ratio (larger : smaller) of

$$\frac{Ct}{Mr^2} : \frac{2Ct}{Mr^2}$$
$$= r^2 : 2R^2$$
$$= r^2 : 2 \times 4r^2 \quad (R = 2r)$$
$$= 1 : 8.$$

The angular speed is reduced for both wheels, but the percentage reduction is 8 times greater for the smaller wheel.

iv) When it rotates with angular speed ω, the kinetic energy of a wheel is $\frac{1}{2}I\omega^2$. The ratio of the kinetic energy of the larger wheel to that of the smaller is

$$\frac{1}{2}(MR^2)\omega^2 : \frac{1}{2}(\frac{1}{2}Mr^2)\omega^2$$
$$= R^2 : \frac{1}{2}r^2$$
$$= 4r^2 : \frac{1}{2}r^2$$
$$= 8 : 1.$$

The solution to part **iii)** of Example 20.12 shows that the use of the larger wheel in the driving mechanism leads to a smaller change in angular speed and so enables the machine to keep working at a steadier rate when there are fluctuations in the power. Such fluctuations are inevitable for many machines. An example is the engine of a car which incorporates a relatively large flywheel, to help smooth out the effects of the intermittent firing in the cylinders. The answer to part **iv)** of the example shows, however, that the amount of energy required to set the larger wheel spinning is much greater than that required to make the smaller wheel spin with the same angular speed. This could be a disadvantage and it is also likely to take longer because the angular acceleration is less for the same torque. When the time required to set a wheel in motion is at a premium, as in the case of a racing car, the flywheel is lighter.

When there is surplus energy in a system, flywheels can be used to store energy. There is a bus design which incorporates a flywheel that is activated when braking takes place. When the bus stops some of its energy is stored in the flywheel and this can be used to boost the power of the engine when the bus starts again. Some toys are designed with flywheels to enable them to go on moving much further by themselves after a child stops pushing.

20.7 The work done by the moment of a force

When a rigid body rotates about its axis through a small angle $\delta\theta$, the point of application of a typical force F moves through an arc of length $r\delta\theta$, as shown in figure 20.31.

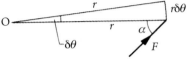

The moment of F about an axis through O is $F \times r \sin \alpha$.

Figure 20.31

The work done by the force is then approximately

$$F \sin \alpha \times r\delta\theta = (F \times r \sin \alpha)\,\delta\theta.$$

This can also be written as (moment about axis) $\times \delta\theta$.

The total work done by this force when the body rotates through an angle θ is thus $\sum \text{moment} \times \delta\theta$.

In the limit as $\delta\theta \to 0$, this becomes $\int (\text{moment})\,d\theta$.

Summing for all the forces, the work done by the total moment, C, is $\int C\,d\theta$.

This again demonstrates the equivalence between the equations for rotation and those for linear motion. It is comparable to $\int F\,dx$ with F replaced by C and x by θ.

When the equation $C = I\ddot{\theta}$ is integrated with respect to θ it gives

$$\int C\,d\theta = \int I\ddot{\theta}\,d\theta$$

$$= I\int \ddot{\theta}\,\frac{d\theta}{dt}\,dt$$

$$= I\int \ddot{\theta}\,\dot{\theta}\,dt$$

But $\qquad \ddot{\theta}\dot{\theta} = \dot{\theta}\ddot{\theta} = \dot{\theta}\dfrac{d\dot{\theta}}{dt} = \dfrac{d}{dt}\left(\tfrac{1}{2}\dot{\theta}^2\right)$ ◄── Compare $v\dfrac{dv}{dt} = \dfrac{d}{dt}\left(\tfrac{1}{2}v^2\right)$.

$\Rightarrow \qquad I\int \ddot{\theta}\dot{\theta}\,dt = \tfrac{1}{2}\dot{\theta}^2$

and $\qquad \int C\,d\theta = \left[\tfrac{1}{2}I\dot{\theta}^2\right].$ ◄── where the brackets mean 'change in'.

This is the work–energy equation for a body rotating about a fixed axis.

Note

This equation has been obtained by integrating the equation of motion. Conversely, the energy equation can be differentiated with respect to t to give the equation of motion. This is done for the compound pendulum on page 260.

20.8 Potential energy change during rotation

The next example illustrates how the use of $\int C \, d\theta$ to find the work done by gravity leads to the expression for the loss in potential energy of the rotating body. The work–energy equation then becomes the equation for the conservation of mechanical energy of the body.

EXAMPLE 20.13   A rigid body of mass M is free to rotate about a horizontal axis through a point A at a distance l from its centre of mass G. Its moment of inertia about the axis is I. The body is displaced through an angle α and let go.

i)   Find, by using $\int C \, d\theta$, the work done by gravity when AG falls into the vertical position and show that this is equal to the loss in potential energy of a particle with the same mass which falls through the same height as G.

ii)   Find an expression for the angular speed when AG is vertical.

SOLUTION

i)   The weight is the force which makes the body rotate about the axis. When it is in the position shown in figure 20.32, the moment of the weight about the axis in the direction of increasing θ is $- Mgl \sin \theta$.

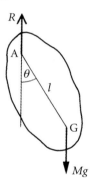

Figure 20.32

The work done by gravity when θ decreases from α to zero is therefore

$$\int_{\alpha}^{0} -Mgl \sin \theta \, d\theta = \Big[Mgl \cos \theta \Big]_{\alpha}^{0}$$
$$= Mgl(1 - \cos \alpha).$$

But the height fallen by G is $h = l(1 - \cos \alpha)$, see figure 20.33. Hence the work done by gravity is equal to Mgh, the loss in potential energy of a particle of mass M which has fallen the same height as G.

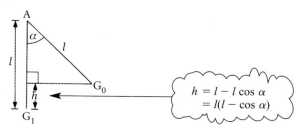

Figure 20.33

ii) By the work–energy equation, or the principle of conservation of energy, the kinetic energy gained when AG is vertical is given by

$$\tfrac{1}{2}I\dot\theta^2 = Mgh$$

The angular speed at this point is then

$$\sqrt{\frac{2Mgh}{I}} = \sqrt{\frac{2Mgl}{I}(1 - \cos\alpha)}.$$

Note

The assumption that the potential energy of a large body is the same as that of a particle of equal mass situated at G follows from the definition of G. If Oy is the co-ordinate axis in the vertical direction, then

$$M\bar y = \sum(m_p y_p) \quad \Rightarrow \quad Mg\bar y = \sum(m_p g y_p).$$

A winch is a useful device for applying a force to a moving object. For example, winches are used for towing gliders into the air and as lifting devices on boats. The photograph below shows a cylinder with a rope wrapped round it, which is used to raise bags of flour in an old working mill.

The well bucket in the next example is raised and lowered using a similar device.

EXAMPLE 20.14 A bucket of mass m for drawing water from a well is attached by a light rope to a cylinder of mass M and radius r. The rope is wound round the cylinder using a light handle and the bucket is then allowed to fall freely from rest. What is its speed when it has fallen a height h
i)   when there is no resistance to its motion
ii) when there is a constant resistive couple of magnitude C?

SOLUTION

i)   Consider the energy of the cylinder and the bucket and let the zero level of potential energy be the initial position of the bucket. The tension T in the rope does no work because it is an internal force.

Note that the axle exerts an upwards force on the cylinder which balances T and the weight of the cylinder. These forces together form a *couple*; the effect is purely rotational.

Figure 20.34

Suppose the speed of the bucket is v when the angular speed of the cylinder is ω. While the cylinder rotates through an angle θ, the bucket falls a height x where

$$x = r\theta.$$

Differentiating $\quad \Rightarrow \quad v = r\omega.$

After falling a height h from rest, the total energy of the bucket is $\frac{1}{2}mv^2 - mgh$.

The cylinder also has a kinetic energy due to its rotation of

$$\frac{1}{2}I\omega^2 = \frac{1}{2}\left(\frac{1}{2}Mr^2\right)\omega^2$$
$$= \frac{1}{4}Mr^2\omega^2.$$

So the total energy of the bucket and cylinder when the bucket has fallen a height h is

$$\frac{1}{4}Mv^2 + \frac{1}{2}mv^2 - mgh.$$

The initial energy is zero so, by the principle of conservation of energy,

$$\frac{1}{4}Mv^2 + \frac{1}{2}mv^2 - mgh = 0$$

$$\Rightarrow \quad \frac{1}{4}(M + 2m)v^2 = mgh$$

$$\Rightarrow \quad v^2 = \sqrt{\frac{4mgh}{2m + M}}$$

$$\Rightarrow \quad v = \sqrt{\frac{4mgh}{2m + M}}.$$

ii) When there is a resistive couple C, the work done against the couple is

$$\int C \, d\theta = C\theta$$

$$= \frac{Ch}{r} \qquad (x = h = r\theta).$$

Then the gain in kinetic energy is the difference between the work done by gravity and the work done against the resistance

$$\Rightarrow \quad \frac{1}{2}mv^2 + \frac{1}{4}Mv^2 = mgh - \frac{Ch}{r}$$

$$\Rightarrow \quad \frac{1}{4}(2m + M)v^2 = \frac{(mgrh - Ch)}{r}$$

$$\Rightarrow \quad v^2 = \frac{4(mgr - C)h}{(2m + M)r}$$

$$\Rightarrow \quad v = \frac{4(mgr - C)h}{(2m + M)r}.$$

The next example illustrates the use of the work–energy principle when the moment of a force about the axis is not constant.

EXAMPLE 20.15   A shed door of mass 30 kg, width 0.76 m and height 1.8 m is standing open at 90° when a gust of wind hits it. The wind is initially perpendicular to the door and does not change its direction as the door shuts. The resultant force on the door due to the wind acts through its centre and is of magnitude 10 N per m² of door 'facing' the wind. A constant frictional couple of 2 Nm opposes the motion of the door.

i)   Assuming the door is a lamina, find its moment of inertia about its axis of rotation.

ii)   Find the moment of the force of the wind about the hinges when the door is open at an angle θ.

iii)   Use the work–energy principle to find the angular speed of the door when it shuts.

SOLUTION

i)   The moment of inertia is given by $\frac{4}{3}Ml^2$. In this case

$$2l = 0.76$$
$$\Rightarrow \quad l = 0.38$$
$$\Rightarrow \quad \text{moment of inertia} = \frac{4}{3} \times 30 \times (0.38)^2 \text{ kg m}^2$$
$$= 6.776 \text{ kg m}^2.$$

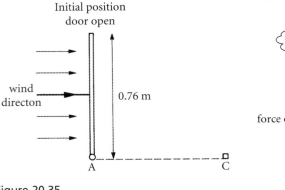

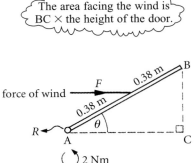

The area facing the wind is BC × the height of the door.

Figure 20.35

ii)   When the door is open at an angle θ, the area facing the wind is

$$0.76 \sin \theta \times 1.8 = 1.368 \sin \theta \text{ m}^2.$$

The force due to the wind is then $13.68 \sin \theta$ N. It acts through the centre, so its moment about the axis of rotation is

$$13.68 \sin \theta \times 0.38 \sin \theta = 5.1984 \sin^2 \theta \text{ Nm}.$$

iii)   There is a frictional couple of 2 Nm resisting the motion, so the resultant moment in the direction of increasing θ is $-(5.1984 \sin2\theta - 2)$ Nm.

The work in joules done in turning the door from $\theta = \dfrac{\pi}{2}$ radians to $\theta = 0$ is

$$\int_{\frac{\pi}{2}}^{0} -(5.1984 \sin^2 \theta - 2)\, d\theta = \int_{0}^{\frac{\pi}{2}} + (5.1984 \sin^2 \theta - 2)\, d\theta$$
$$= \int_{0} (5.1984 \times \tfrac{1}{2}(1 - \cos 2\theta) - 2)\, d\theta$$
$$= \left[2.5992\theta - 1.2996 \sin 2\theta - 2\theta \right]_{0}^{\frac{\pi}{2}}$$
$$= 0.5992 \times \frac{\pi}{2}$$
$$= 0.9412.$$

By the work–energy principle, this is equal to the gain in kinetic energy, $\frac{1}{2}I\dot{\theta}^2$.

From part **i)** the moment of inertia about the axis of rotation is 5.776 kg m².

$$\Rightarrow \qquad 0.9412 = \tfrac{1}{2} \times 5.776 \times \dot{\theta}^2$$

$$\Rightarrow \qquad \dot{\theta}^2 = 0.3259.$$

The angular speed is 0.571 rad s^{-1} (to 3 significant figures).

⚠ Remember angles must be in radians when using calculus.

20.9 The compound pendulum

A compound pendulum is any rigid body which oscillates freely about a fixed horizontal axis. No external forces or couples act on the pendulum apart from its weight and a supporting force at the axis. The next example shows that such a pendulum performs approximate simple harmonic motion when the oscillations are small.

EXAMPLE 20.16

A compound pendulum, of mass M and centre of mass G, is free to rotate about a fixed horizontal axis through a point A. AG is perpendicular to the axis of rotation and of length h. The moments of inertia of the pendulum about the fixed axis and a parallel axis through G are I_A and I_G respectively. Find expressions for

i) the period of small oscillations of the pendulum
ii) the length of a simple pendulum with the same period (called the *simple equivalent pendulum*)
iii) the period in terms of the radius of gyration, k, of the pendulum about the axis through G.

When such a pendulum has a period of 2 seconds, it is called a *seconds pendulum*.

iv) Show that, for a seconds pendulum, k cannot be greater than about $\frac{1}{2}$.
v) Find the value of h for which the period is a minimum.

SOLUTION

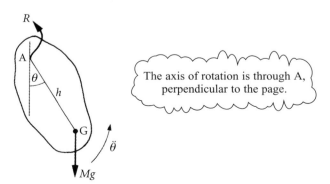

The axis of rotation is through A, perpendicular to the page.

Figure 20.36

i) When AG makes an angle θ with the downward vertical as shown, taking moments about the axis gives

$$Mg \sin \theta \times h = I_A (-\ddot{\theta})$$

> $\ddot{\theta} > 0$ in the direction of increasing θ

where I_A is the moment of inertia about the fixed axis.

$$\Rightarrow \quad \ddot{\theta} = -\frac{Mgh}{I_A} \sin \theta$$

Hence, for small θ, $\ddot{\theta} = -\frac{Mgh}{I_A} \theta$.

This represents simple harmonic motion with period $T = 2\pi \sqrt{\dfrac{I_A}{Mgh}}$.   ①

By the parallel axes theorem, $I_A = I_G + Mh^2$. Therefore, the period for small oscillations is $T = 2\pi \sqrt{\dfrac{I_G + Mh^2}{Mgh}}$.

ii) The period for small oscillations of a simple pendulum of length l is $2\pi \dfrac{l}{g}$ so the length of an equivalent simple pendulum is

$$l = \frac{I_G + Mh^2}{Mh} \quad \text{or} \quad l = \frac{I_A}{Mh} \quad \text{(from ①)}.$$

iii) The radius of gyration about the axis through G is k, so

$$I_G = Mk^2 \quad \Rightarrow \quad I_A = Mk^2 + Mh^2.$$

The period is then

$$T = 2\pi \sqrt{\frac{M(k^2 + h^2)}{Mgh}} \quad \text{(from ①)}$$

$$\Rightarrow \quad T = 2\pi \sqrt{\frac{k^2 + h^2}{gh}}$$

iv) For a given pendulum, the length h, and hence the position of A, can usually be calculated so that it has the correct period. A seconds pendulum has a period of 2 seconds.

In this case, $T = 2$ and squaring both sides of ② gives

$$4 = 4\pi^2 \frac{(k^2 + h^2)}{gh}$$

$$\Rightarrow \quad gh = \pi^2 (k^2 + h^2)$$

$$\Rightarrow \quad h^2 - \left(\frac{g}{\pi^2}\right)h + k^2 = 0.$$

This has real roots for h provided $\left(\dfrac{g}{\pi^2}\right)^2 \geqslant 4k^2$

$$\Rightarrow \quad k \leqslant \frac{1}{2}\left(\frac{g}{\pi^2}\right).$$

But $\dfrac{g}{\pi^2} = 0.994 \approx 1$, so k cannot be greater than about $\frac{1}{2}$.

v) Consider the function

$$f(h) = \frac{(k^2 + h^2)}{h} = \frac{k^2}{h} + h.$$

The period is least when $f(h)$ is a minimum. Differentiating gives

$$f'(h) = -k^2 h^2 + 1$$

so the minimum occurs when $k^2 = h^2$, that is $h = k$.

The period is least when AG is equal to the radius of gyration about the axis through G. (You can check that $f''(k) > 0$ to ensure a *minimum*.)

Deriving the compound pendulum equation using energy methods

It is sometimes easier to derive the equation of motion by differentiating the energy equation. This can be done in the compound pendulum problem as follows.

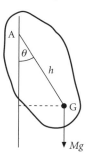

Figure 20.37

The kinetic energy is given by $\frac{1}{2}I_A\dot{\theta}^2$.

Taking A as the base level, the potential energy when the centre of mass is at the position shown in figure 20.37 is

$$-Mgh \cos \theta.$$

The total energy is constant so that

$$\tfrac{1}{2}I_A\dot{\theta}^2 - Mgh \cos \theta = \text{constant}.$$

Differentiating with respect to time gives

$$I_A\dot{\theta}\ddot{\theta} + Mgh \sin \theta \dot{\theta} = 0.$$

Note that $\dfrac{d}{dt}\dot{\theta}^2 = 2\dot{\theta}\dfrac{d\dot{\theta}}{dt}$

Dividing by $\dot{\theta}$ and rearranging gives

$$\ddot{\theta} = -\frac{Mgh}{I_A} \sin \theta$$

as derived previously.

The angular momentum of a rigid body rotating about a fixed axis

When Newton's first two laws are written in terms of the linear momentum of a particle they are equivalent to the following two statements.

- The linear momentum of a particle is constant if no resultant force acts on it.
- The resultant force acting on the particle is equal to the rate of change of its linear momentum.

But how do these laws apply to rotating rigid bodies? Think, for example, of the motion of a rotating space station which has escaped from all external forces. The direction of motion of each particle is continually changing but this does not mean that Newton's laws are contradicted. Although external forces are non-existent, all the particles in the station have forces acting upon them due to their interaction with each other. These internal forces are essential for parts of the space station to retain their rigid shape.

The momentum of a particle is a vector and so it is possible to find its *moment* about the axis of rotation. This moment of momentum is called the *angular momentum of the particle about the axis*. When a rigid body rotates about a fixed axis and the external forces have no resultant moment about the axis, the total angular momentum about the axis is constant. This is shown below.

Figure 20.38 shows a typical particle P of mass m_p rotating with angular speed $\dot\theta_p$ in a circle of radius r_p about a fixed axis. Its momentum is $m_p v_p = m_p r_p \dot\theta_p$ along the tangent so the moment of its momentum about the axis is

$$(m_p v_p)r_p = m_p r_p^2 \dot\theta_p.$$

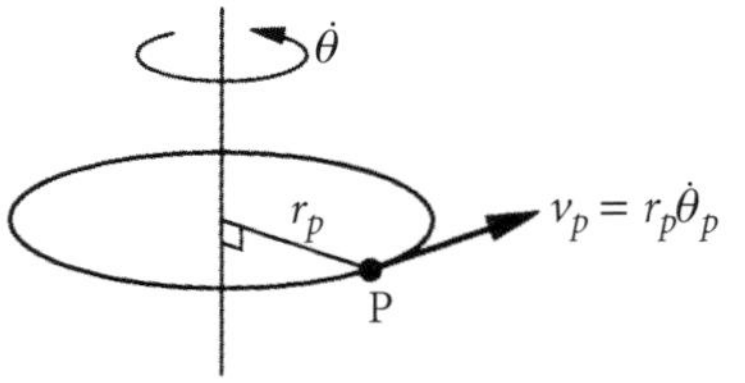

Figure 20.38

Summing for all particles gives the total angular momentum, L, about the axis, namely

$$L = \sum_{\text{all } p} m_p r_p^2 \dot\theta_p.$$

For a rigid body, all the particles have the same angular speed, $\dot\theta$, about the axis, so the expression for angular momentum becomes

$$L = I\dot\theta$$

where $I = \sum m_p r_p^2$ is the moment of inertia of the body about the axis. This expression for angular momentum is analogous to mv for linear motion, as you might expect.

When a rigid body rotates about a fixed axis, the conservation of angular momentum in the absence of external forces follows from the equation of motion

$$C = I\ddot\theta$$

where C is the total moment of forces about the axis. When there are no external forces $C = 0$,

$$\Rightarrow \qquad I\ddot{\theta} = 0$$
$$\Rightarrow \qquad\qquad\qquad \Rightarrow \qquad I\dot{\theta} = \text{constant.}$$

This is the equivalent of Newton's first law of motion. The second law has already been written in the form $C = I\ddot{\theta}$, but it too can be written in terms of the angular momentum of the body:

$$C = \frac{\mathrm{d}}{\mathrm{d}t}(I\dot{\theta}) = \frac{\mathrm{d}L}{\mathrm{d}t}.$$

This is an important result which can be stated as follows.

● For a rigid body rotating about a fixed axis, the resultant moment of forces about the axis of rotation is equal to the rate of change of angular momentum of the body about the axis.

Conservation of angular momentum

An important principle can be deduced from the previous section.

If there is no resultant moment of external forces acting on a body rotating about a fixed axis, its angular momentum remains constant.

It is conservation of angular momentum which explains why a spinning ice-skater draws in her arms to go faster and why a rotating interstellar gas cloud speeds up if it contracts under gravitational attraction.

EXAMPLE 20.17   A turntable which can be modelled as a disc of radius 0.2 m and mass 5 kg is spinning horizontally about an axis through its centre 3 times a second. An object of mass 2 kg is gently placed on it 0.15 m from the centre and remains in place through friction. How fast does the turntable now rotate?

SOLUTION

The angular momentum of the system as a whole is conserved.

The moment of inertia of the turntable is $\frac{1}{2} \times 5 \times 0.2^2 = 0.1$ kg m².

The turntable is rotating at $3 \times 2\pi = 6\pi$ rad s⁻¹.

So the initial angular momentum is $0.1 \times 6\pi = 0.6\pi$ (the object is initially at rest).

If the final angular velocity is ω, the angular momentum of the turntable is 0.1ω.

The speed of the object is 0.15ω so its angular momentum about the axis is

$$\text{mass} \times \text{tangential speed} \times \text{distance from axis} = 2 \times 0.15\omega \times 0.15$$
$$= 0.045\omega.$$

Hence the total final angular momentum is 0.145ω.

$$0.145\omega = 0.6\pi$$
$$\Rightarrow \qquad \omega = 4.14\pi \text{ rad s}^{-1}$$

which is 2.1 rotations per second.

To find the final angular momentum, L, you may prefer to work out the moment of inertia, I, of the turntable plus object and then simply apply $L = I\omega$. This will give the same result.

Note that, although the angular momentum has been conserved, some energy will have been lost. The object was accelerated from rest to its final velocity over a very short period during which some slight movement must have occurred and energy will have been lost overcoming friction.

The impulse of a couple about an axis

Integrating the equation for C gives

$$\int C \, dt = [I\dot{\theta}] = [L].$$

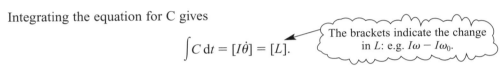
The brackets indicate the change in L: e.g. $I\omega - I\omega_0$.

This is the equivalent of $\int F \, dt = [mv]$ for linear motion. In that case $\int F \, dt$ is called an *impulse*. For rotation $\int C \, dt$ can be regarded as the impulse of a torque or a couple. This gives the following important result.

● The sum of the moments of impulses about the axis of rotation is equal to the change in angular momentum of the body about the same axis.

EXAMPLE 20.18

In a party game a flat board in the shape of a space ship is made to spin round a fixed vertical axis when hit by a small 'meteorite' of mass m. The board is always stationary and facing the player when a meteorite is thrown. Meteorites, which can stick to the board, are not removed between throws. The moment of inertia of the board about its axis is I.

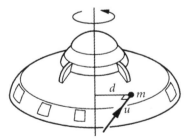

Figure 20.39

i) Assuming that a meteorite is thrown at the empty board and sticks after hitting it at right angles with speed u at a distance d from the axis, find the initial angular velocity of the board.

ii) The board slows down under the action of a constant frictional couple C and the winner in the game is the person who makes it spin for the longest time. Is this a fair game?

iii) Someone suggests that it would be easier to determine the winner by counting the number of revolutions made by the board before coming to rest. Would the game be fair in this case?

SOLUTION

i) During the short period of time taken by the impact, the impulse of the frictional couple is negligible, so the change in total angular momentum about the axis is negligible. In other words, total angular momentum is conserved. You can compare this with conservation of linear momentum when two objects collide.

The angular momentum of the meteorite about the axis just before it hits the board is $mu \times d$. The new moment of inertia of the meteorite and board about the axis after the collision is $I + md^2$, so the angular momentum after the collision is $(I + md^2)\omega$, where ω is the angular speed. Hence

$$(I + md^2)\omega = mud \quad \Rightarrow \quad \omega = \frac{mud}{(I + md^2)}. \qquad \text{①}$$

ii) The impulse of the constant frictional couple in time t is Ct and this is equal to the loss of angular momentum of the board and meteorite during that time. When t is the time for the board to stop rotating:

$$Ct = mud \quad \Rightarrow \quad t = \frac{mud}{C}.$$

This depends only on the speed of the meteorite and the position where it hits the board, and is independent of the moment of inertia of the board when it is hit. So it does not matter how many other meteorites are there already. Provided there is sufficient room on the board, the game is fair.

iii) Suppose that I_n is the moment of inertia of the board about its axis when there are n meteorites attached to it. After another hit this will become $I_n + md^2$. The initial angular speed after this additional meteorite hits the space ship is

$$\omega = \frac{mud}{(I_n + md^2)} \quad \text{(see ①)}.$$

When the board has turned through an angle θ, the work done against the couple is $C\theta$ and this is equal to the loss in kinetic energy. Hence

$$C\theta = \tfrac{1}{2}(I_n + md^2)\omega^2 \quad \Rightarrow \quad \theta = \frac{(mud)^2}{2C(I_n + md^2)}.$$

This is dependent on the value of I_n so it will be more difficult to make the board turn through a given angle as the game progresses. In this case, the game is not fair.

Chapter 20 Exercises

20.1 A drum majorette twirls a baton of mass 0.4 kg and length 0.7 m in a circle around its centre. She says that the end moves at speeds up to 30 mph. Assuming the baton is a thin rod, use the formula $\tfrac{1}{3}Ml^2$ to calculate its moment of inertia and find its kinetic energy in this case.
(Note 1 mile ≈ 1.6 km.)

20.2   A door of mass 35 kg and width 0.8 m is slammed shut with an angular speed of 2.5 rad s^{-1}.

i)   By modelling the door as a lamina and dividing it into vertical strips, or otherwise, find its moment of inertia about the axis through the hinges.

ii)   Find the amount of energy dissipated when the door shuts.

20.3   A potter is throwing clay on a wheel which turns at a constant rate. It starts as a solid cylinder of radius r and height h and gradually changes into a jar with the cross-section shown in the diagram.

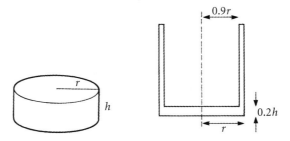

Assume the mass, M kg, and the density of the clay are constant.

i)   Find an expression for the density of the clay.

ii)   Find the height, in terms of h, of the jar when the thickness of the base is $0.2h$ and the inside radius is $0.9r$.

iii)   Assuming the jar is a solid cylinder with another removed from the inside, find its moment of inertia about its axis of rotation in terms of M and r.

iv)   Find the ratio of the kinetic energy of the jar to that of the original cylinder.

20.4   An odd number $(2n + 1)$ of beads each of mass m and negligible size are joined in a straight line by equal light rods of length a. The total mass of the model is M.

i)   Show that the moment of inertia I_M about an axis perpendicular to the line through the middle bead is $\frac{n}{3}(n + 1)Ma^2$.

$$\left(\text{Hint: } \sum_{r=1}^{n} r^2 = \frac{n}{6}(n + 1)(2n + 1) \right)$$

ii)   Use the parallel axes theorem to show that the moment of inertia I_E about the end bead is given by $I_E = I_M + n^2 Ma^2$.
Hence find I_E.

iii)   Verify that the answer to part ii) is the same as the result obtained from first principles.

iv)   Write down the length of the line of beads. The number of rods now increases and the length of each rod decreases in such a way that the total length is always $2l$. Show that in the limit as $n \rightarrow \infty$ and $a \rightarrow 0$, the moments of inertia you found in parts i) and ii) approach the equivalent moments of inertia for a rod of mass M and length $2l$.

20.5   A uniform thin rod AB of mass M and length $2a$ is free to rotate about an axis through its centre O which is inclined at an angle α to the rod AB as shown.

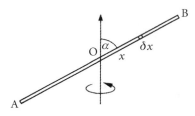

An element of the rod of length δx is situated a distance x from O. ρ is the mass per unit length of the rod.

i)   Write down the mass of the element.

ii) Write down the moment of inertia of the element about the axis.

iii) Form a suitable integral to calculate the moment of inertia of the rod about the axis and evaluate it.

iv) Write M in terms of a and ρ and hence show that the moment of inertia of the rod about the axis is the same as that of a rod of the same mass and of length $2a \sin \alpha$ which is perpendicular to the axis. Check your answer in the cases $\alpha = \dfrac{\pi}{2}$ and $\alpha = 0$.

v)   Use your result to find the moment of inertia of a parallelogram about an axis in its plane which passes through its centre and bisects a side of length $2a$, as shown in the diagram.

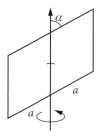

vi) How will the moment of inertia of any lamina be affected by a shearing in the direction of the axis of rotation as shown in the diagram?

shear parallel to axis

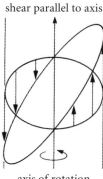

axis of rotation
in plane of lamina

20.6 The ellipse $\dfrac{x^2}{a^2} + \dfrac{y^2}{b^2} = 1$ is rotated completely about the x axis, forming a solid of revolution of mass M.

i) Find the moment of inertia of the solid about the x axis in terms of M.

ii) Write down a similar expression for the moment of inertia of the solid formed when the ellipse is rotated about the y axis.

iii) Comment on your results.

20.7 A point O on a plane lamina is used as the origin in a system of co-ordinates and the lamina is in the xy plane. A typical particle, P, of the lamina has mass m and co-ordinates $(x, y, 0)$. I_z and I_x are the moments of inertia of the lamina about the z axis and the x axis respectively.

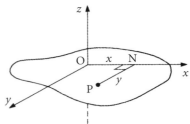

i) Explain why $I_z = \Sigma m(x^2 + y^2)$ summed over all particles and write I_x in terms of another sum.

ii) The lamina is now turned through an angle θ about the x axis.

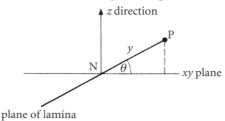

By considering the new moment of inertia of the typical particle, P, about the z axis, show that the moment of inertia about the z axis of the lamina in its new position is given by

$$\sum m(x^2 + y^2 \cos^2 \theta)$$

and so is equal to $I_z - I_x \sin^2 \theta$.

iii) A helicopter blade of mass m is modelled as a rectangle of length a and width b as shown in the photograph. One long edge, AB, is attached to the axle of the rotating mechanism and is a distance c from the axle. The blade can be turned about AB during flight.

Find expressions for the moment of inertia of the blade about the axle when AB is horizontal and the blade is

a) horizontal

b) inclined to the horizontal at an angle θ.

20.8 A solid circular cylinder of radius a and height h is formed by pouring a resin into a mould and allowing it to set. The resin settles unevenly so that the density of the resulting cylinder is 3ρ at the base and decreases uniformly to ρ at the top.

i) Write the density as a function of the distance from the base.

ii) Find the mass, M, of the cylinder.

iii) Find the position of the centre of mass of the cylinder.

iv) Find the moment of inertia of the cylinder about its axis of symmetry.

20.9 As part of an investigation into rotating interstellar dust clouds a simple mathematical model of such a cloud is proposed. The cloud is assumed to have a uniform density ρ and a shape defined by the volume produced by rotating the area between the curves

$$y = \pm be^{-\left(\frac{x}{a}\right)^2}, \quad -\infty < x < \infty,$$

about the y axis.

i) Sketch the above curves and shade in the appropriate area.

ii) By considering thin cylindrical shells of radius x and thickness δx about the y axis, find an expression for the mass M of the dust cloud in terms of ρ, a and b.

iii) Find an expression for the moment of inertia I of the cloud about its axis of symmetry in terms of M and a.

20.10 The flywheel of a car can be modelled as a disc of diameter 0.2 m and thickness 0.02 m from which another concentric disc of diameter 0.07 m has been removed. The density of the wheel is 7800 kg m^{-3}.

i) Find its moment of inertia about its axis.

The starter motor of the engine of the car makes the flywheel rotate at 50 rev s^{-1} in 3 seconds starting from rest. Find

ii) its angular acceleration (assumed constant)

iii) the average couple required to accelerate the flywheel

iv) the kinetic energy of the flywheel after 3 seconds.

20.11 A garden gate can be modelled as a rectangular lamina of width 0.9 m and mass 20 kg. It is kept shut by a spring mechanism which applies a couple equal to $4(1 + \theta)$ Nm when the gate is opened through an angle of θ radians. The gate is opened 1.5 radians and then allowed to shut naturally.

i) Find the moment of inertia of the gate about its hinges.

ii) Find the work done in opening the gate 1.5 radians.

iii) Assuming there is no loss of energy due to friction, find the angular speed of the gate just before it shuts.

20.12 A microwave oven has a turntable which can be modelled as a disc of mass 0.9 kg and radius 0.16 m rotating at $\dfrac{\pi}{6} \text{ rad s}^{-1}$.

i) The turntable has an angular retardation of 0.4 rad s^{-2}. Find the magnitude of the frictional couple acting.

A cylindrical cake of mass 0.6 kg and radius 0.09 m is placed centrally on the turntable and cooked in a light plastic container for 6 minutes at a power of 650 W.

ii) Assuming the same constant frictional couple, find

   a) the kinetic energy, A, given to the cake and turntable

   b) the work done, B, in keeping the turntable rotating at $\dfrac{\pi}{6}\,\sigma\,\text{rad s}^{-1}$

   c) the energy, C, required to cook the cake.

iii) What is the ratio of the energies $A : B : C$?

20.13 A pulley wheel is a disc of mass M and radius a, and can turn freely about a horizontal axis through its centre. Particles of mass M and $\frac{1}{2}M$ hang vertically over the pulley at the ends of a rough, light string. When the system is set in motion, the string does not slip over the pulley.

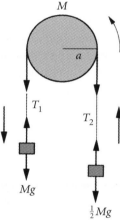

i) By considering the equations of motion of the pulley and the two particles, show that the acceleration of the particles is $\frac{1}{4}g$.

ii) Find the difference in tension between the two portions of the string.

20.14 A cricket bat of mass M kg is pivoted about a horizontal axis through a point A on the handle as shown. The axis is perpendicular to the page, G is the centre of mass and AG $= a$ m.

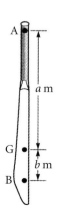

The period of small oscillations about the axis is found to be T seconds.

Show that the moment of inertia of the bat about the axis is $I_A = \dfrac{MgaT^2}{4\pi^2}$.

In order that there should be no impulsive reaction at A when the ball is hit, the impact must be at the 'sweet spot' B given by

$$b = \frac{aI_G}{I_A - I_G}. \text{ Show that } a + b = \frac{gT^2}{4\pi^2}.$$

20.15 Two gear wheels are such that, when they are engaged, their angular speeds are inversely proportional to their radii. One has a radius a and moment of inertia pa^2 about its axis of rotation. The other gear wheel has radius b and moment of inertia qb^2. The first is rotating with angular speed ω when it engages with the second which is initially at rest.

   i)   By considering the change in angular momentum of each wheel separately, find the impulse between the teeth of the gear wheels when they engage.

   ii) Find the angular speed of each wheel.

   iii) Why is the angular momentum not conserved?

   iv) Find the energy lost when the gears engage.

20.16 This question describes a simplified model of a device used to de-spin a satellite. A uniform circular disc of mass $12m$ and radius a lies on a smooth horizontal table and is free to rotate about a fixed vertical axis through its centre. A light wire is attached to a point on the rim of the disc and is wound round this rim. A particle of mass m is attached to the free end of the wire and is initially attached to the rim.

When the disc is rotating with angular speed ω in the opposite sense to that in which the wire is wound the particle is released so that the wire unwinds and remains taut. The length of the wire is chosen so that it is completely unwound at the instant that the disc stops rotating. The particle is then moving at right angles to the wire.

Use the principles of conservation of angular momentum and energy to find the length of the wire.

Personal Tutor

20.1 The racket in the diagram has mass M and is free to rotate about a fixed smooth axis A which is paralel to the plane of the racket and perpendicular to its handle. The moment of inertia of the racket about a parallel axis through its centre of mass, G, is Mk^2.

When the racket is at rest, it is hit at the point P with an impulse J_1 perpendicular to its plane and to the axis of rotation. AG $= h$ and AP $= d$.

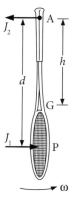

   i)   Find the value of d which ensures that the impulsive reaction, J_2 at A is zero.

   ii) Show that this value of d is equal to the length of a simple pendulum which would perform small oscillations about the axis through A with the same period as the racket.

20.1   The moment of inertia of a body about an axis is $I = \sum Mr^2$ and depends on the position and direction of the axis, which should always be stated.

20.2   Perpendicular axes theorem: $I_z = I_x + I_y$ for a lamina only.

20.3   Parallel axes theorem:
$$I_A = I_G + Md^2$$
$$= I_G + M(AG)^2 \quad \text{for AG perpendicular to the axes.}$$

20.4   To find a moment of inertia by integration follow these steps.
- Divide the body into elements for which the moment of inertia is known.
- Find the mass of a typical element in terms of the density.
- Determine the moment of inertia of the element about the axis.
- Sum for all elements and write the sum as an integral.
- Evaluate the integral.
- Write the density in terms of the total mass and replace it.

20.5   When the density varies, it should be written in terms of a suitable variable and included in the integration.

20.6   The angular speed of a rigid body about an axis is θ, where θ is the angle between a fixed line in the body and a fixed direction in space (both perpendicular to the axis of rotation).

20.7   The kinetic energy of a body rotating about a fixed axis with angular speed θ and moment of inertia I is $\frac{1}{2}I\dot{\theta}^2$ (or $\frac{1}{2}I\omega^2$).

20.8   Angular momentum: $L = I\dot{\theta}$

20.9   Moment of force about axis: $C = \dfrac{\mathrm{d}L}{\mathrm{d}t} = I\ddot{\theta}$

20.10   When there is no resultant moment about the axis, angular momentum is conserved.

20.11   The sum of moments of impulses about an axis is the change in the angular momentum about the same axis.

20.12   Mechanical energy is conserved if no external forces other than gravity do work.

Equivalent quantities

Rotation of a rigid body about a fixed axis		Linear motion	
Moment of inertia about axis:	$I = \sum mr^2$	Mass:	m
Moment of force about axis:	C	Force:	F
Angular displacement:	θ	Displacement:	x or s
Angular velocity:	$\dot{\theta}$ or ω	Velocity:	$\dot{x}$ or v
Angular momentum:	$L = I\dot{\theta}$	Momentum:	$m\dot{x}$
Equation of motion:	$C = I\ddot{\theta} = \dfrac{\mathrm{d}L}{\mathrm{d}t}$	$F = m\ddot{x} = \dfrac{\mathrm{d}(m\dot{x})}{\mathrm{d}t}$	
Kinetic energy:	$\frac{1}{2}I\dot{\theta}^2$ or $\frac{1}{2}I\omega^2$	$\frac{1}{2}m\dot{x}^2$ or $\frac{1}{2}mv^2$	
Work:	$\int C\,\mathrm{d}\theta$	$\int F\,\mathrm{d}x$ or $\int F\,\mathrm{d}s$	
Impulse:	$\int C\,\mathrm{d}t$	$\int F\,\mathrm{d}t$	

21 Stability and small oscillations

A government so situated is in the condition called in mechanics 'unstable equilibrium' like a thing balancing on its smaller end.

John Stuart Mill

What happens when each of these objects is slightly displaced?

The train *can* balance at the high point of a roller coaster but a slight disturbance causes it to move rapidly away from its equilibrium position. The train is said to be in *unstable equilibrium*. Compare this with the swing boat. When this is disturbed, the resulting net forces will tend to restore it to its equilibrium position. The swing is in *stable equilibrium*.

You would like a garage door to be in equilibrium in any position so that a breath of wind does not bring it crashing on to your car. It is safest when the forces acting are always balanced so that it is in *neutral equilibrium*.

Think of other objects which can be placed in positions of stable or unstable equilibrium. What happens to the centre of mass when each is displaced by a small amount?

When you balance a coin on its edge, its centre of mass is at its highest point and displacing the coin lowers the centre of mass. The potential energy is at a maximum in the upright position. Conversely, the potential energy of a swing is at a minimum when it is hanging in equilibrium.

In this chapter energy principles are used to find positions of equilibrium and determine their nature and also to analyse the oscillations about positions of stable equilibrium which are caused by small displacements. It is assumed that all forces are conservative so that total energy is conserved. This means that forces such as friction are assumed to be negligible.

You might be surprised to learn that energy methods often provide an easier way to find equilibrium positions than analysing resultant forces and taking moments.

21.1 Potential energy at equilibrium

Consider a body modelled as a particle which is able to move in one dimension, such as a ball hanging on the end of an elastic string. Assume that its displacement from a fixed position is denoted by x and also that all forces acting are conservative, so that the energy equation applies.

$$\text{kinetic energy} + \text{potential energy} = \text{constant}$$
$$\tfrac{1}{2}mv^2 + V(x) = E$$

where the potential energy $V(x)$ is the total of *all* types of energy due to position x. In the case of the ball this is the total of gravitational and elastic energy. It is often denoted by $V(x)$ to indicate that it is simply a function of the position of the body, but abbreviated to V in equations. Differentiating the energy equation with respect to x gives

$$mv\frac{dv}{dx} + \frac{dV}{dx} = 0.$$

Now $mv\dfrac{dv}{dx} = ma = F$, by Newton's second law where F is the resultant force acting on the particle. So

$$F + \frac{dV}{dx} = 0$$

$$\Rightarrow \qquad F = \frac{dV}{dx}.$$

But a particle is in equilibrium if and only if the resultant force is zero. Hence $\dfrac{dV}{dx} = 0$ in an equilibrium position.

The potential energy has a stationary value if and only if the body is in an equilibrium position.

Sometimes the position of a body is specified not by its displacement from a given position but by some other variable. For example, the position of a pendulum is usually expressed by giving the angle θ it makes with the vertical (see figure 21.1).

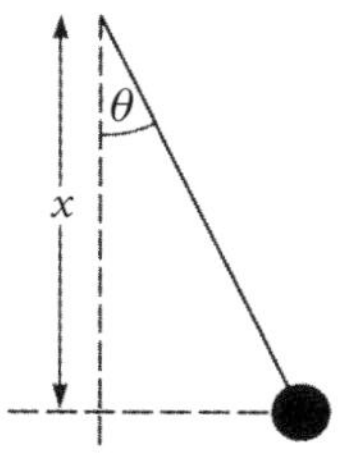

Then the derivative with respect to *this* variable is still zero at equilibrium, since by the chain rule

$$\frac{dV}{d\theta} = \frac{dV}{dx}\frac{dx}{d\theta} = 0 \quad \text{when} \quad \frac{dV}{dx} = 0.$$

Figure 21.1

This argument can be applied not just to a particle but to a more general system, provided that it is constrained to move in such a way that its position, and therefore its potential energy, can be specified by the value of a single variable: normally an angle or a displacement. The conclusion is

> when a body acted on by conservative forces is free to move so that its potential energy can be given as a function $V(x)$ of a single variable x, its equilibrium positions are given by the stationary values of V with respect to this variable.

EXAMPLE 21.1

One end of an elastic string of natural length l and modulus λ is fixed to the ceiling and a ball of mass m is attached to the other end so that the string hangs vertically. Show, *using energy methods*, that in equilibrium the string is stretched an amount $\dfrac{mgl}{\lambda}$.

SOLUTION

In figure 21.2, OA represents the position of the unstretched string. Consider the potential energy of the system when the mass is displaced x below A. The elastic energy is $\dfrac{\lambda x^2}{2l}$. The gravitational energy, relative to the level of point A, is $-mgx$: negative because the ball is below A. So the total potential energy is

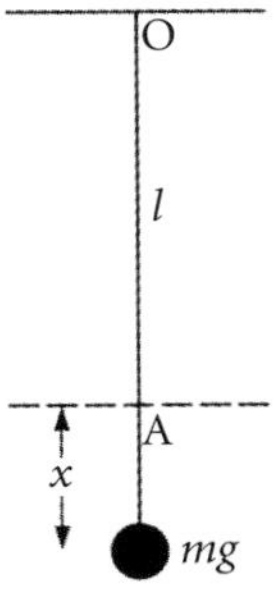

$$V = \frac{\lambda x^2}{2l} - mgx$$

$$\Rightarrow \quad \frac{dV}{dx} = \frac{lx}{l} - mg.$$

When

$$\frac{dV}{dx} = 0, \frac{\lambda x}{l} - mg = 0$$

$$\Rightarrow \quad x = \frac{mgl}{\lambda}. \qquad \qquad \textcircled{1}$$

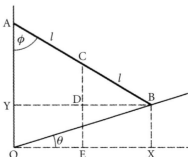

Figure 21.2

Note

The choice of the horizontal through A as the zero level of gravitational potential energy is arbitrary. Any fixed point will do. Changing V by a constant does not affect the result when $\dfrac{dV}{dx}$ is found.

This familiar example shows the energy method works, even though in this case simply balancing the forces $\left(mg = T = \dfrac{\lambda x}{l} \right)$ gives equation ① immediately and is rather easier. The following examples show that energy methods can often be a quicker way of locating equilibrium positions than considering forces.

EXAMPLE 21.2

Figure 21.3 shows a uniform ladder AB of mass m and length $2l$ resting in equilibrium with its upper end A against a smooth vertical wall and its lower end B on a smooth inclined plane. The inclined plane makes an angle θ with the horizontal and the ladder makes an angle ϕ with the wall.

What is the relationship between θ and ϕ?

Figure 21.3   O

SOLUTION

There is no friction at A and B, so the reaction forces do no work when the ladder moves. The equilibrium position can thus be found using the energy method. The only contribution to the potential energy of the system is the gravitational force, which may be considered to act at the centre of the ladder, C. Note that ϕ varies as the ladder moves, but θ, the inclination of the plane to the ground, is constant. The energy can be expressed in terms of a single variable, ϕ.

Take the ground level OX as the zero level of gravitational energy. The height of C above this level is

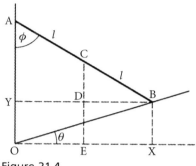

$$CD = CD + DE = CD + BX$$
$$= l \cos \phi + OX \tan \theta$$
$$= l \cos \phi + 2l \sin \phi \tan \theta$$

$(OX = YB = 2l \sin\phi)$.

Hence the potential energy is

$$V = mg(l \cos \phi + 2l \sin \phi \tan \theta)$$

$$\frac{dV}{d\phi} = mgl(-\sin \phi + 2 \cos \phi \tan \theta).$$

Figure 21.4

Equilibrium occurs when $\dfrac{dV}{d\phi} = 0$, that is when

$$-\sin \phi + 2 \cos \phi \tan \theta = 0.$$

Hence $\tan \phi = 2 \tan \theta$ at equilibrium.

 Any fixed level can be taken as the reference for zero energy; OX is the most convenient here. It would be wrong to choose the zero energy level through A or B because A and B are not fixed.

The next example shows a situation with more than one equilibrium position.

EXAMPLE 21.3 A uniform rod OQ of mass m and length a is smoothly jointed to a fixed point at O, so that it can rotate in a vertical plane. P is a fixed point vertically above O so that OP = a. The ends of an elastic string of length a and modulus $2mg$ are connected to P and Q.

i) Find the potential energy of the system when the rod makes an angle θ with the upward vertical, assuming the string is stretched.

   Hence find positions of equilibrium of the rod.

ii) Are there any other equilibrium positions when the string is unstretched?

(You may assume that the string is attached in such a way that the rod can rotate completely in the vertical plane without being caught up by the string.)

Solution

i) Figure 21.5 shows the position when the rod makes a general angle θ clockwise from the upward vertical with $0 \leqslant \theta \leqslant \pi$. The string is just taut when triangle OPQ is equilateral, that is $\theta = \dfrac{\pi}{3}$.

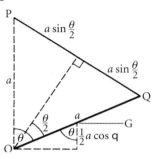

Figure 21.5

Consider the rod when in the general position *with the string taut*, that is $\theta > \dfrac{\pi}{3}$. Then

$$PQ = 2a \sin \frac{\theta}{2}$$

so the extension is $x = 2a \sin \dfrac{\theta}{2} - a$. The elastic energy is then

$$\frac{\lambda x^2}{2a} = \frac{2mg\left(2a \sin \dfrac{\theta}{2} - a\right)^2}{2a}$$

$$= mga\left(2 \sin \frac{\theta}{2} - 1\right)^2.$$

The centre of mass of the rod is at the mid-point G, so the gravitational potential energy, referred to zero level through O, is $mg \times$ height of G above O, that is $mg(\tfrac{1}{2}a \cos \theta)$. This remains correct when $\theta > \dfrac{\pi}{2}$, as $\cos \theta$ is then negative *and* G is *below* O.

The total potential energy is

$$V = mga\left(\frac{1}{2} \cos \theta\right) + mga\left(2 \sin \frac{\theta}{2} - 1\right)^2.$$

Differentiating to find the equilibrium points gives

$$\frac{dV}{d\theta} = -mga\left(\frac{1}{2} \sin \theta\right) + 2mga\left(2 \sin \frac{\theta}{2} - 1\right) \cos \frac{\theta}{2}$$

$$= -mga\left(\sin \frac{\theta}{2} \cos \frac{\theta}{2}\right) + 2mga\left(2 \sin \frac{\theta}{2} - 1\right) \cos \frac{\theta}{2}$$

$$= mga \cos \frac{\theta}{2}\left(-\sin \frac{\theta}{2} + 4 \sin \frac{\theta}{2} - 2\right)$$

$$= mga \cos \frac{\theta}{2}\left(3 \sin \frac{\theta}{2} - 2\right).$$

Equilibrium points are given when $\dfrac{dV}{d\theta} = 0$, that is when

$$\cos\frac{\theta}{2} = 0 \quad \text{or} \quad \sin\frac{\theta}{2} = \frac{2}{3}$$

$$\Rightarrow \quad \frac{\theta}{2} = \frac{\pi}{2} \quad \text{or} \quad \frac{\theta}{2} = \arcsin\frac{2}{3} = 0.730 \text{ (3 significant figures)} \quad \text{or} \quad \pi - 0.730$$

$$\Rightarrow \quad \theta = \pi \quad \text{or} \quad \theta = 1.46 \quad \text{(or } 2\pi - 1.46 \text{ giving the position left of OP).}$$

The three equilibrium positions with the string taut, are shown in Figure 21.6.

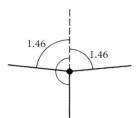

Figure 21.6

ii) It remains to look at the case when the string is *unstretched*, $u < \dfrac{\pi}{3}$. In this case, the equilibrium position is the one where the rod is balanced vertically upwards. The energy method confirms this: only the gravitational term $mga\left(\frac{1}{2}\cos\theta\right)$ contributes to the potential energy and this has a stationary value at $\theta = 0$.

 Remember when using calculus that the angle θ must be in radians.

QUESTION 21.1 Consider a bead threaded on a smooth wire bent as shown in figure 21.7 and held in a vertical plane.

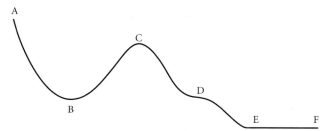

Figure 21.7

At which points on the curve can the bead be in equilibrium?

Think about what happens when the bead is slightly displaced from each of these points.

Which are positions of stable equilibrium?

QUESTION 21.2 In Example 21.3 there are three equilibrium positions: $\theta = 0$, π and 1.46 (ignoring the symmetrical case). But not all of these are stable.

Which do you think are stable and which unstable?

The potential energy can be written

$$V = mga\left[\tfrac{1}{2}\cos\theta + \left(2\sin\frac{\theta}{2} - 1\right)^2\right] \quad \text{when } \frac{\pi}{3} \leqslant \theta \leqslant \pi$$

$$V = mga(\tfrac{1}{2}\cos\theta) \qquad\qquad\qquad \text{when } 0 < \theta \leqslant \frac{\pi}{3} \text{ (string not taut).}$$

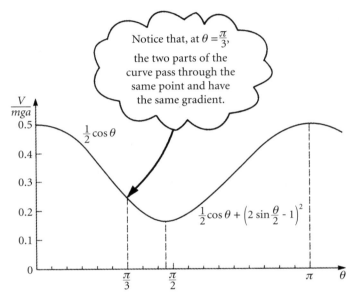

Notice that, at $\theta = \frac{\pi}{3}$, the two parts of the curve pass through the same point and have the same gradient.

Figure 21.8

The graph in figure 21.8 shows how the potential energy varies with θ.

Think about what happens when the bead is slightly displaced from each of these points.

Can you infer any relationship between the type of stationary point of a potential energy curve and the stability of the equilibrium position?

Condition for stability of equilibrium

Suppose that the potential energy V has a *minimum*, V_0, at the equilibrium position x_0. Displacing the body from its equilibrium position is equivalent to supplying a small amount of kinetic energy K_0 in the form of a velocity taking it away from the equilibrium position. Assume this is in the direction of increasing x (see figure 21.9).

During the subsequent motion

$$V + K = \text{constant } E = V_0 + K_0$$

where K denotes the kinetic energy.

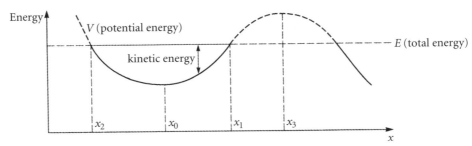

Figure 21.9

The graph shows the potential energy $V(x)$ and total energy E of the body at points near x_0. The difference $E - V(x)$ is the kinetic energy. You can see that at the point x_1, the kinetic energy is zero. It cannot become negative so the body cannot move past x_1 (dotted region of curve). Furthermore, since x_1 is *not* a position of equilibrium, the body begins to return to its equilibrium position. When it returns to x_0, it will have some velocity so will continue to a point x_2 and oscillate about the position of equilibrium. This is what is meant by a position of *stable equilibrium*: the body returns after a sm displacement.

Note, if K_0 is greater, leading to a *large* displacement past the point x_3, thi the body into a new energy region and it would not return to x_0.

Activity

21.1 Show that when V is a maximum at the equilib given a small displacement, the kinetic energy further from the equilibrium point. Draw figure 21.9 (remember that $E \geqslant V(x)$). Show th unstable.

0

Summary of stability criter

The results can be summarised as follows.

● The potential energy is a minimum $\Leftrightarrow$ a stable on a horizontal
● The potential energy is a maximum $\Leftrightarrow$ an u any force. This is
   $V'(x) = 0$, $V''(x) < 0$ graph has a point of

QUESTION 21.3   There are other types of equilibrium position. stable or unstable?
surface. $V = $ constant, $\dfrac{dV}{dx} = 0$ and a displa

known as neutral equilibrium. Suppose
inflexion. What could this represent? Wou

EXAMPLE 21.4

The potential energy of a particle is given by the equation $V = \dfrac{Ae^{kx}}{x}$.

Show that it has one position of equilibrium and that this is stable.

SOLUTION

$$\frac{dV}{dx} = \frac{A(xke^{kx} - e^{kx})}{x^2}$$

$$= \frac{Ae^{kx}}{x^2}(kx - 1)$$

$e^{kx} \neq 0$, so $\dfrac{dV}{dx} = 0$ when $kx - 1 = 0$, i.e. when $x = \dfrac{1}{k}$.

When x is just less than $\dfrac{1}{k}$, $\qquad \dfrac{dV}{dx} < 0$.

When x is just greater than $\dfrac{1}{k}$, $\qquad \dfrac{dV}{dx} > 0$.

> Sometimes this method of determining the nature of a stationary point is easier than differentiating twice.

So $x = \dfrac{1}{k}$ gives minimum potential energy and the equilibrium is stable.

EXAMPLE 21.5

A smooth circular hoop of radius a is fixed in a vertical plane. A small smooth ring of mass m is threaded on the hoop and is joined to the highest point of the hoop by a light elastic spring of natural length d and stiffness k, where $ka^2 - mga > 0$. At time t, the angle between the spring and the downward vertical is θ where $-\dfrac{\pi}{2} < \theta < \dfrac{\pi}{2}$.

i) Show that the potential energy, V, relative to the highest point of the hoop, can be written in the form

$$V = 2(ka^2 - mga)\cos^2\theta - 2kad\cos\theta + \tfrac{1}{2}kd^2.$$

ii) Given that $d < 2\left(a - \dfrac{mg}{k}\right)$, show that there are three positions of equilibrium. Discuss the stability of each position.

iii) In the case when $d > 2\left(a - \dfrac{mg}{k}\right)$, show that there is just one equilibrium position and that this position is stable.

Discuss briefly the case when $d = 2\left(a - \dfrac{mg}{k}\right)$.

the spring has extension x

$-mg(d + x)\cos\theta + \tfrac{1}{2}kx^2$

$mg \times 2a\cos\theta\cos\theta + \tfrac{1}{2}k(2a\cos\theta - d)^2$

$^2 - mga)\cos^2\theta - 2kad\cos\theta + \tfrac{1}{2}kd^2.$

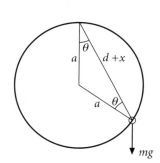

Figure 21.10

ii) $$V'(\theta) = -4(ka^2 - mga)\cos\theta\sin\theta + 2kad\sin\theta$$

$$= 2a\sin\theta[kd - 2(ka - mg)\cos\theta]$$

$$= 4a(ka - mg)\sin\theta\left[\frac{kd}{2(ka - mg)} - \cos\theta\right] \qquad ①$$

Let $\dfrac{kd}{2(ka - mg)} = c.$

> $c > 0$ as $ka^2 > mga$

When $d < 2\left(a - \dfrac{mg}{k}\right)$, $c < 1$, so there are three solutions to $V'(\theta) = 0$ (when $\sin\theta = 0$

or $\cos\theta = c$), giving three positions of equilibrium.

These are $\theta = 0$ and $\theta = \pm\alpha$, where $\cos\alpha = c$.

> differentiate
> ①

$$V''(\theta) = 4a(ka - mg)\cos\theta\,(c - \cos\theta) + 4a(ka - mg)\sin^2\theta$$

> maximum
> P.E.

$$\Rightarrow \qquad V''(0) = 4a(ka - mg)\,(c - 1) < 0 \qquad \text{so } \theta = 0 \text{ is unstable}$$

$$V''(\pm\alpha) = 4a(ka - mg)\sin^2\alpha > 0 \qquad \text{so } \theta = \pm\alpha \text{ are stable}$$

> min⋅

iii) When $d > 2\left(a - \dfrac{mg}{k}\right)$, $c > 1$ so $\cos\theta = c$ is impossible.

There is just one position of equilibrium when $\theta = 0$.

$V''(0) = 4a(ka - mg)\,(c - 1) > 0$ so this position is stabl⎺

iv) When $d = 2\left(a - \dfrac{mg}{k}\right)$, $c = 1$ so $\alpha = 0$ and there is o⎺

Now $V''(0) = 0$, so investigate signs of $V'(\theta)$ near ⎼

$$V'(\theta) = 4a(ka - mg)\sin\theta\,(1 - \cos\theta)$$

$$= 4a(ka - mg)\sin\theta \times 2\sin^2\frac{\theta}{2}$$

This changes from negative to positive as θ in⎺
$\theta = 0$ gives a minimum for $V(\theta)$ and hence a p⎼ when

⎼ this

Note

Potential energy as a function of ⎼ds to systems

The discussion of stationary values of potential ener⎼ variables.
V can be expressed as a function of a *single* vario⎼inciple in use in such
case, the system is said to have 'one degree of fr⎼ ⎼al energy can be
equilibrium positions coincide with stationary val⎼ ⎼ave to be considered
with more than one degree of freedom, i.e. whe⎼
Investigation 21.1 on the accompanying CD de⎼
a case. Although the problem has two degrees ⎼
expressed as a function of two variables), this ⎼
in the investigation.

21.2 Small oscillations about stable equilibrium

When a system is displaced from a position of stable equilibrium it usually oscillates about this position. You have already met systems like this such as the spring–mass oscillator and simple pendulum, in which small oscillations have been modelled by simple harmonic motion. It is possible to use the energy equation to obtain an approximation to simple harmonic motion in the more complex systems you have met in this chapter.

In general the energy equation is:

$$\tfrac{1}{2}m\dot{x}^2 + V(x) = \text{constant}.$$

Differentiating this with respect to t gives:

$$m\dot{x}\ddot{x} + V'(x)\dot{x} = 0$$

$$\Rightarrow \qquad \ddot{x} = -\frac{1}{m}V'(x). \qquad\qquad ①$$

When $V'(x)$ is a linear function of x, this gives the simple harmonic motion equation directly, but see the comment on page 400.

The potential energy of a ball of mass m hanging on the end of an elastic string of stiffness k and with extension x, for example, can be written as

$$V(x) = \tfrac{1}{2}kx^2 - mgx$$

$$\Rightarrow \qquad V'(x) = kx - mg.$$

Substituting in ① gives

$$\ddot{x} = -\frac{1}{m}(kx - mg).$$

The equilibrium position is when $x = \dfrac{mg}{k} = x_0$ so

$$\ddot{x} = -\frac{k}{m}(x - x_0)$$

which is simple harmonic motion about $x = x_0$.

Setting $y = x - x_0$ gives the standard equation

$$\ddot{y} = -\omega^2 y$$

$$\omega^2 = \frac{k}{m}.$$

When the potential energy is a function of an angle, θ, the kinetic energy can usually be written in the form $K\dot{\theta}^2$ and you can differentiate the energy equation with respect to t to get a similar result with x replaced by θ.

It is often the case, however, that $V'(x)$ or $V'(\theta)$ is not a linear function of x or θ and then you need to find a linear approximation for it. You have met an example of this in the case of the simple pendulum, where $\sin\theta$ is written as θ when θ is small. You can often make this approximation using series expansions or trigonometrical methods, but the technique of using the second derivative of the potential energy at the equilibrium position is simpler.

Figure 21.11 shows the graph of V'(x).

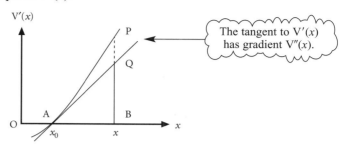

Figure 21.11

The gradient of $V'(x)$ at x_0 is

$$V''(x_0) = \frac{QB}{AB} = \frac{QB}{x - x_0}.$$

When $(x - x_0)$ is small, $QB \approx PB = V'(x)$

$$\Rightarrow \qquad V''(x_0) \approx \frac{V'(x)}{x - x_0}.$$

$$\Rightarrow \qquad V'(x) \approx (x - x_0)V''(x_0)$$

This is the linear approximation for $V'(x)$ in terms of x.

The simple harmonic motion approximation for small oscillations about the equilibrium position x_0 can then be found using equation ①.

$$m\ddot{x} = -V'(x)$$

so $\qquad m\ddot{x} = -(x - x_0)V''(x_0)$

$$\Rightarrow \qquad \ddot{x} = -\frac{V''(x_0)}{m}(x - x_0).$$

Replacing $(x - x_0)$ by y gives

$$\ddot{y} = -\omega^2 y$$

with $\qquad \omega^2 = \frac{V''(x_0)}{m}.$

Notice that this motion is simple harmonic motion only if $V''(x_0)$ is positive, which is the condition for stable equilibrium.

EXAMPLE 21.6

Figure 21.12 shows a spring controlled flap. The flap can be modelled as a uniform rectangular lamina of length AB = $2l$, smoothly hinged about an axis through B. The spring BD is horizontal and is connected by a cord passing over a smooth pulley C and attached to the flap at A. BC = $2l$. When the flap is horizontal (A is at C), the spring has its natural length l. The modulus λ of the spring is $\frac{1}{4}mg$.

i) Write down the potential energy of the system when the flap makes an angle θ with the horizontal.

Hence show that $\theta = \dfrac{\pi}{4}$ is an equilibrium position.

ii) Differentiate the energy equation to give the equation of motion and express this in terms of $\phi = \theta - \dfrac{\pi}{4}$.

Hence show that the period of small oscillations about the equilibrium position is $\dfrac{2\pi}{\omega}$, where $\omega^2 = \dfrac{3\sqrt{2}g}{4l}$.

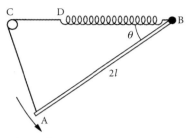

Figure 21.12

SOLUTION

i) When the flap makes an angle θ with the horizontal, the spring is extended from its natural length by $CA = 4l \sin\left(\dfrac{\theta}{2}\right)$.

Elastic P.E. $= \dfrac{\lambda}{2l}\left(4l\sin\dfrac{\theta}{2}\right)^2 = 2mgl\sin^2\dfrac{\theta}{2}$ (since $\lambda = \tfrac{1}{4}mg$).

Gravitational P.E. $= -mgl\sin\theta$ when BC is the zero level.

Total P.E. $= V = 2mgl\sin^2\dfrac{\theta}{2} - mgl\sin\theta$.

Differentiate to find the equilibrium position:

$$\dfrac{dV}{d\theta} = 2mgl\sin\dfrac{\theta}{2}\cos\dfrac{\theta}{2} - mgl\cos\theta$$

$$= mgl\sin\theta - mgl\cos\theta. \qquad ①$$

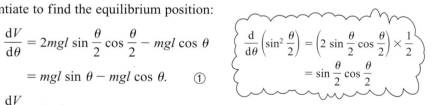

When $\dfrac{dV}{d\theta} = 0$, $\sin\theta = \cos\theta$

$\Rightarrow \qquad \theta = \dfrac{\pi}{4}$.

ii) When the flap is moving, P.E. + K.E. is constant.

The moment of inertia of the flap, rotating about a pivot through one end is $\tfrac{4}{3}ml^2$. So the kinetic energy of the flap is

$$\tfrac{1}{2}\left(\tfrac{4}{3}ml^2\right)\dot{\theta}^2 = \tfrac{2}{3}ml^2\dot{\theta}^2.$$

Therefore

$$\tfrac{2}{3}ml^2\dot{\theta}^2 + V = \text{constant}.$$

Differentiate with respect to t to give the equation of motion

$$\tfrac{2}{3}ml^2 \times 2\dot\theta\ddot\theta + \frac{dV}{d\theta}\dot\theta = 0$$

$$\Rightarrow \quad \left(\frac{4ml^2}{3}\right)\ddot\theta = -V'(\theta) \qquad\qquad ②$$

$$= -mgl\,(\sin\theta - \cos\theta)$$

$$V''(\theta) = +mgl\,(\cos\theta - \sin\theta)$$

$$\Rightarrow \quad V''\!\left(\frac{\pi}{4}\right) = \sqrt{2}\,mgl.$$

Using the linear approximation $V'(\theta) = \left(\theta - \dfrac{\pi}{4}\right)V''\!\left(\dfrac{\pi}{4}\right)$ in equation ② gives

$$\left(\frac{4ml^2}{3}\right)\ddot\theta \approx -\left(\theta - \frac{\pi}{4}\right) \times \sqrt{2}\,mgl$$

$$\Rightarrow \quad \ddot\theta = \frac{-3\sqrt{2}}{4l}g\left(\theta - \frac{\pi}{4}\right)$$

which is simple harmonic motion about $\theta = \dfrac{\pi}{4}$ with period $\dfrac{2\pi}{\omega}$ where $\omega^2 = \dfrac{3\sqrt{2}}{4l}g$.

Note

You can also use trigonometry to solve this problem, by replacing θ with $\dfrac{\pi}{4} + \phi$ and writing $\sin\phi \approx \phi$.

Chapter 21 Exercises

21.1 A spring of modulus λ and natural length a is attached to a point P on the ceiling and an identical spring is attached to a point Q on the floor a distance $2a$ vertically below P. The other end of each spring is attached to a ball of mass m, so that it is free to oscillate in a vertical line.

   i) Write down the gravitational potential energy and the elastic energy when the ball is displaced x below O, the mid-point of PQ. Hence deduce there is an equilibrium position at $x = \dfrac{mga}{2\lambda}$.

   ii) Differentiate the energy equation to show the equation of motion is

$$\ddot x = -\frac{2\lambda}{ma}\left(x - \frac{mga}{2\lambda}\right).$$

   Hence write down the period of the oscillation about the equilibrium position.

21.2 A small lamp of mass m is at the end A of a light rod AB of length $2a$ attached at B to a vertical wall in such way that the rod can rotate freely about B in a vertical plane perpendicular to the wall. A spring CD of natural length a and modulus of elasticity λ is joined to the rod at its mid-point C and to the wall at a point D a distance a vertically above B.

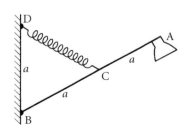

Show that if $\lambda > 4\,mg$ the lamp can hang in equilibrium away from the wall and find the angle DBA.

21.3   The diagram shows a smooth wire bent into the form of a circle in a vertical plane. A ring P is threaded on the circle and tied to a light inextensible string which passes over a pulley O at the highest point of the circle. A particle of the same mass as the ring hangs at the other end of the string. Use the energy method to find two positions of equilibrium, one stable and one unstable.

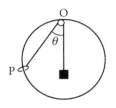

21.4   In a diatomic molecule, one atom is very much heavier than the other and the lighter atom is free to move on a straight line through the centre of the heavier atom. The potential energy for the force between two atoms in a diatomic molecule can be represented approximately by

$$V(x) = -ax^{-6} + bx^{-12}$$

where x is the atomic separation and a and b are positive constants.
   i)   Locate any positions of stable equilibrium of the lighter atom.
   ii) Find the period of small oscillations about such positions, assuming the mass of the lighter atom is m.
   iii) Sketch the potential energy as a function of x. How does the force vary as the atoms approach each other?

21.5   A uniform circular cylinder of mass M and radius a lies at rest inside a fixed hollow cylinder of radius $3a$. The axes of both cylinders are horizontal. A particle of mass m is fixed to a point at the top of the inner cylinder which can roll without slipping inside the hollow cylinder. The system is subject to a vertical gravitational field g.
   i)   Show that $\phi = \frac{1}{2}\theta$, where ϕ and θ are as shown in the diagram.
   ii) Find an expression for the gravitational potential energy of the system as a function of θ.
   iii) Locate the positions of equilibrium and determine the relationship between M and m for the initial position to be one of stable equilibrium. Show that in this case it is the only equilibrium position.

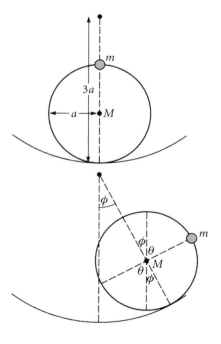

21.6   A light plane uniform square lamina of side $2a$ is mounted in a vertical plane and is free to rotate about an axis through its centre and perpendicular to its plane. Initially two of the edges of the lamina are horizontal. A particle of mass km, $k > 0$, is attached to the centre of the upper edge and particles of mass m are attached to each of the two bottom corners.

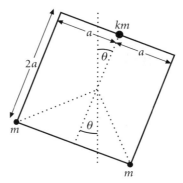

i) By considering a rotation through an angle θ of the lamina show that the potential energy of the system relative to the centre of the lamina is $mga(k-2)\cos\theta$.

ii) Locate any positions of equilibrium.

iii) Discuss the dependence of the nature of the stability of any equilibrium positions on the value of k.

iv) Given that $k = 1$ and the mass of the lamina can be neglected write down the energy equation. By differentiating this equation with respect to time find the period of small oscillations of the lamina about the position of stable equilibrium.

Personal Tutor

21.1 Four light rods each of length $2a$ are freely hinged at their ends to form a rhombus ABCD which is suspended at A from a fixed point. A light spring of natural length $2a$ and stiffness $\dfrac{3mg}{a}$ connects the points A and C. A particle of mass $\dfrac{3m}{2}$ is attached at the point C and each of the rods AB and AD carries a particle of mass m at its mid-point. The arrangement is shown in the diagram.

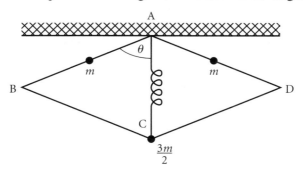

i) Show that the potential energy V of the system relative to the zero level through A is given by

$$V = mga(24\cos^2\theta - 32\cos\theta + 6).$$

ii) Deduce that there are two positions of equilibrium and show that only one of these is stable.

The system now performs small oscillations about the position of stable equilibrium, when $\theta = \alpha$. Writing $\theta = \alpha + \phi$ and assuming that both ϕ and $\dot{\phi}$ remain very small, the kinetic energy T is given by $T = \frac{23}{3}ma^2\dot{\phi}^2$ and the potential energy $V = \dfrac{mga}{3}(-14 + 40\phi^2)$.

iii) Find the approximate period of small oscillations about the position of stable equilibrium.

KEY POINTS

21.1 When a body acted on by *conservative* forces is free to move so that its potential energy can be given as a function V(x) of a single variable x, its equilibrium positions are given by the stationary values of V(x) with respect to this variable.

21.2 A *minimum* value of potential energy corresponds to a position of *stable* equilibrium; a *maximum* value to a position of *unstable* equilibrium.

21.3 Small oscillations about positions of stable equilibrium can generally be modelled by simple harmonic motion. The equation of motion can be derived by differentiating the energy equation with respect to time.

21.4 The linear approximation

$$V'(x) = (x - x_0)V''(x_0)$$

can be used when $x \approx x_0$. It might be convenient to transform the variable representing displacement so that it has a zero level at the equilibrium position.

Index